Cytoskeleton: signalling and cell regulation

The Practical Approach Series

SERIES EDITOR

B. D. HAMES
Department of Biochemistry and Molecular Biology
University of Leeds, Leeds LS2 9JT, UK

See also the Practical Approach web site at **http://www.oup.co.uk/PAS**

★ indicates new and forthcoming titles

Affinity Chromatography
Affinity Separations
Anaerobic Microbiology
Animal Cell Culture (2nd edition)
Animal Virus Pathogenesis
Antibodies I and II
Antibody Engineering
Antisense Technology
★ Apoptosis
Applied Microbial Physiology
Basic Cell Culture
Behavioural Neuroscience
Bioenergetics
Biological Data Analysis
Biomechanics – Materials
Biomechanics – Structures and Systems
Biosensors
★ Caenorhabditis Elegans
Carbohydrate Analysis (2nd edition)
Cell-Cell Interactions
The Cell Cycle
Cell Growth and Apoptosis
★ Cell Growth, Differentiation and Senescence
★ Cell Separation
Cellular Calcium
Cellular Interactions in Development
Cellular Neurobiology
Chromatin
★ Chromosome Structural Analysis
Clinical Immunology
Complement
★ Crystallization of Nucleic Acids and Proteins (2nd edition)
Cytokines (2nd edition)
The Cytoskeleton
Diagnostic Molecular Pathology I and II
DNA and Protein Sequence Analysis
DNA Cloning 1: Core Techniques (2nd edition)
DNA Cloning 2: Expression Systems (2nd edition)

DNA Cloning 3: Complex Genomes (2nd edition)

DNA Cloning 4: Mammalian Systems (2nd edition)

★ DNA Microarrays

★ DNA Viruses

Drosophila (2nd edition)

Electron Microscopy in Biology

Electron Microscopy in Molecular Biology

Electrophysiology

Enzyme Assays

Epithelial Cell Culture

Essential Developmental Biology

Essential Molecular Biology I and II

★ Eukaryotic DNA Replication

Experimental Neuroanatomy

Extracellular Matrix

Flow Cytometry (2nd edition)

Free Radicals

Gas Chromatography

Gel Electrophoresis of Nucleic Acids (2nd edition)

★ Gel Electrophoresis of Proteins (3rd edition)

Gene Probes 1 and 2

Gene Targeting (2nd edition)

Gene Transcription

Genome Mapping

Glycobiology

Growth Factors and Receptors

Haemopoiesis

★ High Resolution Chromatography

Histocompatibility Testing

HIV Volumes 1 and 2

★ HPLC of Macromolecules (2nd edition)

Human Cytogenetics I and II (2nd edition)

Human Genetic Disease Analysis

★ Immobilized Biomolecules in Analysis

Immunochemistry 1

Immunochemistry 2

Immunocytochemistry

★ *In Situ* Hybridization (2nd edition)

Iodinated Density Gradient Media

Ion Channels

★ Light Microscopy (2nd edition)

Lipid Modification of Proteins

Lipoprotein Analysis

Liposomes

Mammalian Cell Biotechnology

Medical Parasitology

Medical Virology

MHC Volumes 1 and 2

★ Molecular Genetic Analysis of Populations (2nd edition)

Molecular Genetics of Yeast

Molecular Imaging in Neuroscience

Molecular Neurobiology

Molecular Plant Pathology I and II

Molecular Virology

Monitoring Neuronal Activity

★ Mouse Genetics and Transgenics
Mutagenicity Testing
Mutation Detection
Neural Cell Culture
Neural Transplantation
Neurochemistry (2nd edition)
Neuronal Cell Lines
NMR of Biological Macromolecules
Non-isotopic Methods in Molecular Biology
Nucleic Acid Hybridisation
★ Nuclear Receptors
Oligonucleotides and Analogues
Oligonucleotide Synthesis
PCR 1
PCR 2
★ PCR 3: PCR In Situ Hybridization
Peptide Antigens
Photosynthesis: Energy Transduction
Plant Cell Biology
Plant Cell Culture (2nd edition)
Plant Molecular Biology
Plasmids (2nd edition)
Platelets
Postimplantation Mammalian Embryos
★ Post-translational Processing
Preparative Centrifugation
Protein Blotting
★ Protein Expression
Protein Engineering
Protein Function (2nd edition)
Protein Phosphorylation (2nd edition)
Protein Purification Applications
Protein Purification Methods
Protein Sequencing
Protein Structure (2nd edition)
Protein Structure Prediction
Protein Targeting
Proteolytic Enzymes
Pulsed Field Gel Electrophoresis
RNA Processing I and II
RNA-Protein Interactions
Signalling by Inositides
★ Signal Transduction (2nd edition)
Subcellular Fractionation
Signal Transduction
★ Transcription Factors (2nd edition)
Tumour Immunobiology
★ Virus Culture

Cytoskeleton: signalling and cell regulation

A Practical Approach

Edited by

KERMIT L. CARRAWAY

and

CAROLIE A. CAROTHERS CARRAWAY

Departments of Cell Biology and Anatomy (R-124) and
Biochemistry and Molecular Biology
University of Miami School of Medicine
Miami, FL 33101
USA

This book has been printed digitally in order to ensure its continuing availability

OXFORD
UNIVERSITY PRESS

Great Clarendon Street, Oxford OX2 6DP

Oxford University Press is a department of the University of Oxford.
It furthers the University's objective of excellence in research, scholarship,
and education by publishing worldwide in

Oxford New York

Auckland Bangkok Buenos Aires Cape Town Chennai
Dar es Salaam Delhi Hong Kong Istanbul Karachi Kolkata
Kuala Lumpur Madrid Melbourne Mexico City Mumbai Nairobi
São Paulo Shanghai Singapore Taipei Tokyo Toronto

with an associated company in Berlin

Published in the United States
by Oxford University Press Inc., New York

Reprinted 2001

A catalogue record for this book is available from the British Library

Library of Congress Cataloging in Publication Data
Cytoskeleton : Signalling and cell regulation : a practical approach /
edited by Kermit L. Carraway and Corolie A. Carothers Carraway.
p. cm. – (The practical approach series ; 221)
Includes bibliographical references and index.
1. Cytoskeleton. 2. Cellular control mechanisms. 3. Cellular
signal transduction. I. Caraway, Kermit L. II. Carraway, C. A.
C. (Corolie C. Carothers) III. Series.
QH603.C96C976 2000 571.6'54—dc21 99-41249

ISBN 0-19-963782-2 (Hbk)
0-19-963781-4 (Pbk)

Preface

The impetus for this volume was the confluence of two events. First, the editors had recently finished a monograph entitled *Signaling and the cytoskeleton* and were suffering the relief and withdrawal symptoms associated with such an event. Secondly, we were contacted by Dr Hames about updating our previous *Practical approach* volume, *The cytoskeleton: a practical approach*. Since the earlier volume was published, there has been a tremendous expansion in interest in the broad area of the relationship of the cytoskeleton to signalling and cell regulatory events. The emphasis for this volume was dictated by the fact that research on cellular signalling is one of the most fertile areas of biological and biomedical sciences and promises to become even more active in the near future. By the same token, there are important implications of this area of research for understanding the dynamics of the cytoskeleton. Finally, one of the emerging principles of signalling research is that signalling pathways are organized on scaffolding matrixes, which are often associated with the cellular cytoskeleton. It is becoming clear that one cannot, or at least should not, study these two areas in isolation. Thus, we were intrigued with the idea of contributing to that needed integration by providing a book of experimental protocols bridging the two areas.

One of the comments in the preface of our previous *Practical approach* volume was that we could provide only a fraction of the approaches and protocols which would be interesting and relevant in one volume. In dealing with the two broad areas of research and facing an exponentially growing literature on signalling alone, the problem was even further amplified. We have attempted in this volume to provide some of both the flavour and meat of the subject. Our hope is that this volume will afford perspective as well as insights into approaches and techniques and will serve at least as an illuminating guide to this exciting area of research.

Miami, Florida K. L. C.
May, 1999 C. A. C. C.

Contents

Contributors

JOSÉ LUIS ALONSO
Children's Hospital, Departments of Surgery and Pathology, 300 Longwood Avenue, Boston, MA 02115, USA.

CHRISTOPHE AMPE
Department of Biochemistry, University Ghent and Department of Medical Protein Chemistry, Flanders Interuniversity Institute for Biotechnology, Belgium.

YUN BAO
Department of Biological Sciences, University of Notre Dame, Notre Dame, IN 46556–0369 USA.

GIOVANNA BERMANO
Intracellular Targeting Group, Rowett Research Institute, Bucksburn, Aberdeen AB21 9SB, Scotland.

KRZYSZTOF BOJANOWSKI
University of California at Berkeley LBNL, 91 Bolivar Drive, Berkeley, CA 94710, USA.

CLIFFORD BRANGWYNNE
Children's Hospital, Departments of Surgery and Pathology, 300 Longwood Avenue, Boston, MA 02115, USA.

CORALIE A. CAROTHERS CARRAWAY
Department of Biochemistry and Molecular Biology, University of Miami School of Medicine, PO Box 016960, Miami, FL 33101, USA.

KERMIT L. CARRAWAY
Department of Cell Biology and Anatomy, University of Miami School of Medicine, PO Box 016960, Miami, FL 33101, USA.

CHRISTOPHER S. CHEN
Department of Biomedical Engineering, Johns Hopkins University, Traylor 710, 720 Rutland Avenue, Baltimore, MD 21205, USA.

MARINA E. CHICUREL
216 Koshland Way, Santa Cruz, CA 95064, USA.

JOHN A. COOPER
Department of Cell Biology, Washington University School of Medicine, 300 Euclid Avenue, Box 8228, St. Louis, MO 63110, USA.

LAURA DIKE
BD ViaSante, 2 Oak Park, Bedford, MA 01730, USA.

KENNETH H. DOWNING
Lawrence Berkeley National Laboratory, Donner Laboratory, 1–326, Berkeley, CA 97420, USA.

DOUGLAS J. FISHKIND
Department of Biological Sciences, University of Notre Dame, Notre Dame, IN 46556–0369, USA.

HELEN M. FLINN
Ludwig Institute for Cancer Research, 91 Riding House Street, London W1P 8BP, UK.

WOLFGANG H. GOLDMANN
Children's Hospital, Departments of Surgery and Pathology, 300 Longwood Avenue, Boston, MA 02115, USA.

HIDEMASA GOTO
Laboratory of Biochemistry, Aichi Cancer Centre Research Institute, Chikusa-ku, Nagoya, Aichi 464–8681, Japan.

JOHN E. HESKETH
Intracellular Targeting Group, Rowett Research Institute, Bucksburn, Aberdeen AB21 9SB, Scotland.

SUI HUANG
Children's Hospital, Departments of Surgery and Pathology, 300 Longwood Avenue, Boston, MA 02115, USA.

HIROYASU INADA
Laboratory of Biochemistry, Aichi Cancer Centre Research Institute, Chikusa-ku, Nagoya, Aichi 464–8681, Japan.

MASAKI INAGAKI
Laboratory of Biochemistry, Aichi Cancer Centre Research Institute, Chikusa-ku, Nagoya, Aichi 464–8681, Japan.

DONALD E. INGBER
Children's Hospital, Departments of Surgery and Pathology, 300 Longwood Avenue, Boston, MA 02115, USA.

TATIANA S. KARPOVA
Department of Cell Biology, Washington University School of Medicine, 300 Euclid Avenue, Box 8228, St. Louis, MO 63110, USA.

KYUNG-MI LEE
Ben May Institute, Room S305, University of Chicago, 5841 South Maryland, Chicago, IL 60637, USA.

ANDREW MANIOTIS
Department of Anatomy and Cell Biology, Iowa Cancer Center, College of Medicine, University of Iowa, 51 Newton Road, Iowa City, Iowa 52242–1109, USA.

ROBERT MANNIX
Children's Hospital, Departments of Surgery and Pathology, 300 Longwood Avenue, Boston, MA 02115, USA.

CHRISTINE M. MANUBAY
Department of Biological Sciences, University of Notre Dame, Notre Dame, IN 46556–0369, USA.

HELEN MCNAMEE
Department of Cancer Biology, Dana Farber Cancer Institute, 44 Binney Street, Boston, MA 02115, USA.

CHRISTIAN J. MEYER
Children's Hospital, Departments of Surgery and Pathology, 300 Longwood Avenue, Boston, MA 02115, USA.

KOH-ICHI NAGATA
Laboratory of Biochemistry, Aichi Cancer Centre Research Institute, Chikusa-ku, Nagoya, Aichi 464–8681, Japan.

KEIJI NARUSE
Children's Hospital, Departments of Surgery and Pathology, 300 Longwood Avenue, Boston, MA 02115, USA.

KEVIN KIT PARKER
Children's Hospital, Departments of Surgery and Pathology, 300 Longwood Avenue, Boston, MA 02115, USA.

GEORGE PLOPPER
Department of Biological Sciences, University of Nevada, Las Vegas, 4505 Maryland Parkway, Las Vegas, NV 89154–4004, USA.

THOMAS POLTE
Children's Hospital, Departments of Surgery and Pathology, 300 Longwood Avenue, Boston, MA 02115, USA.

D. L. PURICH
Department of Biochemistry and Molecular Biology, University of Florida, Gainesville, FL 32610–0245, USA.

ANNE J. RIDLEY
Ludwig Institute for Cancer Research, 91 Riding House Street, London W1P 8BT, UK.

IRINA ROMENSKAIA
Department of Biological Sciences, University of Notre Dame, Notre Dame, IN 46556–0369, USA.

F. S. SOUTHWICK
Department of Medicine, J277, University of Florida College of Medicine, Gainesville, FL 32610–0277, USA.

JOËL VANDEKERCKHOVE
Department of Biochemistry, University Ghent and Department of Medical Protein Chemistry, Flanders Interuniversity Institute for Biotechnology, Belgium.

NING WANG
Physiology Program, Harvard School of Public Health, Boston, MA 02115, USA.

LI YAN
Children's Hospital, Departments of Surgery and Pathology, 300 Longwood Avenue, Boston, MA 02115, USA.

W. L. ZEILE
Department of Medicine, University of Florida College of Medicine, Gainesville, FL 32610–0277, USA.

Abbreviations

ABTS	2′,2′-azino-bis(3-ethylbenzathiazoline)-6-sulfonic acid
AMCA	amino-methylcoumarin-acetic acid
β-Gal	β-galactosidase protein
BCIP	5-bromo-4-chloro-3-indolyl-phosphate
BHI	brain heart infusion
BSA	bovine serum albumin
CBP	cytoskeletal-bound polysomes
CCD	charge-coupled device
CCE	complete cytopathic effect
c.f.u.	colony-forming units
CHO	Chinese hamster ovary
CMV	cytomegalovirus
Con A	concanavalin A
2D	two-dimensional
3D	three-dimensional
DAB	diaminobenzidine
DEPC	diethylpyrocarbonate
DIC	differential interference contrast
DIG	digoxigenin
DMEM	Dulbecco's modified Eagle's media
DMF	dimethylformamide
DMSO	dimethyl sulfoxide
DTT	dithiothreitol
EDC	1-ethyl-3-(3-dimethylaminopropyl)carbodiimide
EDTA	ethylenediaminetetraacetic acid
EGFR	epidermal growth factor receptor
EGTA	ethylene glycol-bis(β-aminoethyl ether)*N*,*N*,*N*′,*N*′-tetraacetic acid
ELISA	enzyme-linked immunosorbent assay
EM	electron microscopy
FBS	fetal bovine serum
FCS	fetal calf serum
FDLD	fluorescence detected linear dichroism
FITC	fluorescein isothiocyanate
FP	free polysomes
G418	geneticin antibiotic
GFP	green fluorescent protein
GLUT 1	glucose transporter 1
GM-CSF	granulocyte-macrophage colony-stimulating factor

GST	glutathione *S*-transferase
HBS	Hepes-buffered saline
Hepes	*N*-(2-hydroxyethyl)piperazine-*N*′(2-ethanesulfonic acid)
h.p.i.	hours post-infection
IEF	isoelectric focusing
IF	intermediate filaments
IgG	immunoglobulin
ISH	*in situ* hybridization
kPa	kilo Pascals
MALDI-TOF	matrix-assisted laser desorption ionization-time of flight
MAPK	mitogen-activated kinase
MBP	membrane-bound polysomes
m.o.i.	multiplicity of infection
MT	microtubule
NBT	nitroblue tetrazolium
NCI	National Cancer Institute
NRK	normal rat kidney cell
PAGE	polyacrylamide gel electrophoresis
PBS	phosphate-buffered saline
PCR	polymerase chain reaction
p.f.u.	plaque-forming units
PIP_2	phosphatidylinositol 4,5-bisphosphate
PipesBS	Pipes-buffered saline
PMSF	phenylmethylsulfonyl fluoride
PSD	post-source decay
PTH	phenylthiohydantoin
REF	rat embryo fibroblast
RER	rough endoplasmic reticulum
rhph	rhodamine phalloidin
RPAS	recombinant phage antibody system
r.p.m.	rounds per minute
RT	room temperature
RT-PCR	reverse transcription polymerase chain reaction
SDS	sodium dodecyl sulfate
SDS–PAGE	sodium dodecyl sulfate–polyacrylamide electrophoresis
SH	Src homology
SNR	signal-to-noise ratio
STE	saline, Tris, EDTA solution
sulfo-NHS	*N*-hydroxysulfosuccinimide
SV40	simian virus 40
TBS	Tris-buffered saline
TMC	transmembrane complex
TRITC	tetramethylrhodamine isothiocyanate
3′ UTR	3′ untranslated region

1

The cytoskeleton in the transduction of signal and regulation of cellular function

CORALIE A. CAROTHERS CARRAWAY

1. Introduction

Cells are constantly bombarded with signals from their environments, and their responses to these cues determine all aspects of their function. Each type of cell responds to these environmental stimuli in a sensitive, selective, and temporally ordered manner. Each signal is perceived and integrated into a carefully regulated cellular response which reflects a hierarchy imposed by each cell. Responses to stimuli can range from relatively simple ones, such as activation of a housekeeping metabolic pathway, to global reorganizations which encompass multiple metabolic pathways and elicit massive morphological alterations. In many cases there is at least minimal involvement of portions of the intracellular cytoskeletal network. In some instances, such as mitogenesis, complete disassembly and reassembly of all the cytoskeletal networks is required. In these cases morphology becomes destiny, and the pathways which link cell activation to cytoskeletal changes are key to the transmission of signals for the pleiotypic effects necessary for appropriate cellular response.

The concept that interactions between membrane proteins and the cytoskeleton play a central role in cellular responses is an old one. The classical methods of microscopy and cell perturbation studies, as well as biochemical fractionations and reconstitution studies, were key to the development of these basic concepts, first in simple systems such as the erythrocyte and the platelet, and then in more complex cells. However, the more recent elucidation at the molecular level of signalling pathways involved in the transduction of extracellular signals has renewed interest in roles for the cytoskeleton in mediating cellular alterations. The diversity of the systems being used to answering these basic questions broadens perspectives. Similarly, the array of tools and approaches available for the study of the molecular mechanisms involved in the development and transmittal of signalling responses expands the number of questions which can be asked.

Two broad classes of cellular proteins comprise the major players in the mediation of signal to the appropriate pathways to be modified. The first group are the components of the signalling pathways, including ligands, receptors, several classes of intracellular transducing enzymes and structural proteins, and transcription factors. The second major group is the proteins which comprise the cytoskeletal networks: microfilaments, which are actin polymers; intermediate filaments, assembled from one or more of several classes of cell-specific intermediate filament subunits; and microtubules, which are assemblies of polymerized tubulin. Microfilaments and microtubules are dynamic structures whose plasticity is necessary for cellular function; intermediate filaments are less dynamic in most cell types and tend to serve as a buffer against major cellular deformation. Co-ordination of the regulation of morphology is integrated via interactions among the three cytoskeletal arrays, resulting in 'cross-talk' between the different structural elements.

Numerous proteins which bind each of these elements in either their monomeric or polymeric forms regulate assembly–disassembly of the filament structures as well as interactions between the structures. Among these, and perhaps most important to the transmission of pleiotypic effects, are the proteins which link signalling elements to these cytoskeletal networks. An extensive array of membrane skeletal proteins, some ubiquitous and some more cell-specific, have been identified and implicated in the mediation of these interactions. Others will likely be identified by a diversity of approaches, including co-purification with signalling complexes, characterization of kinase or phosphatase substrate proteins, identification of regions of sequence homology to known domains important in protein–protein interactions, and characterization of genes and proteins involved in mediating developmental processes.

The last decade has seen the development of a major area, domain identification and characterization, which should contribute significantly to the understanding of the roles of specific proteins in the dissemination of signalling information throughout the cell. A growing number of domains and motifs which are involved in catalysis or which mediate protein–protein interactions in signalling complexes are being characterized. The presence of these sequences is beginning to allow the prediction of a newly discovered protein's function, or a portion of its function, based on the presence of homologous domains or motifs. This foot in the door then permits the rational design of experiments to test assumptions based on sequence similarities.

Research on the involvement of the cytoskeleton in cellular functions has contributed for over thirty years to our conceptual understanding of many aspects of cell behaviour. Components of the cytoskeletal network are major regulators of processes as diverse as establishment and maintenance of gross cell morphology, polarity, transduction of force, motility, adhesion to both matrix components and cells, cell division, and secretion (for a comprehensive review see ref. 1). Further, the cytoskeleton has long been implicated in the organization and reorganization of receptors in the plasma membrane and is therefore critical to cell recognition mechanisms for many types of associations, ranging from tissue formation to the immune killing of foreign cells. Thus the association of cytoskeletal elements with membrane components became an early paradigm for the transduction of signals to the cytoplasm from the cell surface and, conversely, from the cell interior to the membrane. Interaction sites for membrane proteins with the interior of the cell are also key integration sites for the transmission of signals to several pathways, eliciting pleiotypic responses of cells to signals. Membrane–cytoskeletal complexes are mediators of cross-talk between receptors. Cell surface receptors for diverse ligands, including growth factors and hormones, cell–matrix and cell–cell adhesion proteins are linked transmembranally to microfilaments, which in turn, interact with both microtubules and intermediate filaments. These interactive systems of membranes with all of the cytoskeletal arrays are capable of eliciting the global responses of cells to ligands such as mitogens, which elicit major morphological perturbations.

2. Model systems

The types of model systems employed in the study of membrane–cytoskeleton interactions and their roles in cellular processes are many and varied, as illustrated by the examples given in *Table 1*. From these studies major concepts relating to the regulation of cell function by environmental cues were developed. Studies on the aberrant behaviour of cancer cells and mechanisms for the bypass of normal regulatory processes were major contributors to our conceptual understanding of the importance of the roles of the cytoskeleton and cytoskeletal plasticity in cell function. Tumour cells also provided the impetus for the convergence of the studies of cytoskeleton-mediated cellular functions with those on (proto)oncogene and oncogene products, our entry into modern signal transduction studies. Finally, genetics studies in diverse developmental model systems have afforded key insights into biological functions of gene products which are homologues of known signalling and cytoskeletal proteins.

3. Approaches and techniques

A spectrum of techniques has played important roles in identifying and characterizing the involvement of the cytoskeleton in cellular function and regulation. These can be divided into three broad categories; microscopy, biochemical methods, and genetics. Basic methods for some of the most common classical cell biological and biochemical approaches have been described in some detail in the previous volume on the cytoskeleton in this series (2). The present volume elaborates on all three of these approaches,

Table 1. Model systems and cell processes in cytoskeleton research

Cell type	Force generation/ maintenance	Morphology/ polarity	Motility/ chemotaxis	Adhesion	Activation/ signal transduction	Other
Muscle	X	X				Differentiation, development
Erythrocyte		X				
Fibroblast	X	X	X	X	X	
Platelet		X		X	X	Secretion
Neural		X	X	X	X	Secretion
Lymphoid				X	X	
Epithelial		X		X	X	Differentiation, development, patterning
Tumour cells		X	X	X	X	Mechanisms for malignancy, metastasis
Developmental systems	X	X		X		Differentiation, development, patterning
Dictyostelium						
Drosophila						
C. elegans oocyte						
Yeast		X				Mating, cell division

Table 2. General approaches to the study of the cytoskeleton

Type of approach	Methods, tools	References[a]
Cellular localization	Microscopy, electron microscopy, immunolocalization, epitope tagging, microinjection, *in situ* hybridization	Ref. 3 A: Chap. 3, 5–7, 9–11 B: Chap. 1, 2
Structure and organization	Electron microscopy; X-ray crystallography; electron crystallography	Refs 4, 5 A: Chap. 8
Cell fractionation	Centrifugation, immunoprecipitation, affinity-based precipitation	Refs 6, 7 A: Chap. 4, 10, 11
Reconstitution	Filament sedimentation assays, detergent solubilization/reconstitution, chaotrope disruption/reconstitution, perturbation methods	A: Chap. 2–4, 9 B: Chap. 6, 8, 10
Specificity of interaction	Yeast two-hybrid, phage display, ELISA-type assays, fusion protein assays, epitope tags, competition assays, recombinant domains/ motifs, cross-linking, gel overlays	A: Chap. 2, 4, 8–10 B: Chap. 3–5
Biological function	Antibody blocking, antisense, deletions, site-directed mutations, domain swaps, chimeric proteins, knock-outs, knock-ins	A: Chap. 2, 5, 8, 9 B: Chap. 5

[a] A: chapters in this volume. B: chapters in previous volume (ref. 2).

using protocols which illustrate modern refinements in technology and designer reagents. A compendium of methodologies and locations of their descriptions is listed in *Table 2*.

The first approach, a direct visualization of the components of interest *in situ*, is critical to a conceptual understanding of the interactions. Immunolocalization microscopy, a time-honoured approach for this type of study, was described in detail in ref. 2 (Chapter 1). The present volume elaborates on special applications of this basic technique in addition to describing state-of-the-art advances in video imaging and three-dimensional reconstructions from microscopy. Once cellular co-localization is established, the second approach is to determine whether molecular complexes containing the components can be isolated. Basic biochemical techniques are used for the partial disassembly of the cell and isolation of fractions containing both cytoskeleton and signalling protein(s). A wide selection of assay methods, discussed in Chapter 4, can then be used to identify both general and specific components in the complexes. A complementary approach is a test of the hypothesis of direct association by reconstitution of complex proteins. This may be done using purified constituent proteins, constituent proteins expressed in a host cell, or a combination of these. The final approach addresses the question of the role of the protein in loss- or gain-of-function studies. Genetics studies in amenable organisms have led the way in much of this research, but other methods, including blocking antibodies and antisense or ribozyme inhibitors of function, are now being used widely. More recently, the application of gene modulation technology by knock-out or knock-in studies in the whole organism promises to provide new insights into specific functions, particularly when applied to specific tissues and cell types.

Each of these approaches has its inherent problems. Co-localization by immunomicroscopic approaches cannot provide information on intermolecular interactions involved in the cellular organization of the assemblies. Co-fractionation is subject to introduction of artefactual associations of extraneous materials or release of components which would normally be present (for a discussion see ref. 7). However, if both microscopic and biochemical isolation approaches are suggestive of these interactions, the final *in vitro* test is reconstitution of the complexes from their components. Difficulties may be envisioned on several fronts in reconstitution studies. The first stems from our inability to adequately reconstruct the optimal (*in vivo*) conditions for binding, even when using 'physiological' buffers. A second may entail use of the appropriately modified protein for binding assays. For example, a specifically phosphorylated form of a protein, domain, or motif may be required for its active binding conformation or stability. Use of the unmodified form, as one frequently obtains, for example, in *E. coli* expression systems or from chemical syntheses, will lead to a false negative result in a binding assay. On the positive side, the reliability of the reconstitution approach derives from the specificity and stability of the interactions between many proteins in signalling complexes, mediated by specific binding domains and motifs (8–10).

The ultimate test for a biological role of these interactions, of course, can be performed only *in vivo*. Genetic approaches for gene mutation for alteration of function, gene knock-outs and knock-ins, and loss-of-function experiments and can provide information on the overall role of a molecule in the development and normal function of an organism. However, even with genetic studies one sometimes encounters examples of biochemical redundancies which mask the importance of the function of a specific protein. A classic example is knock-out experiments on Src family members. Disruption of the genes of individual family members showed relatively mild phenotypic effects on mice except in specific tissues. Knock-outs of multiple family members were required for extensive effects (reviewed in refs 11 and 12). These back-up mechanisms probably argue for, rather than against, the importance of the protein in normal function (11). An important aspect of redundancy is likely that the redundant protein(s) share conserved sequences in family member binding domains and motifs as well as catalytic domains.

Collectively, these approaches, particularly when used in combination, provide powerful tools for the accumulation, assembly, and integration of information at the cellular and molecular levels on the cytoskeleton and its numerous roles in nearly every aspect of cell function. However, the field is still relatively young, and major technological advances can still be anticipated. Forecasting is always risky, but we foresee progress along several fronts. One will certainly be the increasing sophistication of biophysical methods for examining properties of cells and cell fractions. A second is likely to come from new methods for investigating macromolecular and supramolecular complexes. A third will surely develop from computer-based modelling studies of protein interactions. And increasingly powerful genetic systems and technologies for gene transfer will permit more sophisticated *in vivo* analyses. Among these gene analyses is the potential for measuring gene expression changes of hundreds of genes using chip technology. Finally, just as the important feature in the cell is the integration of multiple signals and structural elements, we predict that advances will come most rapidly from those who can apply, integrate, and interpret results from multiple methods directed toward answering specific questions.

References

1. Carraway, K. L., Carraway, C. A. C., and Carraway, K. L. III. (1998). *Signaling and the cytoskeleton.* Springer, Berlin.
2. Carraway, K. L. and Carraway, C. A. C. (ed.) (1992). *The cytoskeleton: a practical approach.* IRL Press, Oxford.
3. Lacey, A. J. (ed.) (1999). *Light microscopy in biology: a practical approach.* IRL Press, Oxford.
4. Steinmetz, M. O., Hoenger, A., Tittmann, P., Fuchs, K. H., Gross, H., and Aebi, U. (1998). *J. Mol. Biol.*, **278**, 703.
5. Mendelson, R. A. and Morris, E. (1994). *J. Mol. Biol.*, **240**, 138.
6. Graham, J. M. and Rickwood, D. (ed.) (1997). *Subcellular fractionation: a practical approach.* IRL Press, Oxford.
7. Carraway, K. L. and Carraway, C. A. C. (1982). In *Antibody as a tool: the application of immunochemistry* (ed. J. J. Marchalonis and G. W. Warr), p. 509. John Wiley, Chichester, England.
8. Lin, D. and Pawson, T. (1997). *Trends Cell Biol.*, **7**, Centerfold.
9. Bork, P., Schultz, J., and Ponting, C. P. (1997). *Trends Biochem. Sci.*, **22**, 296.
10. Sudol, M. (1998). *Oncogene*, **17**, 1469.
11. Brown, M. T. and Cooper, J. A. (1996). *Biochim. Biophys. Acta*, **1287**, 121.
12. Thomas, S. M. and Brugge, J. S. (1997). *Annu. Rev. Cell Dev. Biol.*, **13**, 513.

2

Assaying binding and covalent modifications of cytoskeletal proteins

CHRISTOPHE AMPE and JOËL VANDEKERCKHOVE

1. Introduction

Cytoskeletal dynamics are essential for cell survival. The microfilament system is one of the major players in cell motility processes. It exerts force for movement either by the action of molecular motors on relatively stable filaments or by dynamic turnover of actin filaments. In the latter case processes such as filament polymerization, depolymerization, and bundling are regulated in a temporal and spatial manner, governed by actin-binding proteins. The complexity of the actin system arises from the multitude of actin-binding proteins, often with partly overlapping activities, displaying multiple functions or effects on actin polymerization, depending on the intracellular environment. Therefore, detailed biochemical studies of individual actin-binding proteins and their interaction with actin or with regulatory ligands are necessary to understand cellular motility phenomena. The field is now slowly progressing towards *in vitro* studies with more than one actin-binding protein, because it is clear that in cells these proteins either compete with each other or have additive or co-operative effects on actin reorganization.

In the previous edition of this series the delineation of actin-binding domains in proteins was a major topic (1). In this chapter we will extend this analysis by describing experiments to identify actin-binding residues in proteins together with the analysis of the actin interactions. We also present two assays for phosphatidylinositide binding, since it is thought that in a number of cases PIP_2-binding sites partially overlap actin-binding sites. In addition, we discuss analysis of post-translational modifications.

2. How to identify actin-binding regions in known actin-binding domains or proteins

2.1 Information on actin-binding residues can be derived from sequence comparison

Once an actin-binding protein has been identified and sequenced, one is confronted with the question of which part of the molecule interacts with actin and whether one can alter the effect on actin binding or eliminate actin binding by mutagenesis. Identifying these segments in proteins can be straightforward if the sequences of different members of a protein family are known. Often, but certainly not always, the region that is making the contact with actin is highly conserved (for examples see ref. 2). This conservation can be found using multiple alignment algorithms if the actin-binding region is in a sufficiently large linear stretch of amino acids. As a rule of thumb it is better to compare distantly related actin-binding proteins. One should, of course, interpret these data with caution; sequences may be conserved for other reasons, e.g. involved in binding of other ligands or in the scaffolding of the protein. A distinction between scaffolding sequences (in the interior of the protein) and surface regions can be found using other algorithms predicting hydropathy (3). In exceptional cases similar actin-binding motifs may be present in otherwise unrelated proteins. For example, protein kinase C-epsilon has a functional actin-binding sequence also present in the small actin sequestering protein thymosin β4 (4).

2.2 Crystal structures of actin-binding proteins and calculated models help to design mutants

We found it very useful to guide our mutagenesis experiments by knowledge of the three-dimensional structure of actin-binding proteins, several of which have been solved. The choice of where to make the mutation is greatly simplified, since one can avoid those mutations that will likely disrupt the proper folding of the protein. Of even greater value are those rare cases where the structure of an actin-binding protein complexed to actin is known. Using the profilin I–actin structure (5) as an example, inspection of the interface indicates that 16 residues are involved in actin interaction. One may then model a mutation in one of those side chains and perform an energy minimization for optimal side chain positioning. Increased or reduced contact or steric hindrance at the interface can often be seen immediately. In addition, one may calculate the altered electrostatic surface potentials for the mutant. This exercise is very informative when making charge reversal mutants (see also below) and can predict disruption of salt bridges successfully. Indeed Arg74 (*Figure 1*) and Lys88 from human profilin I are involved in salt bridge formation with actin. Mutation of either of these to glutamic acid leads to a

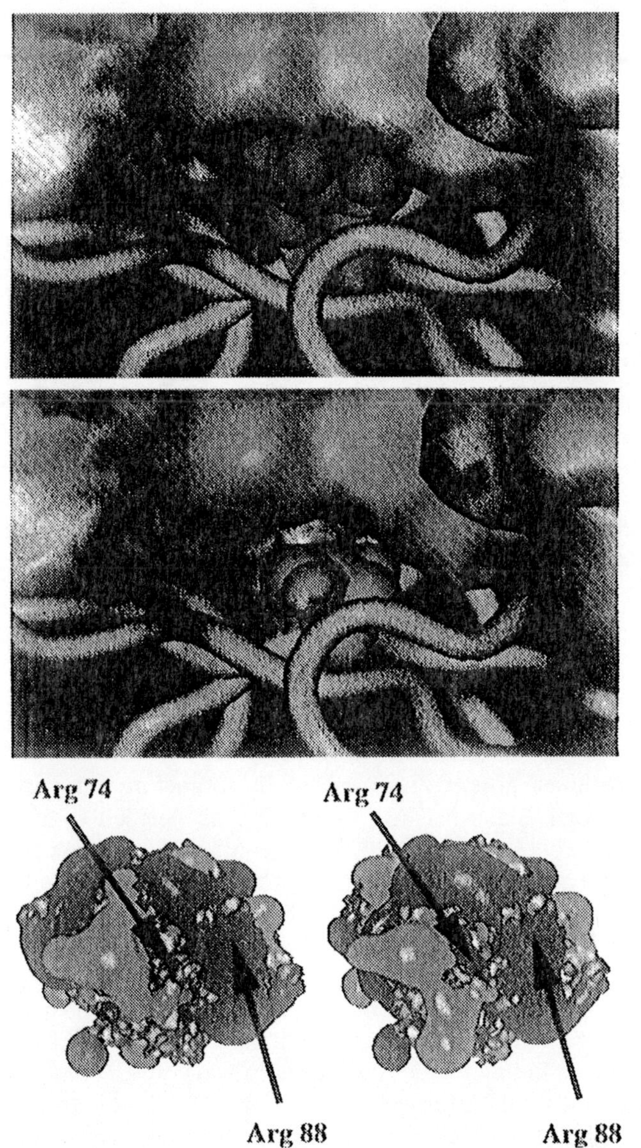

Figure 1. *Upper* and *middle* panel: electrostatic interaction of profilin I residue Arg74 with actin. The backbone of profilin I is rendered as a blue ribbon. The van der Waals surfaces of actin and of residue Arg74 are shown (*upper*). Red represents surfaces with negative potential, blue regions with positive potential. Mutation to Glu results in repelling of surfaces (*middle*). *Lower* panel: solvent accessible surfaces of bovine profilin I (*left*) and II (*right*) coloured according to electrostatic potential (red negative, blue positive). The three-dimensional structure of profilin II was calculated with *HOMOLOGY* (Molecular Simulations) using the known profilin I structure (5) as a template. The surface potential was calculated using *DELPHI*. Settings for Lys and Arg were +1, and for His +0.5. A charge of −1 was assigned for Asp and Glu. The positions of Arg74 and Arg88 are indicated.

dramatic decrease in affinity for actin of the recombinant mutants (6) (Lambrechts *et al.*, unpublished data). Additionally, if sequence similarity is sufficiently high, one can theoretically calculate a structure using a known structure as a template. For instance, the mammalian profilin isoforms share 67% identity. Using the known profilin I three-dimensional structure (5), we modelled the structure of bovine profilin II. If we put the calculated profilin II structure back into the complex with actin, the resulting interface is very similar to the original interface with profilin I, consistent with a similar affinity of both profilins for actin (7). Additionally, we calculated the electrostatic surface potential of both isoforms to pin-point residues involved in PIP_2 binding, as both isoforms display a different affinity for this ligand (8) (*Figure 1*). A similar strategy was successful for profilin isoforms from *Acanthamoeba* (9).

Once regions and/or residues of interest (e.g. for actin binding, for phospholipid binding) have been delineated, their significance must be confirmed by mutagenesis. One may opt for either of the two strategies outlined below: the use of chemically synthesized peptide mimetics or of recombinant proteins. Obviously, the actin binding capacity of these mutants must then be compared with the wild-type activity.

3. Actin peptide mimetics

Relatively small peptides can mimic an activity of intact proteins. The approach we used previously is to create a set of chemically synthesized peptide mutants in which residues potentially involved in contacting actin were modified, either alone or in pairs (11–13). Generally, we decided to make charge reversal mutants because, if involved in a contact, these analogues result in a strong phenotype. However, other substitutions can be as informative (12). The main advantage of using peptides lies in the ease and speed of synthesis and purification. For a 25 amino acid peptide the synthesis (with an Applied Biosystems 431A synthesizer) takes about one day. Further processing and purification require an additional three days (a second synthesis can be started at day two). Problems with solubility are rarely encountered when analysing actin-binding peptides. Often it is also convenient to include, either at the NH_2- or COOH-terminal, a unique cysteine residue (internal, naturally occurring cysteines in the sequence are then usually converted to serines). This thiol group can be modified with iodoacetamide derivatives carrying a biotin moiety, a radioactive tracer, or a fluorescent group for detection or isolation. Alternatively, biotin can be directly coupled to the amino-terminus as the last step in the synthesis process. One drawback to the use of peptides is the fact that their affinities are usually one or two orders of magnitude lower than those of the intact proteins (14). Also, a loss of peptide activity may occur due to structural constraints rather than abolition of a specific contact between the peptide and actin. Circular dichroism experiments can provide information on structural characteristics of the peptide (13). Microinjection of such peptides into cells has sometimes been shown to cause cellular effects (15); however, applications *in vivo* have been limited.

In principle, one can use all of the classical actin-binding assays for analyses of the effects of the peptides, including polymerization assays with pyrene–actin. The methods and protocols for these have been covered in the previous edition of this book (1, 10). We therefore focus on additional protocols, especially those suited for analysis of actin-binding peptide mimetics. One involves cross-linking (*Protocol 1*); the other is a competitive cross-linking experiment coupled to co-sedimentation (*Protocol 2*). In both assays it is very

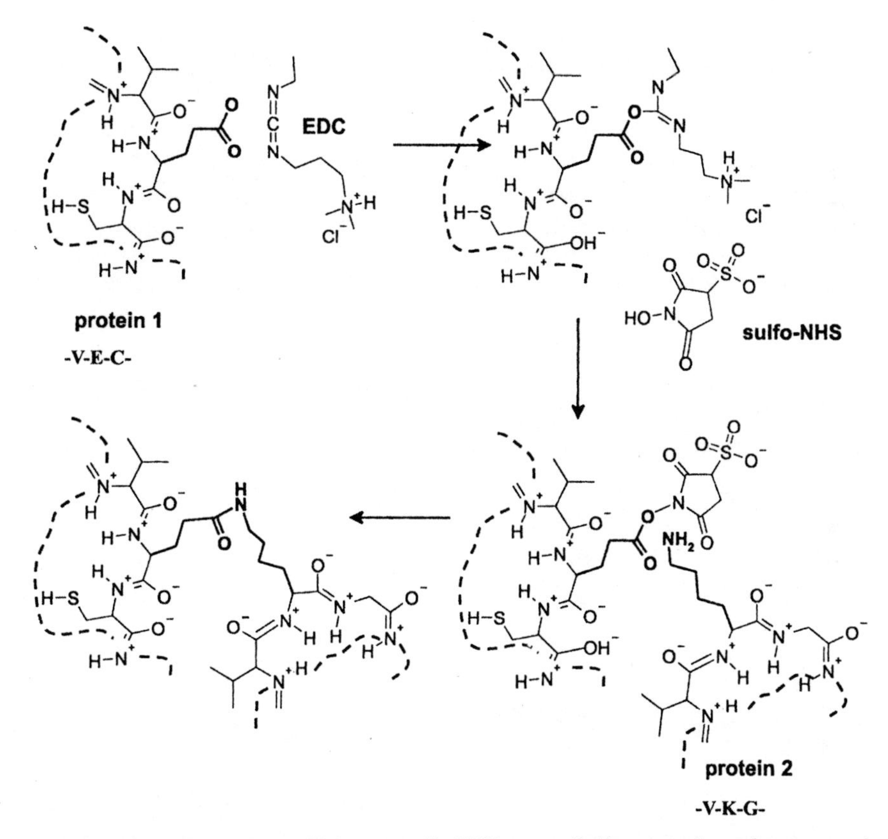

Figure 2. Mechanism of the EDC and sulfo-NHS cross-linking reaction of two proteins (the reacting side chains: Glu in protein 1 and Lys in protein 2, are in bold). EDÇ will react with a carboxyl group of a protein forming an *O*-acylurea derivative (*top right*) that is relatively unstable because it may hydrolyse. This derivative may react with sulfo-NHS to form a more stable intermediate that further reacts with amines. If working in amine-free buffers the amino group comes from a lysine residue in close spatial proximity, resulting in an isopeptide bond. As no atoms of the coupling reagent are incorporated, EDC is designated a zero-length cross-linker. The surface of the proteins is represented by a dashed line. The amino acid sequences depicted are indicated by their one letter code.

important to analyse constant amounts of actin and different concentrations of peptides, ranging from substoichiometric amounts to a large excess over actin.

We typically use the zero-length cross-linker EDC in combination with sulfo-NHS, covalently coupling glutamic acid (in theory also aspartic acid, although rarely observed) and lysine residues and creating an isopeptide bond (*Figure 2*). Obviously, other bifunctional reagents can be applied. As cross-linking efficiency varies from one protein complex to another, one often needs to adjust the cross-linker concentration. Ideally, one needs as little cross-linker as possible, while still maintaining high coupling efficiency. Samples should be in amine-free buffers, such as phosphate or Hepes buffers. Tris should not be used, since it is a reactive amine. *Protocol 2* works for both G- and F-actin. Obvious controls include incubation of the cross-linking reagent with actin (especially relevant when analysing binding to F-actin) or with the peptide alone at the highest concentrations used. The products analysed by SDS–PAGE typically show the covalent complex as a band with a molecular weight that is approximately the sum of the molecular weights of actin and the peptide (*Figure 3A*). The specificity from the peptide mimetic binding can often be assessed by comparing the behaviour of the wild-type and mutant peptides. When working with charge reversal mutants (K to E or vice versa), elimination of a cross-linking site may occur, causing the mutant to show a dramatically reduced binding. In addition, it is useful to quantify the cross-linked product formed using densitometric scans for Coomassie-stained gels or PhosphorImager data with radioactively labelled peptides. When the amount of product is plotted against the concentration of the peptide, a

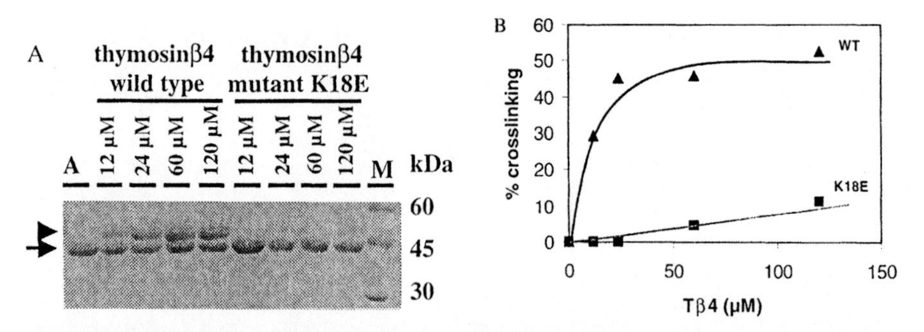

Figure 3. Cross-linking of wild-type and mutant K18E thymosin β4 to actin. (A) Increasing concentrations of thymosin β4 are incubated with a constant amount of 12 μM actin and treated with EDC and sulfo-NHS. In the case of wild-type this results in a covalently coupled complex with an apparent molecular weight of 47 kDa (42 kDa from actin, 5 kDa from thymosin β4); in the case of the mutant there is only marginal cross-linking at 120 μM. M indicates the molecular weight markers. (B) Densitogram of the Coomassie-stained gel in (A) showing the amount of complex formed as a function of the thymosin β4 concentration (WT, filled triangle, thymosin β4; K18E, filled square, mutant thymosin β4 at position 18).

hyperbolic curve is obtained (*Figure 3B*). This analysis is an important quality control for the data. Curves not reaching plateau values are indicative of non-specific reactions or very low affinity reactions.

Protocol 1. The interaction of actin-binding proteins with G- or F-actin using chemical cross-linking (12, 13)[a]

Reagents

- Actin: prepared according to Spudich and Watt (29), as modified by others (e.g. ref. 12)
- G-ATP-Ca-actin in phosphate buffer: 5 mM K-phosphate pH 7.6, 0.2 mM ATP, 0.1 mM $CaCl_2$, 0.2 mM DTT
- Actin-binding protein, also in an amine-free buffer (e.g. phosphate, Hepes); buffer ions containing primary amines (e.g. Tris) quench the cross-linking reaction
- EDC: 20 mg/ml stock solution of 1-ethyl-3-(3-dimethylaminopropyl)carbodiimide (Sigma) in H_2O, freshly dissolved prior to addition to the protein mixture
- Sulfo-NHS: 20 mg/ml stock solution of *N*-hydroxysulfosuccinimide (Pierce) in H_2O, freshly dissolved prior to addition to the protein mixture

Method

1. Depending on analysis of G- or F-actin binding properties of the particular actin-binding protein, start with either G-ATP-Ca-actin in phosphate buffer or pre-polymerized actin in phosphate buffer to which KCl and $MgCl_2$ are added (100 mM and 2 mM final concentrations, respectively). The final actin concentration after addition of all components (after step 2) is typically 12 μM (0.5 mg/ml).
2. Add the actin-binding protein (preferably over a concentration range from substoichiometric to several fold in excess over the actin concentration) and adjust to the end volume with phosphate buffer (G-actin interaction) or phosphate buffer + KCl and $MgCl_2$ (F-actin conditions). Incubate at RT for 20–30 min (longer incubation times at 4°C may be favourable when working at lower total protein concentrations).
3. Dilute the freshly dissolved EDC and sulfo-NHS stock solutions 25-fold in the protein mixtures (final concentration 4 mM). Incubate for an additional 45 min at RT.
4. Quench the cross-linking reaction with an excess of a primary amine (e.g. Tris–HCl buffer) or immediately analyse on SDS–PAGE followed by Coomassie staining.

[a] Complexes of actin with short peptides often cannot be fully separated from actin alone. In this case one can use radioactively labelled peptides and a PhosphorImager to quantify the products.

In a more sophisticated version of this protocol one may assay competing reactions. This can be done by incubating constant amounts of actin and constant amounts of one actin-binding protein (or one actin-binding peptide mimetic) with increasing concentrations of a competitor peptide. In this case, the competitors must be differentiated from the original binding protein by molecular weight or a suitable tag for recognition. If competition occurs, the amount of cross-linked product of the actin-binding partner present in constant amounts will decrease, whereas the cross-linked product from the competitor will increase. If one uses an actin-binding protein and a competing WT peptide derived from it, then this assay also proves the specificity of a peptide mimetic. The fixed concentration of the actin-binding partner should be around the K_d value if using a protein or below the saturating concentration (as determined above) when using an actin-binding peptide mimetic. In addition, one can discriminate between G- and F-actin binding, because this cross-linking assay can be easily followed by a sedimentation assay (*Protocol 2*) prior to analysis by SDS–PAGE.

Protocol 2. Competitive cross-linking of actin-binding proteins to F-actin using a sedimentation assay

Equipment

- Beckman airfuge (allows reaction volumes as low as 25 μl), alternatively TLA-100 (reaction volumes up to 1 ml) and polypropylene tubes

Method

1. Perform the experiment using F-actin as described in *Protocol 1*, except that the filaments are pre-incubated for 30 min with a constant concentration of one of the actin-binding partners prior to executing *Protocol 1*, step 2 with the other actin-binding partner. Controls are actin and each of the two actin-binding components alone, and each partner with F-actin.
2. After executing *Protocol 1*, step 3 centrifuge at 100 000 *g* for 15 min.
3. Carefully remove the supernatant, rinse the F-actin pellet with 20 μl F-buffer, and resuspend it in SDS–PAGE sample loading buffer.
4. Analyse equal proportions of supernatant and pellet on SDS–PAGE followed by Coomassie staining.

A similar approach involves a competitive sedimentation assay, also known as a continuous variation experiment (13, 16) (*Protocol 3*). This assay also uses sedimentation, but is performed without cross-linking. Thus, it is not suited for analysing F-actin-binding peptide mimetics, since F-actin–peptide complexes are sometimes insufficiently stable during sedimentation or become non-specifically trapped in the pellets due to the use of high concentrations. Although more difficult to perform, this assay estimates the ratio of

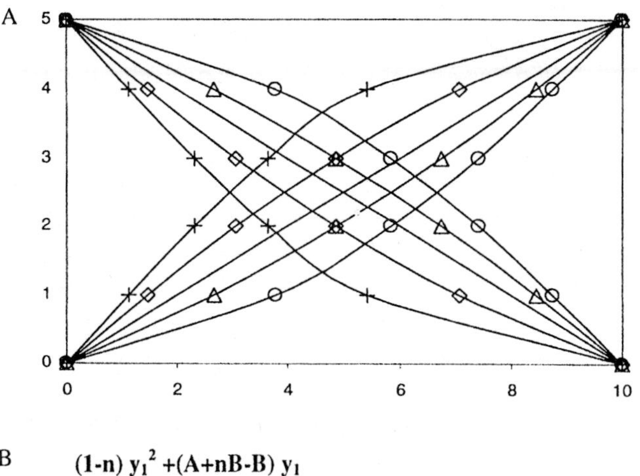

B

$$x = \frac{(1-n)\,y_1^2 + (A+nB-B)\,y_1}{nB-(n-1)y_1}$$

$$x = \frac{(1-n)\,y_2^2 + (nB-B-A)y_2 + AB}{B+(n-1)y_2}$$

A = [Ligand$_1$] +[Ligand$_2$] = [Ligand$_{tot}$]
B = [Actin-Ligand$_1$] + [Actin-Ligand$_2$] = [F-actin-binding sites]
y_1 = [Actin-Ligand$_1$]
y_2 = [Actin-Ligand$_2$]
x = [Ligand$_1$]
$n = K_{d2}/K_{d1}$

Figure 4. (A) Hypothetical binding curves in continuous variation experiments of two competing proteins for different n values (the ratio of the two K_d values). Shown are n = 0.2 (o), 0.4 (Δ), 1 (solid lines), 2 (◇), and 10 (+). The amounts of bound protein 1 (increasing Y values) or competing protein 2 (decreasing Y values) are plotted versus the total concentration of protein 1. (B) Equations used to generate the upper and lower curve.

the K_d of two competing actin-binding proteins. The concentration of both actin-binding partners is varied inversely, but the total concentration of actin-binding units is kept constant. After analysis and quantification, the graphical presentation yields two curves (*Figure 4*) (examples of such analyses can be found in refs 13 and16). Note that if the K_ds are equal, the graphs are straight lines intersecting at the mid-point, if a symmetrical series is used.

Protocol 3. Competition between two F-actin-binding proteins using co-sedimentation with a continuous variation protocol (13, 16)

Equipment and reagents

- G-ATP-Ca-actin in G-buffer
- Airfuge (see *Protocol 2*)

Method

1. Prepare F-actin and perform two separate co-sedimentation experiments (without cross-linking) with varying concentrations of each of the two actin-binding proteins, as described in *Protocol 2*, steps 2–4. Use the same actin concentration for both and determine for each ligand the concentration that is needed to saturate the actin filaments, using densitometry of Coomassie-stained SDS–PAGE gels.
2. Make a series of samples of polymerized actin with the same actin concentration as in step 1 in F-buffer, but varying the concentrations of the two ligands: decreasing concentrations throughout the series for the first ligand and increasing for the second. This is done in such a way that:
   (a) Their highest concentrations added are sufficient to saturate the actin filaments (based on experiments in step 1).
   (b) The sum of their concentrations is constant.

   Preferably make a symmetrical series (e.g. vary from 10 μM of ligand 1 and no ligand 2 to the reverse: no ligand 1 and 10 μM of ligand 2). Incubate these samples at RT.
3. Spin at 100 000 *g* for 15 min and proceed as described in *Protocol 2*, steps 3 and 4. Analyse the resuspended pellets by densitometry of Coomassie-stained gels for quantification for each combination of concentrations, i.e. the amounts of ligand 1 and 2 that bind to F-actin.
4. Bring together in one graph:
   (a) [ligand 1–actin complex] versus $[\text{ligand 1}]_{total}$ from experiment in step 1.
   (b) [ligand 1–actin complex] versus $[\text{ligand 1}]_{total}$ from experiment in step 2.
   (c) [ligand 2–actin complex] versus $[\text{ligand 2}]_{total}$ from experiment in step 1.
   (d) [ligand 2–actin complex] versus $[\text{ligand 2}]_{total}$ from experiment in step 2.

4. Creation of mutants using recombinant technology

As more and more actin-binding proteins are produced by recombinant methods, site-directed mutagenesis has become an attractive tool for altering the activity of actin-binding proteins. Since the procedure for the creation and production of such proteins is more complicated than chemical peptide synthesis, one must be more astute in choosing and designing the mutants, unless one adopts the phage display strategy that allows screening of libraries (described below).

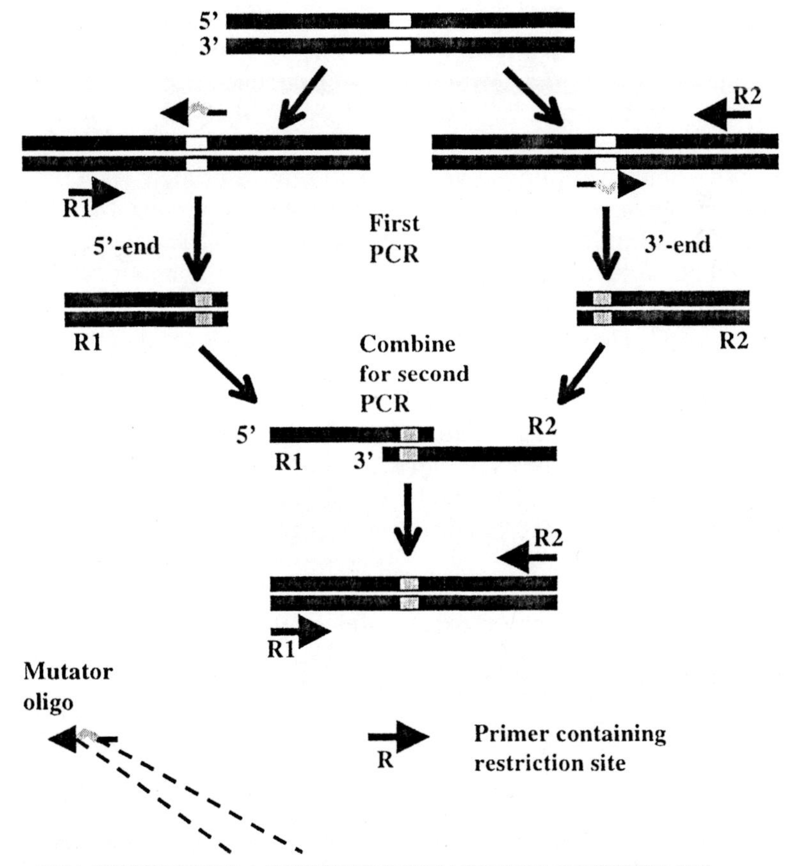

Figure 5. Polymerase chain reaction by splice overlap extension. The two DNA strands of the gene are shown as black bars with the position of the site to be mutagenized in white. In a first PCR, in two separate reactions, the 5′ and 3′ region are amplified with an oligo containing a restriction site (arrows with R1 or R2) and an oligo containing mutated codons (arrow with grey). The amplified products are mixed, and a new PCR reaction is initiated yielding the desired mutant. A hypothetical mutator oligo is shown below; N is any mixture of nucleotides. Flanking regions need to have sufficient length.

4.1 Mutagenesis using the polymerase chain reaction

Most mutagenesis experiments presently employ the polymerase chain reaction. Outlined below is a simple protocol that works very well and is versatile for creating site-directed mutants in actin-binding domains (or proteins). It is based on the splice overlap primer extension method (17). Recent PCR-based commercial systems, such as the pAlter (Promega) or Quick-change (Stratagene) system, work as well, but the method described in *Protocol 4* offers the advantage that new start and stop codons can be added simultaneously (important when one wants to analyse a single domain of a multidomain actin-binding protein). The reaction requires four different primers. Two primers are complementary to the portion of the gene into which one wants to insert a new start and stop codon (these will be referred to as, respectively, the forward and reverse primers). Both of these primers contain a different restriction site suitable for cloning into an expression vector. The other two primers are the mutagenesis primers (mutator oligo) and are complementary to the site at which one wants to create the mutation(s), except for the desired mutant codon, which should be positioned in the middle of the primer (see *Figure 5*). If one uses a mutator oligo with a degenerate codon, one may obtain multiple mutants at one position in a single PCR reaction (18) (Rossenu *et al.*, submitted). Also, two to six adjacent codons may be altered in a single reaction, which is useful when creating mutant libraries (Rossenu *et al.*, unpublished data). Two separate PCR reactions are required for creation of the 5′ and 3′ ends of the gene with the mutation (see *Figure 5*). After amplification, both reaction products are combined, forward and reverse primers are again added, and a new PCR reaction is initiated. The final product is purified and cloned into an expression vector.

Protocol 4. PCR mutagenesis by splice overlap extension (17)

Equipment and reagents

- Cycling heating block
- Wizard™ minipreps (Promega): plasmid DNA containing the wild-type gene is isolated using a Wizard™ miniprep and used as template
- Four different primers: a forward and reverse primer (each containing a different restriction site and a start and stop codon) and two mutator primers[a] (see *Figure 5*)
- Agarose gel electrophoresis tank
- Gene clean DNA purification kit (Bio 101, La Jolla)
- PCR pre-mix: 2 mM $MgCl_2$, PCR buffer, 20 mM Tris–HCl pH 8.4, 50 mM KCl, and 2.5 mM of each dNTP mix in H_2O (Molecular Bio-products)
- *Taq* polymerase (Goldstar, Eurogentec) or *Pfu* polymerase (Stratagene)

Method

1. Mix, in two separate reactions (one for the 5′ and one for the 3′ end of the gene) 10–20 ng of template DNA, with the PCR pre-mix and two primers (0.1- to 0.5-fold molar excess), the forward primer and the

reverse mutator primer in one reaction, the forward mutator and reverse primer in the other. The total volume of each reaction is 100 μl (approx. 90 μl pre-mix).

2. Place on a cycling heating block for 2 min at 94 °C.
3. Decrease the temperature to 80 °C for 2 min and add 2.5 U polymerase.
4. Amplify fragments using 25 cycles of 94 °C for 30 sec, 52 °C for 30 sec, and 72 °C for 30 sec, followed by 72 °C for 10 min (for *Pfu* longer extension times are required).
5. Mix 1 μl of each of the amplified DNA fragments, add 90 μl of PCR pre-mix, and fresh forward and reverse primers.
6. Amplify the fragment using steps 2–4.
7. Purify the amplified fragment on an agarose gel, excise the two bands, and use Gene clean to extract the DNA.
8. Perform a digest on the restriction sites, incorporated via the forward and reverse primers in the PCR product, and ligate into a suitable vector.

[a] Using degenerate mutator oligos at one or more codons (up to six adjacent ones), one can create libraries of mutants. Non-mutated 5′ and 3′ flanking sequences are usually 15–18 nucleotides long, depending on the GC content.

4.2 A few considerations when working with recombinant proteins

Several commercial vectors are available for cloning and expression of recombinant proteins in *E. coli*. However, in general, *E. coli* does not perform post-translational modifications on the recombinant proteins. As lack of these modifications may influence actin binding activity an initial careful comparison between eukaryotic and recombinant wild-type is useful. One may also opt to use different sequence tags to allow easy purification. However for quantitative biochemistry one should keep in mind that the tag sequence may modulate or interfere with the activity of the protein. We also advise including a control for the structural integrity of the mutants. Indeed, for some mutants of profilin, actophorin, and gelsolin-S2, we observed decreased stability, often characterized by aberrant elution on gel filtration columns, increased aggregation, a changed urea denaturation profile, increased susceptibility to proteases, or a change of secondary structure by circular dichroism analysis.

4.3 Phage display of an actin-binding protein

We have designed a novel way to analyse the actin binding capacity of mutants based on a combination of phage display and ELISA. We chose the M13 phage display system (commercially available from Pharmacia), designed to be monovalent, to present human thymosin β4 mutants on the surface of the phage (18) (*Protocol 5*). The first step is the creation of the library using the PCR protocol described above. If not present, *Sfi*I and *Not*I restriction sites are added using the forward and reverse primers, respectively. The library of genes is inserted into the M13-derived phagemid: pCANTAB5E, which contains in-frame the coding information for:

- a signal sequence
- the (actin-binding) mutant protein
- the E-tag sequence recognized by an antibody
- the M13 gIII protein

The phagemid is propagated in *E. coli*. Addition of a helper phage produces recombinant phages. The important point here is that a phage displaying a particular protein or mutant on its surface contains the genetic information for that particular protein inside. The system works very well with thymosin β4, and we have also displayed the villin headpiece (18, and unpublished data). Having a library of mutant plasmids in *E. coli*, one may take one of two approaches, depending on the size of the library (*Figure 6*). Either one can

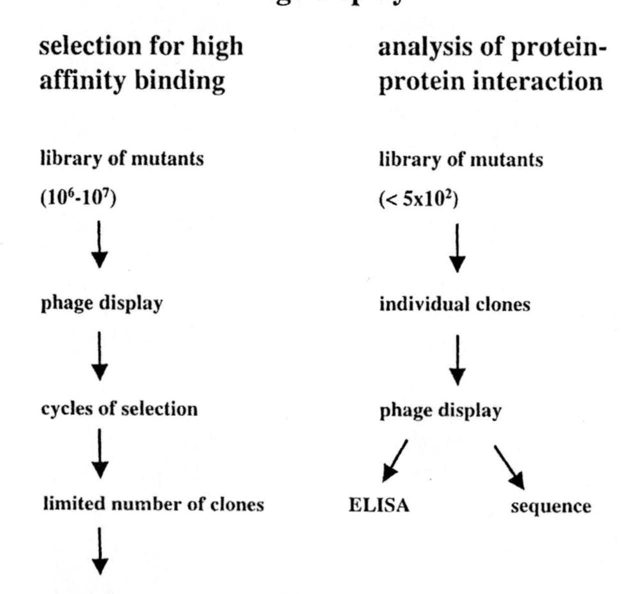

Figure 6. Different strategies for phage display of proteins. (*Left*) Classical phage display for the selection of high affinity interactions starting from a large library. (*Right*) Use of phage display for systematic analysis of a protein–protein interaction. For details see *Protocol 6*; for the ELISA see *Protocol 7* and *Figure 7*.

clone, pick individual colonies, and produce selected phages or one can create a population of phages immediately for selection purposes.

Protocol 5. Phage display (18) (Rossenu *et al.*, unpublished data)

Equipment and reagents

- Gene clean DNA purification kit (Bio 101, La Jolla) and Wizard™ minipreps (Promega)
- Bio-Rad gene pulser and electroporation cuvettes (or equivalent)
- Benchtop centrifuge with variable speed
- Electroporation-competent *E. coli* TG1 cells
- Ampicillin (200 mg/ml in H_2O), kanamycin (25 mg/ml in H_2O), 2 M glucose (all filter sterilized)
- SOC medium: the same as SOBAG medium but without ampicillin and with 250 mM KCl and only 27.8 ml of 2 M glucose
- M13KO7 helper phage (Pharmacia)
- SOBAG medium: to 20 g Bacto tryptone, 5 g Bacto yeast extract, 0.5 g NaCl, add distilled water to approx. 900 ml and autoclave. Cool to 50–60°C, add the following: 10 ml of sterile 1 M $MgCl_2$, 55.6 ml of sterile 2 M glucose, and 5 ml of a sterile 20 mg/ml ampicillin solution. For plates add 15 g Bacto agar prior to autoclaving.
- 2 × YT media: 17 g tryptone, 10 g yeast extract, 5 g NaCl per litre, adjust to pH 7.0 with NaOH, and sterilize by autoclaving

Method

1. Create a library of mutants as described in *Protocol 4* using degenerate mutator oligos, a forward primer containing an *Sfi*I restriction site, and a reverse primer containing a *Not*I site such that the coding region of the gene will be in-frame with the coding sequences for the signal peptide and the gIII protein when ligated into the vector. The PCR product is ligated in *Sfi*I and *Not*I restricted phagemid pCANTAB5E (Pharmacia).
2. Electroporate 100 ng of salt-free ligation mixture in 40 μl electroporation-competent TG1 cells. Dilute immediately in 1 ml SOC medium.
3. Grow cells for 30 min at 37 °C.
4. Plate cells on SOBAG plates and incubate overnight at 30 °C.
5. Pick as many individual transformants as one can subsequently handle and grow in 1 ml SOBAG medium.
6. Use part of the culture to isolate the DNA with Wizard™ minipreps for DNA sequencing, the other part for rescue. The recombinant phages are produced by superinfection of the TG1 cells containing the phagemid with M13KO7 helper phage.
7. Transfer 10 μl of the culture to 990 μl of 2 × YT medium containing 100 μg/ml of ampicillin and 62.5 μl of 2 M glucose. Incubate at 37 °C with shaking at 250 r.p.m. for 3 h.
8. Add 2.5×10^9 p.f.u. of M13KO7 to the cell suspension.
9. Incubate for 30 min at 37 °C with shaking at 150 r.p.m., followed by 30 min with shaking at 250 r.p.m.

Protocol 5. *Continued*

10. Transfer to Eppendorf tubes and centrifuge for 10 min at 800 *g* and carefully remove the supernatant.
11. Resuspend the pellet in 1 ml of 2 × YT medium containing 100 μg/ml of ampicillin and 50 μg/ml of kanamycin.
12. Transfer to a culture tube containing 4 ml of 2 × YT medium with 100 μg/ml of ampicillin and 50 μg/ml of kanamycin and incubate overnight at 37 °C.
13. Spin at 1000 *g* for 10 min.
14. Filter the supernatant (containing the recombinant phages) through a 0.45 μm filter and transfer to a sterile culture tube. Store at 4 °C or use in the ELISA (*Protocol 6*).

4.3.1 Selection of stronger binding variants

If the library is large, one may select for stronger ligand-binding mutants. Given the diversity of proteins, no general detailed protocol is available that will consistently be successful. Here we provide a few guidelines. A selection cycle consists of one incubation, several washes, and an elution. Ideally, one incubates the selector ligand (on a solid support, e.g. magnetic beads, a plastic tube) in amounts substoichiometric to the total number of recombinant phages in the population. It is important to select incubation and washing conditions under which both the ligand and the displayed protein are stable. The number of washes and the stringency depends on the K_d of the interaction and on the k_{off}. For interactions with a natural high affinity, selection in combination with free WT ligand may be included. Elution of the retained phages requires more stringent conditions; however, these need to be chosen so that the selected phages remain infective. Four to six selection cycles are usually required. Knowledge of the input number of phages for each cycle enables calculation of the enrichment, often 10- to 30-fold in the first three cycles and then decreasing.

4.3.2 Systematic analysis of libraries

If one has small libraries (< 400 mutants), one may systematically analyse the interaction of several isolated phages displaying a particular mutant. Then no selection is applied, rather, mutants are purified by cloning. Clones are grown, and part is used to isolate the phagemid to sequence the relevant portion of the mutant gene. The remainder is used for rescue to produce recombinant phages that are also assayed for their capability to interact with the ligand in a modified ELISA. We have used *Protocols 5* and *6* successfully to analyse over 100 mutants of thymosin β4 for binding to actin (Rossenu *et al.*, submitted). Although one can coat actin directly to the ELISA plates in good yield, only a small fraction remains functional. We have good experience with coating

A

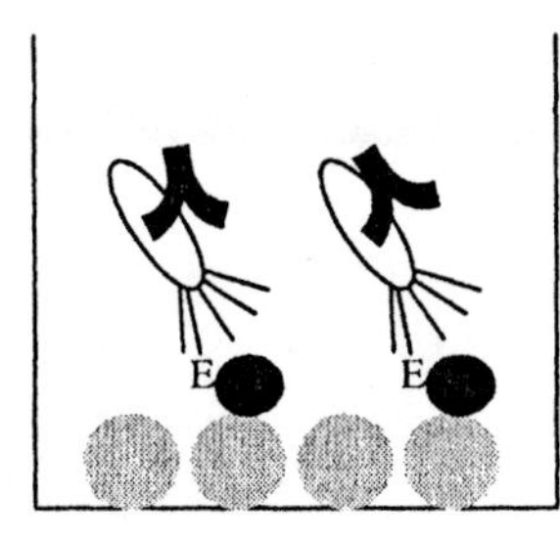

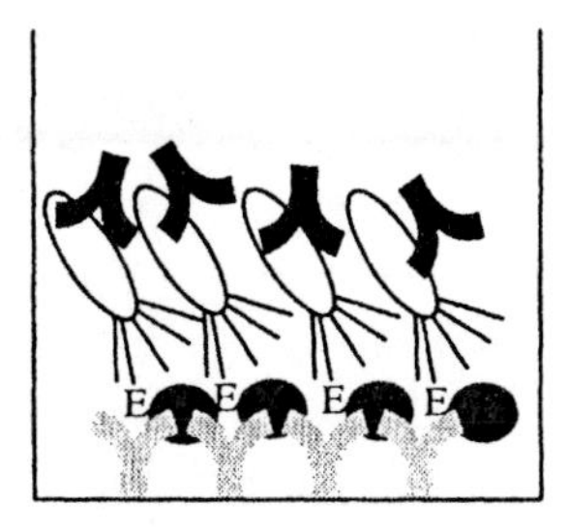

absorption equivalent to protein-protein complex concentration

A

absorption equivalent to total protein concentration

$\mathbf{A_0}$

B $$Kd/[Act_{tot}])+1= A_0/A$$

$$\text{Rel. affinity} = A_{mut}/A_{0mut} \times A_{0WT}/A_{WT}$$

Figure 7. (A) Schematic of the ELISA used for analysis of a phage displayed actin-binding protein. In the well on the left, coated actin (grey circles) (or neutravidin with biotinylated actin) will interact with the actin-binding protein (black circles) displayed on the surface of the recombinant phages. In the well on the right the coated anti-E-tag antibody (**Y**) recognizes the E-tag sequence with high affinity. In both cases the amount of phages can be measured using the same anti-M13 antibody (**Y**) conjugated to an enzyme. The absorbance (A or A_0) can be used in the expressions shown below. (B) The upper expression shows there is a linear correlation between the K_d and the ratio of the absorbances. The lower expression allows assessing a relative binding strength.

actin, biotinylated either on Cys374 or Gln41, using iodoacetyl-LC-biotin or cadaverine-biotin (*Protocol 7*) on neutravidin plates. As these actin residues are on opposite faces of the protein molecule, chances are high that at least one of the modifications will not interfere with the actin-binding protein–actin interaction under investigation.

In parallel with actin, the E-tag antibody is coated in different wells (*Figure 7*). The recombinant proteins carry not only the mutant protein but also an E-tag sequence COOH-terminal from it. This epitope is recognized with very high affinity by the E-tag antibody. Both the interaction with a particular thymosin β4 mutant and the total concentration of recombinant phages present (via the E-tag) are measured in one ELISA. Thus, with one detection module (*Protocol 6*), one may determine relative affinities in this system. Indeed, the ratio of the absorption proportional to the amount of displayed protein (measured via the E-tag) over the absorption proportional to the amount of ligand–target protein complex exhibits a linear correlation with the K_d (Rossenu *et al.*, submitted). One should keep in mind that this is a solid phase assay. The advantage of this system lies in the number of mutants one can analyse in a relatively short time. In addition, the system yields data on the relative importance and the tolerance at each of the mutated sites.

Protocol 6. ELISA-based assay for testing binding of recombinant phages (Rossenu *et al.*, unpublished data)

Equipment and reagents

- 96-well microtitre plates (Nunc, Immunosorp)
- ELISA plate reader
- PBS buffer: 10 mM phosphate buffer pH 7.4, 150 mM NaCl
- PBS/Tween: PBS buffer with 0.05% Tween 20
- Blocking buffer: PBS with 2% skimmed milk powder
- ABTS substrate: add 100 mg ABTS to 450 ml of 0.05 M citric acid pH 4.0, filter sterilize, and store at 4°C
- RPAS detection module including positive control phage, positive control antigen, ABTS, and an anti-M13 IgG conjugated to horseradish peroxidase (Pharmacia)
- Immunopure neutravidin (Pierce)
- E-tag antibody (Pharmacia)

Method

1. Coat microtitre plates with 400 ng/100 μl Immunopure neutravidin or 1 μg/100 μl E-tag antibody in PBS at 4°C overnight. Coat for each as many wells as one has different recombinant phages plus negative and positive controls.[a]
2. Block excess binding sites with 200 μl blocking buffer.
3. Add biotinylated target protein (see *Protocol 7* for biotinylation of actin) to the neutravidin-coated wells for 1 h at RT.
4. Wash microtitre plate three times with PBS/Tween.
5. Add 100 μl/well of each of the recombinant phage (when analysing high affinity interactions dilute phage 1 in 10 or 1 in 100 in PBS), including the phage carrying the wild-type protein, in blocking buffer to the neutravidin-coated wells or 1:100 dilution of phage to the E-tag antibody-coated wells. Incubate for 2 h at RT.
6. Wash the plate four times with PBS/Tween.
7. Add anti-M13 (IgG diluted 1:2500), conjugated with horseradish peroxidase, for 1 h at RT.
8. Wash the wells three times with PBS/Tween.
9. Add 200 μl ABTS substrate to the wells and incubate at RT.
10. Measure the absorbance at 405 nm.

11. Correct for blanks and calculate the relative affinity using the expression (see *Figure 7*):

$$Abs_{mut} / Abs_{mutE\text{-}tag} \times Abs_{WTE\text{-}tag} / Absw_{WT}$$

[a] The positive control is a coated control antigen incubated with the control phage from the RPAS-module. Negative controls include neutravidin and the E-tag antibody with subsequent incubation of recombinant phage, and neutravidin plus biotinylated actin without incubation with phage.

Protocol 7. Biotinylation of actin

Equipment and reagents

- NAP10 desalting column (Pharmacia)
- Transglutaminase: 2 U in 200 μl water (Sigma)
- Iodo-acetyl-LC-biotin (Pierce)
- Biotin-cadaverine: 2 mg in 100 μl 50% DMF (Sigma)

A. *Biotinylation on Cys374*

1. Apply 1 ml of actin (2 mg/ml) to a NAP10 column previously equilibrated in G-buffer (see *Protocol 1*) without DTT.
2. Elute with successive 1 ml aliquots of G-buffer without DTT.
3. Add to the second ml of the eluate, containing the actin, a fivefold molar excess of iodo-acetyl-LC-biotin.
4. Incubate 2 h at RT. Dialyse four times against 1 litre of G-buffer containing DTT at 4°C.

B. *Biotinylation on Gln41*

1. Apply 1 ml of actin (2 mg/ml) to a NAP10 column previously equilibrated in transglutaminase buffer (5 mM Tris, 1 mM $CaCl_2$, 1 mM DTT, 0.4 mM ATP pH 8.0).
2. Elute with successive 1 ml aliquots of this buffer.
3. Add to the second ml of the eluate, containing the actin, 75 μl of enzyme and 75 μl of biotin-cadaverine.
4. Incubate overnight at 4°C.
5. Add 15 μl of 1 M KCl, 3 μl of 1 M $MgCl_2$, and 15 μl of 100 mM EGTA. Allow actin to polymerize at RT for 2 h.
6. Pellet the filaments in an ultracentrifuge at 80 000 *g* at 4°C for 2 h.
7. Remove supernatant and wash pellet twice with F-buffer (*Protocol 1*) containing 10% DMF.
8. Resuspend the pellet in regular G- or F-buffer, as desired, and dialyse twice overnight against this buffer.

5. Analysis of actin-binding proteins for PIP_2 interaction

PIP_2 is a ligand for many actin-binding proteins, modulating their activities in a positive or negative manner with respect to actin binding. For those cases where inhibition is observed, the PIP_2-binding site is thought to partly overlap with the actin-binding site. Therefore, it is useful to analyse proteins mutated in their actin-binding sites, especially those with a charge reversal, for PIP_2 binding. We present two simple assays for PIP_2 binding using micelles, a model for interaction with phosphoinositides in membranes that have a more complex composition. The information one obtains is a relative binding strength. In both assays the amount of actin-binding protein is kept constant and the amount of PIP_2 varies. A control without PIP_2 is included.

The microfiltration assay (*Protocol 8*), which consumes more PIP_2, takes only about half a day, and all samples can be treated simultaneously. The assay consists of centrifuging a relatively small actin-binding protein in the presence of PIP_2–micelles through a membrane with a cut-off larger than the molecular weight of the proteins and smaller than the molecular weight of the PIP_2–micelle (with associated protein) (8, 19). The fraction of the protein passing through the filter is analysed by SDS–PAGE and represents that not associated with the micelle (see *Figure 1* in ref. 8, and *Figure 5* in ref. 19).

Protocol 8. Analysis of PIP_2 binding using microfiltration (8, 19)

Equipment and reagents

- Millipore Ultrafree-MC microfiltration units with a suitable cut-off (e.g. 30 000 Da for profilin, cofilin, and actophorin)
- Aliquots of a stock of 1 mM of PIP_2 (Sigma) in distilled water: sonicate for 30 min and store at –80°C

Method

1. Prior to use sonicate the PIP_2–micelles stock for 5 min.
2. Incubate a fixed amount of protein (final concentration usually around 5 μM) with an increasing amount of PIP_2 to several fold excess in 150 μl of an appropriate binding buffer (e.g. 75 mM KCl, 10 mM Tris–HCl pH 7.5, 0.5 mM DTT) for 30 min on ice.[a,b]
3. Load the samples into the microfiltration units and spin for 1 min at 2000 *g*.
4. Analyse the flow-through on SDS–PAGE.
5. Quantify by densitometry.

[a] One usually requires up to a 100-fold excess of PIP_2 to have the protein completely micelle bound.
[b] Some buffer components, such as Ca^{2+}, precipitate with PIP_2. Prior to analysis of the protein always check buffer compatibility with the highest concentration of PIP_2 used. In some rare cases mutants also precipitate.

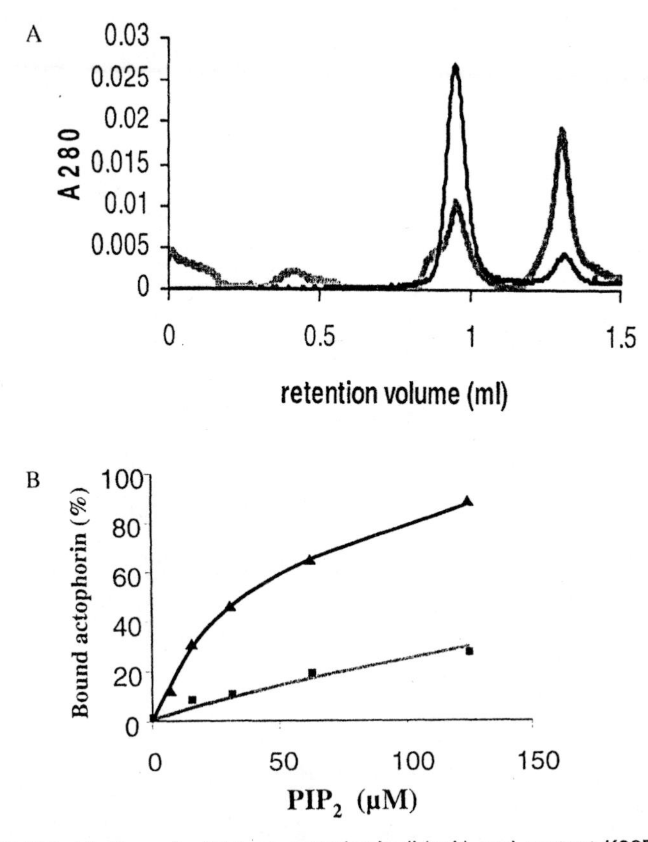

Figure 8. (A) PIP_2 binding of wild-type actophorin (black) and mutant K92E (grey). Gel filtration on Superdex 75 of 5 μM actophorin with 125 μM micellar PIP_2. The absorption is monitored at 280 nm. The peak of free actophorin elutes at 1.3 ml, the bound form is shifted to the exclusion volume (0.95 ml). Note that at similar concentrations the mutant hardly binds. (B) Plot of amount of bound wild-type actophorin (filled triangle, black) or mutant K92E (filled square, grey) as a function of the PIP_2 concentrations (M. Van Troys, D. Dewitte, J. Vandekerckhove, and C. Ampe, unpublished data).

An alternative method involves gel filtration (*Protocol 9*), which requires less PIP_2 and is suitable for larger proteins, but is more time-consuming (typically six or seven consecutive runs). After incubation of the protein with the PIP_2–micelles, the complexes are passed over a suitable gel filtration column. The free protein is separated from the PIP_2-bound (often in the void volume) form, and the elution is monitored by the UV absorption at 280 nm. For quantification the peak height of the free form can be integrated, and fractions can also be analysed by SDS–PAGE. With increasing concentrations of PIP_2 the peak intensity of the free form will decrease. One usually calculates the amount of bound form from this decrease (*Figure 8*).

Protocol 9. Analysis of PIP_2 binding using gel filtration

Equipment

- A SMART system with a Superdex 75 or a Superdex 200 (Pharmacia) (or equivalent systems), depending on the size of the protein

Method

1. Equilibrate the column in the binding buffer.
2. Repeat *Protocol 8*, steps 1 and 2 for one sample at a time.[a]
3. Load the sample onto the SMART system and elute isocratically, monitor at 280 nm.
4. Integrate peak height and/or analyse samples on SDS–PAGE. Calculate the amount of the bound form from the decrease of the free form.[b]
5. Plot amount of bound form versus the total PIP_2 concentration.

[a] One usually requires less PIP_2 compared to the amounts in *Protocol 8* to shift the protein completely to the bound form.
[b] The increase in peak height for the bound form is due both to an increase in absorption and a change in the refractive index. Therefore, it is not appropriate for quantitation.

6. Post-translational modifications and identification of phosphorylation sites using mass spectrometry

Most eukaryote proteins are post-translationally modified. Removal of the initiator methionine and subsequent acetylation of the new amino-terminal acid is very common. Actin and some actin-binding proteins from *Acanthamoeba* (profilins and actobindin) are methylated on histidine and lysine, respectively (20–22). Actin is also ADP-ribosylated by toxins (23). The physiological significance of these modifications is often not clear, but they may affect the function of the protein. Since recombinant forms expressed in *E. coli* usually lack these modifications, they may have an altered or reduced activity. In addition, the instability of the protein may be enhanced. Whereas acetylation and methylation are stable modifications, phosphorylation is transient. Numerous actin-binding proteins are phosphorylated, including actin in some species. A well-characterized example is phosphorylation on threonine 203 of *Physarum* actin by the actin fragmin kinase (24, 25). Phosphorylation of this residue interferes with association of other actin molecules and with DNase I (24).

One can identify the type and position of a stable modification in a peptide by combining Edman degradation and mass spectrometry, especially when the gene sequence is known. Indeed, PTH-amino acids have a characteristic elution time in reversed-phase systems. A shift in the retention time of a

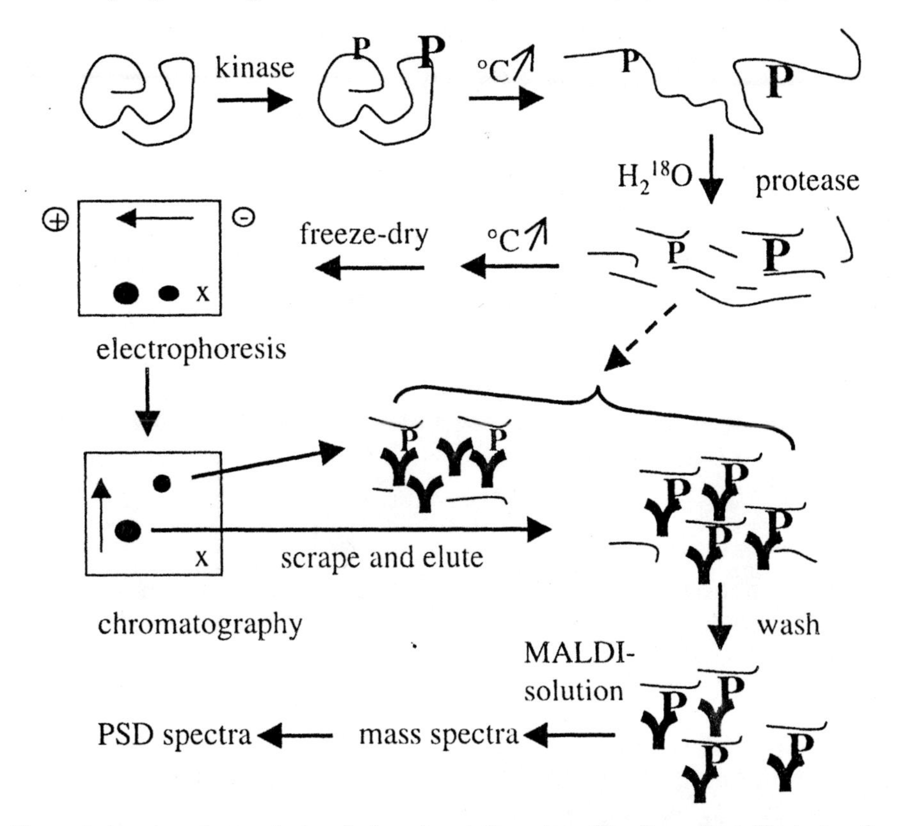

Figure 9. Strategy for analysis of phosphorylation sites. The flow chart illustrates the different steps of *Protocol 10*. The protein is phosphorylated by a kinase on a minor (small **P**) and a major site (large **P**). The running direction during thin-layer electrophoresis and chromatography are indicated by *arrows*, x denotes the site of application. The dashed *arrow* indicates a short cut, but with loss of quantitative information as major and minor phosphopeptides, recognized by the anti-phosphotyrosine antibodies (Y), are present in the same sample.

particular PTH-derivative (compared to that predicted from the gene sequence) suggests that this amino acid is modified. Subtracting the theoretically calculated mass from the measured mass identifies or yields clues on the type of modification. However, identification of phosphorylation site(s) is not always so straightforward, often because the phosphoform is a minor species in the population and is unstable. Recently, our laboratory successfully applied a combination of two-dimensional thin-layer chromatography, affinity precipitation with anti-pTyr-antibody, and mass spectrometry to identify major and minor phosphorylation sites in gelsolin (26) (*Protocol 10*). The method is relatively simple and fast, but requires access to a MALDI-TOF instrument capable of generating PSD spectra (*Figure 9*). Currently, this analysis is applicable only to phosphotyrosine-containing proteins, since reliable anti-phosphothreonine or phosphoserine antibodies are not yet available.

Protocol 10. Identification of tyrosine phosphorylation sites (25, 30, 31)

One should have some estimate to what extent the protein is phosphorylated (e.g. 2%, 60%, 150%, etc.) and the number of expected major and minor sites (determined in an initial two-dimensional electrophoresis/chromatography experiment, predicted from consensus sequences, information on related proteins, etc.).

Equipment and reagents

- Cellulose thin-layer plates, 20 × 20 cm (Merck)
- Multiphor II (Pharmacia) and a chromatography tank
- MALDI-TOF mass spectrometer with reflectron (Reflex III, Bruker-Franzen Analytic Gmbh, Bremen) or equivalent
- Anti-phosphotyrosine–agarose (clone PT66) (Sigma)
- A MALDI-matrix stock solution: 40 mg of α-cyano-4-hydroxycinnamic acid and 8 mg of 2,5-dihydroxybenzoic acid (both from Sigma) in 1 ml of 0.1% trifluoroacetic acid, 50% acetonitrile
- $H_2{}^{18}O$ (Arc Laboratories, Netherlands)
- Proteases with narrow specificity: trypsin, EndoLys-C, EndoAsp-N, etc., sequencing grade (Boehringer)

Method

1. Phosphorylate 100–500 pmol of protein using the appropriate kinase conditions.[a]
2. Inactivate the kinase by boiling for 3 min, and cool the sample on ice.
3. Add protease (1:40, w/w) and 50% $H_2{}^{18}O$, appropriate 10 × protease buffer, if necessary, and incubate at 37 °C for 4 h.[b]
4. Inactivate the protease by boiling for 5 min with or without addition of inhibitors.
5. Lyophilize the sample and dissolve in a minimal amount of acetic acid:pyridine:water (50:5:945, by vol.) pH 3.5 and apply to cellulose plate.[c]
6. Perform horizontal electrophoresis in one dimension at 400 V for 3 h and allow the plate to dry.
7. Perform chromatography in second dimension (perpendicular to the first) with 1-butanol:pyridine:acetic acid:water (75:50:15:60, by vol.); allow the plate to dry.
8. Visualize peptides by autoradiography.[d]
9. Scrape radioactive spots from the plate and put the powder in an Eppendorf tube. Elute with 20 mM Tris–HCl pH 7.5, by continuous shaking at RT for 30 min. Centrifuge at 12 000 r.p.m. for 5 min.

10. Incubate the supernatant with 5 μl of anti-phosphotyrosine–agarose suspension with constant end-over-end rotation at 4 °C overnight; centrifuge briefly at 2200 r.p.m. Wash the beads with 1 ml of 200 mM NaCl, 10 mM Tris–HCl pH 7.5. Repeat this four times. Wash twice with 1 ml distilled water.
11. Dilute the MALDI-matrix stock fivefold in 0.1% trifluoroacetic acid in 50% acetonitrile, add 2 μl to the beads, and mix. Transfer 0.5 μl of the suspension to the MALDI-target and air dry.
12. Record mass spectra in the reflectron mode; optimal running parameters are different for each sample. For details see ref. 25.

[a] The amount of protein to be taken depends on the amount of phosphate incorporated.
[b] The use of $H_2{}^{18}O$ is optional but greatly simplifies interpretation of PSD spectra. Only the peptide fragments containing the COOH-terminal amino acids contain the heavy isotope, giving rise to a doublet with a mass difference of 2 Da (for some enzymes also a triplet with 4 Da difference).
[c] The thin-layer step is optional, as the immunoadsorption will also work directly on the digest. However one loses quantitative information, i.e. which phosphopeptide is major and which minor.
[d] Smearing of the radioactivity indicates that the phosphopeptide has limited solubility under the electrophoresis or chromatography conditions used. One may then opt to digest the protein with a protease with different specificity or a combination of two proteases.

7. Conclusions

Mutagenesis has proven to be a powerful tool to study protein–protein or protein–ligand interactions. Although the use of chemically synthesized peptides and derived mutants has limitations, it is an attractive technique, primarily for a fast initial screening of which region to mutate. In this respect, multiple peptide synthesis (such as SPOT synthesis) (27) may become important in the future. Peptides will remain useful for those proteins that are difficult to produce as recombinants in *E. coli*. In these cases the use of baculovirus expression systems also may offer a solution, but this approach requires a more specialized set-up. Obtaining purified functional recombinant proteins is often tedious, but since the scaffold of site-directed mutants is often largely preserved, this method is advantageous over the use of peptides, which sometimes adopt a functional structure only upon binding actin. In addition, by shuttling the genes to an appropriate vector, the recombinant actin-binding proteins may be transiently or stably expressed in eukaryotic cells. Using phage display, one may facilitate purifying binding proteins and, for this reason alone, this technique merits attention. One can easily analyse the interaction of many mutants in a reasonable amount of time. Quantitative biochemistry remains difficult because the concentration of proteins is usually below nM.

If the three-dimensional structure of the actin-binding protein is known, modelling experiments can be helpful in predicting which residues contact actin. It is rather exceptional that solving the crystal or NMR structure precedes detailed biochemical analysis of a protein. However, modelling techniques capable of predicting folds using solved structural modules as templates will become increasingly important in the future. This should be particularly true for actin-binding proteins, since many of these share similar structural modules despite having low sequence similarity and different functions (2, 28).

Acknowledgements

C. A. is a research associate of the Fund for Scientific Research—Flanders (FWO) (Belgium).

References

1. Matsudaira, P. (1992). In *The cytoskeleton: a practical approach* (ed. K. L. Carraway and C. A. C. Carraway), pp. 73–98. Oxford University Press/IRL.
2. Van Troys, M., Vandekerckhove, J., and Ampe, C. (1999). *Biochim. Biophys. Acta*, **1448**, 323.
3. Kyte, J. and Doolittle, R. F. (1982). *J. Mol. Biol.*, **157**, 105.
4. Prekeris, R., Hernandez, R. M., Mayhew, M. W., White, M. K., and Terrian, D. M. (1998). *J. Biol. Chem.*, **273**, 26790.
5. Schutt, C., Myslik, J., Rozycki, M., Goonesekere, N., and Lindberg, U. (1993). *Nature*, **365**, 810.
6. Sohn, R., Chen, J., Kobland, K., Bray, P., and Goldschmidt-Clermont, P. (1995). *J. Biol. Chem.*, **270**, 21114.
7. Lambrechts, A., Van Damme, J., Goethals, M., Vandekerckhove, J., and Ampe, C. (1995). *Eur. J. Biochem.*, **230**, 281.
8. Lambrechts, A., Verschelde, J.-L., Jonckheere, V., Goethals, M., Vandekerckhove, J., and Ampe, C. (1997). *EMBO J.*, **16**, 484.
9. Fedorov, A. A., Magnus, K. A., Graupe, M. H., Lattman, E. E., Pollard, T. D., and Almo, S. C. (1994). *Proc. Natl. Acad. Sci. USA*, **91**, 8636.
10. Cooper, J. (1992). In *The cytoskeleton: a practical approach* (ed. K. L. Carraway and C. A. C. Caraway), pp. 47–71. Oxford University Press/IRL.
11. Vancompernolle, K., Goethals, M., Huet, C., Louvard, D., and Vandekerckhove, J. (1992). *EMBO J.*, **13**, 4739.
12. Van Troys, M., Dewitte, D., Goethals, M., Carlier, M.-F., Vandekerckhove, J., and Ampe, C. (1996). *EMBO J.*, **15**, 201.
13. Van Troys, M., Dewitte, D., Verschelde, J. L., Goethals, M., Vandekerckhove, J., and Ampe, C. (1997). *J. Biol. Chem.*, **272**, 32750.
14. Van Troys, M., Dewitte, D., Goethals, M., Vandekerckhove, J., and Ampe, C. (1996). *FEBS Lett.*, **397**, 191.
15. Friederich, E., Vancompernolle, K., Huet, C., Goethals, M., Finidori, J., Vandekerckhove, J., *et al.* (1992). *Cell*, **70**, 81.
16. Way, M., Pope, B., and Weeds, A. G. (1992). *J. Cell Biol.*, **119**, 835.
17. Ho, S. N., Hunt, H. D., Horton, R. M., Pullen, J. K., and Pease, L. R. (1989). *Gene*, **77**, 51.

18. Rossenu, S., Dewitte, D., Vandekerckhove, J., and Ampe, C. (1997). *J. Protein Chem.*, **16**, 499.
19. Haarer, B. K., Petzold, A. S., and Brown, S. (1993). *Mol. Cell. Biol.*, **13**, 7864.
20. Weihing, R. R. and Korn, E. D. (1972). *Biochemistry*, **11**, 1538.
21. Ampe, C., Vandekerckhove, J., Brenner, S. L., Tobacman, L., and Korn, E. D. (1985). *J. Biol. Chem.*, **260**, 834.
22. Vandekerckhove, J., Van Damme, J., Vancompernolle, K., Bubb, M. R., Lambooy, P. K., and Korn, E. D. (1990). *J. Biol. Chem.*, **265**, 12801.
23. Aktories, K. (1994). *Mol. Cell. Biochem.*, **138**, 167.
24. Gettemans, J., De Ville, Y., Vandekerckhove, J., and Waelkens, E. (1992). *EMBO J.*, **11**, 3185.
25. Gettemans, J., De Ville, Y., Waelkens, E., and Vandekerckhove, J. (1995). *J. Biol. Chem.*, **270**, 2644.
26. De Corte, V., Demol, H., Goethals, M., Van Damme J., Gettemans, J., and Vandekerckhove, J. (1999). *Protein Sci.*, **8**, 234.
27. Frank, R. and Overwin, H. (1996). *Methods Mol. Biol.*, **66**, 149.
28. Puius, Y. A., Mahoney, N. M., and Almo, S. C. (1998). *Curr. Opin. Cell Biol.*, **10**, 23.
29. Spudich, J. A. and Watt, S. (1971). *J. Biol. Chem.*, **246**, 4866.
30. Zhao, Y. and Chait, B. (1994). *Anal. Chem.*, **66**, 3723.
31. Gevaert, K., Demol, H., Verschelde, J.-L., Van Damme, J., De Boeck, S., and Vandekerckhove, J. (1997). *J. Protein Chem.*, **16**, 335.

3

Methods to study actin assembly and dynamics in yeast

TATIANA S. KARPOVA and JOHN A. COOPER

1. Introduction: yeast as a model organism for studies of the actin cytoskeleton

Budding yeast (*Saccharomyces cerevisiae*) is a good model organism for studies of the actin cytoskeleton (reviewed in ref. 1). Actin and many components of the yeast actin cytoskeleton are highly homologous to cytoskeletal proteins of higher eukaryotes. Another advantage of yeast is the relative simplicity of the actin cytoskeleton. For example, actin and most of the actin-binding proteins are represented by a single isoform. The genetics of yeast is well developed. The genome of yeast has been sequenced, which simplifies the search for new actin-binding proteins, either by homology searches or by testing the phenotypes of mutants with targeted ORF disruptions.

In yeast, genetics can be easily combined with biochemistry. Yeast cells can be grown in large quantities, which permits one to purify large amounts of proteins. Biochemical properties of purified proteins may be tested in *in vitro* assays. New components of purified protein complexes may be identified by gel separation and subsequent mass spectrometry, using the complete database of yeast protein sequences. Cytology of yeast is also well developed. Proteins may be localized in the yeast cell by immunostaining or by GFP-tagging *in vivo*. Tagging of intracellular structures by GFP-fused proteins permits one to monitor those structures in the cell cycle by means of video microscopy. The combination of genetics, biochemistry, and cytology methods enables powerful studies of actin cytoskeletal function and regulation.

Genetic manipulations of yeast and yeast immunostaining are beyond the scope of this review. We recommend excellent reviews instead (1, 2). Here we describe only basic biochemical and microscopic assays, which permit one to observe and quantify yeast F-actin polymerization and function *in vivo* and *in vitro*. We will discuss F-actin assembly and disassembly assays *in vitro* and *in vivo*, measurement of F-actin and G-actin in yeast cells, and observations of F-actin in live cells. These assays may be applied to characterize phenotypes of actin cytoskeleton mutants and to study functions of actin-binding proteins and actin cytoskeleton regulation.

2. Measurements of F-actin and G-actin *in vivo*

In contrast to higher eukaryotes, yeast maintains a low level of G-actin (3). Therefore it may be expected that actin filaments require stabilization by actin-binding proteins. This is indeed the case. Mutants were found in the genes that stabilize F-actin (3). *Protocols 1–3* may be used for studies of actin polymerization in different yeast cytoskeletal mutants.

The F-actin and G-actin assays are different approaches and are complementary. Different results with these two assays are not unexpected.

The G-actin assay includes unambiguous quantitation of actin by immunoblots. The disadvantage of this assay is that permeabilized cells are not fixed; therefore, permeabilization may cause the solubilization of some F-actin. In addition, the assay assumes that all unpolymerized actin is released upon permeabilization. While it is clear that actin monomers are freely permeable (4), one cannot exclude the possibility that unpolymerized actin also exists in some other forms, such as in large complexes that are not soluble and readily released.

The F-actin assays are based on rhodamine phalloidin staining of fixed cells. In fixed cells the actin structures are well preserved. None the less, some of these actin structures might display reduced affinity for phalloidin due to the presence of wild-type or mutant actin-binding proteins.

Actin may exist in a form detected in both or neither assays. For example, some actin filaments may depolymerize rapidly and become soluble on permeabilization, so this pool would be measured in both assays. Alternatively, denatured and precipitated actin would neither bind rhodamine phalloidin nor be soluble.

Protocol 1. G-actin level in unfixed permeabilized cells

This version of the assay was used to measure the amount of soluble actin released from yeast cells (3). The permeabilization procedure was modified from ref. 4.

Equipment and reagents

- Microcentrifuge
- Reagents for Bradford assay
- Equipment and reagents for Western blots
- Protease inhibitors
- 1 × MKEI: 2 mM $MgCl_2$, 100 mM KCl, 1 mM EGTA, 20 mM imidazole–HCl pH 7.0
- 5 mg/ml saponin

Method

1. Grow a 25 ml yeast culture to ~ 5 × 10^6 cells/ml.
2. Collect the cells by centrifugation. Wash the pellet with 1.0 ml of cold 1 × MKEI with protease inhibitors.

3. Pellet the cells in a microcentrifuge for 2 sec, aspirate the supernatant, and freeze immediately in liquid nitrogen.
4. Thaw the frozen pellet at RT, add an equal volume of 1 × MKEI (about 100 μl) with protease inhibitors, and freeze again.
5. Thaw the sample once again at RT in the presence of 0.5 mg/ml saponin (0.1 vol. of 5 mg/ml saponin) in MKEI buffer with freshly added protease inhibitors.
6. Sediment the cells in a microcentrifuge at top speed for 15 min.
7. Use the supernatant for Bradford assay and for Western blotting with mouse C4 monoclonal anti-actin and secondary [^{125}I] goat anti-mouse. Use serial twofold yeast actin dilutions as a standard.
8. Quantitate bands on Western blots with a PhosphorImager. Normalize the actin content in the supernatant to total protein in the supernatant, determined by Bradford assay.

Protocol 2. Fluorescent phalloidin-binding assay by methanol extraction from fixed cells

This version of the assay was used to measure the fluorescence of rhodamine phalloidin extracted by methanol from cell populations of wild-type strains and mutant strains defective in F-actin polymerization (3). This procedure was modified from ref. 5.

Equipment and reagents

- Fluorimeter
- Reagents for Bradford assay
- 37% formaldehyde
- PBS
- PBS with 0.2% Triton X-100
- 0.33 μM rhodamine phalloidin (Molecular Probes Inc.)

Method

1. Fix a logarithmically growing cell culture (10^6 cells/ml) for 3 h in 3.7% formaldehyde by adding 10 × formaldehyde directly to the medium. Before fixation remove an aliquot containing 10^7 cells to determine the total protein content by Bradford assay.
2. Collect the fixed cells by centrifugation, and divide into equal aliquots containing 2 × 10^7 cells each. Treat these five samples of fixed cells independently.
3. Wash the cell samples in PBS and resuspend the pellets in a final volume of 100 μl of PBS with 0.2% Triton X-100. Triton lowers non-specifically bound rhodamine phalloidin.
4. Use half of each aliquot (10^7 cells) to make serial twofold 50 μl

Protocol 2. *Continued*

dilutions in PBS + 0.2% Triton. Incubate the remaining half of each aliquot and its serial twofold dilutions with 0.33 μM rhodamine phalloidin for 30 min on ice. Incubate one of the five samples and its twofold dilutions with 300 μM non-labelled phalloidin. Wash twice in 1 ml of PBS.

5. Resuspend the pellet carefully in 100 μl of PBS and mix with 1 ml of methanol. Extract the rhodamine phalloidin bound to F-actin by incubation for 2 h. Sediment the cells by centrifugation.
6. Assay the fluorescence of the supernatant with a fluorometer.
7. Subtract the values obtained for the non-labelled phalloidin-containing control sample from those obtained for other samples to correct for non-specific binding of labelled phalloidin. If the cells are properly washed after staining, then non-specific binding accounts for 8–10% of total binding. Plot the values, corrected for non-specific binding, versus total protein content to normalize the F-actin to total protein.
8. Derive an expected fluorescence for each mutant strain from the standard wild-type curve depending on the total protein content of the mutant strain. Represent the fluorescence obtained in the experiment as the percentage of expected. Treat the four samples and their dilutions as independent measurements to calculate the mean and standard error for each strain.

Protocol 3. Fluorescent phalloidin-binding assay by imaging of individual cells

This version of the assay was used to measure F-actin levels from digital fluorescence images of cells stained with rhodamine phalloidin. The advantage of this assay is that the cells at different stages of the cell cycle may be assayed without synchronization of the culture (6). In our experiments images were acquired with a cooled CCD 300T-RC camera (DAGE-MTI), a frame grabber LG-3 (Scion Corporation, Frederick, MD), and a Power Macintosh. Image capturing and measurements of fluorescence intensity and length of cell axes were performed with *NIH Image 1.62*. The amount of non-specific binding was insignificant, and no correction for non-specific binding was necessary.

Reagents

- 37% formaldehyde
- PBS
- PBS with 0.05% Triton X-100
- 0.165 μM rhodamine phalloidin (Molecular Probes Inc.)

Method

1. Fix 5 ml of a logarithmically growing cell culture (10^6 cells/ml) for 1 h in 3.7% formaldehyde by adding 10 × formaldehyde directly to the medium.
2. This stage and subsequent stages should be performed in glass tubes to avoid loss of cells. Collect the fixed cells by centrifugation; wash in 5 ml of PBS.
3. Resuspend the pellets in a final volume of 100 μl of PBS with 0.05% Triton X-100.
4. Incubate the sample with 0.165 μM rhodamine phalloidin for 30 min on ice. Wash twice in 1 ml of PBS.
5. Mount a small volume of the culture on a glass slide, so that cells are immobilized by contact with the coverslip.
6. Acquire digital fluorescence images. Image the middle focal plane of the cell. In wide-field microscopy each focal plane collects the light from the whole specimen, so fluorescence from the specimen at one focal plane adequately represents the fluorescence of the whole specimen. Obtain the background fluorescence from a nearby area of the same size and subtract it from the cell fluorescence.
7. Measure fluorescence intensity of the cells. Measure the major and minor axes of the mother and the bud. Calculate the volume from the axes assuming that the mother and the bud are prolate spheroids. Normalize the fluorescence to cell volume.

3. Measurements of assembly and disassembly of F-actin *in vitro* and *in vivo*

Yeast actin can be purified from wild-type strains and *act1* mutants (3, 7). Although the purification procedure includes rounds of polymerization–depolymerization, proteins with partial polymerization defects may be also purified with partial loss of the mutant protein. An easy sedimentation assay (*Protocol 4, Figure 1*) permits one to determine the efficiency of purified actin polymerization (3).

A rhodamine-labelled actin assembly assay was also developed for actin polymerization in permeabilized cells. Cells are permeabilized by rapid freezing. Foci of actin assembly are revealed by this assay. This test may be applied to studies of actin polymerization and of distribution of polymerization foci in different actin-binding protein mutants (4).

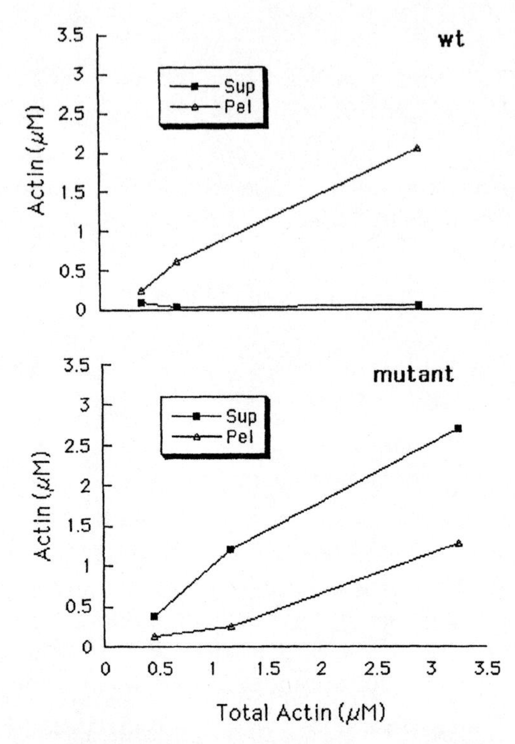

Figure 1. Assay of actin sedimentation by polymerization. Graphs of the concentration of actin in the supernatant (filled squares) and pellet (open triangles) fractions versus total actin in wild-type strain and mutant. Adapted from ref. 3.

Protocol 4. Sedimentation assay of actin polymerization

Equipment and reagents

- Equipment and reagents for SDS–PAGE gels
- 1 × MKEI: 2 mM $MgCl_2$, 100 mM KCl, 1 mM EGTA, 20 mM imidazole–HCl pH 7.0

Method

1. Polymerize purified actin at 6 μM for 2 h at room temperature in 1 × MKEI. Periodically mix the samples gently to accelerate the approach to steady state.
2. Dilute the actin solution to 3 μM, 1 μM, and 0.5 μM in 1 × MKEI and incubate 100 μl samples for an additional 2 h. Periodically mix the samples gently to accelerate the approach to steady state.
3. Remove 25 μl of 100 μl for the sample 'Total' and add to 25 μl of 2 × SDS–PAGE sample loading buffer.

4. Centrifuge the remaining 75 μl at 150 000 *g* for 30 min at 22 °C (70 000 r.p.m. in a TLA 100.1 rotor in a TL100 centrifuge) (Beckman Instruments, Inc.).
5. Remove 25 μl from the meniscus for the sample 'Supernatant', and add to 25 μl of 2 × SDS sample buffer.
6. Remove the remaining liquid from the tube, avoiding the pellet. Dissolve the pellet in 150 μl of 1 × SDS sample buffer.
7. Boil all the three samples. Separate the proteins on 10% SDS–polyacrylamide gels. Stain the gels with Coomassie blue.
8. Scan the gels and quantitate the intensities of the relevant bands with *NIH Image 1.61*. Intensities of actin bands may be converted to absolute values by comparison with the internal standards.

4. Actin structures in live cells

4.1 Video microscopy of GFP-tagged actin and actin-binding proteins

GFP-tagging permits one to observe protein location in live cells. Video microscopy is widely used to track GFP-marked cellular structures in yeast. Here we describe basic methods of GFP-tagging and observation of actin and actin-binding proteins in yeast. Further descriptions of video microscopy in yeast may be found in refs 8 and 9.

Actin and actin-binding proteins were tagged with GFP in yeast (10, 11). GFP-tagged proteins may be introduced into the cell on autonomously replicating plasmids (10). However, to obtain a strain stable for the GFP-marker, one may replace a chromosomal wild-type copy with a tagged version of the gene (11), or insert a GFP-tag, generated by PCR, at the C- or N-terminus of the gene (12, 13).

C- and N-terminal tags may differ in their effect on protein function (14, 15), therefore if one type of fusion does not complement the gene disruption, it is worth trying another one. It is always preferable to have a GFP-tagged construct that rescues the disruption phenotype, to be sure that protein functions are not perturbed by tagging. A less desirable alternative if the GFP-tagged protein does not rescue the null mutant is to use a cell with a wild-type copy of a relevant gene. One then must show that GFP-tagged protein has the same distribution as untagged protein, localized by an independent approach (10).

The yeast actin cytoskeleton includes cables and patches, which can be detected by rhodamine phalloidin and immunostaining (16). Actin and many actin-binding proteins tagged with GFP may also be used as a patch label (10, 11). Unfortunately, no vital label for cables has been found yet. A C-terminal

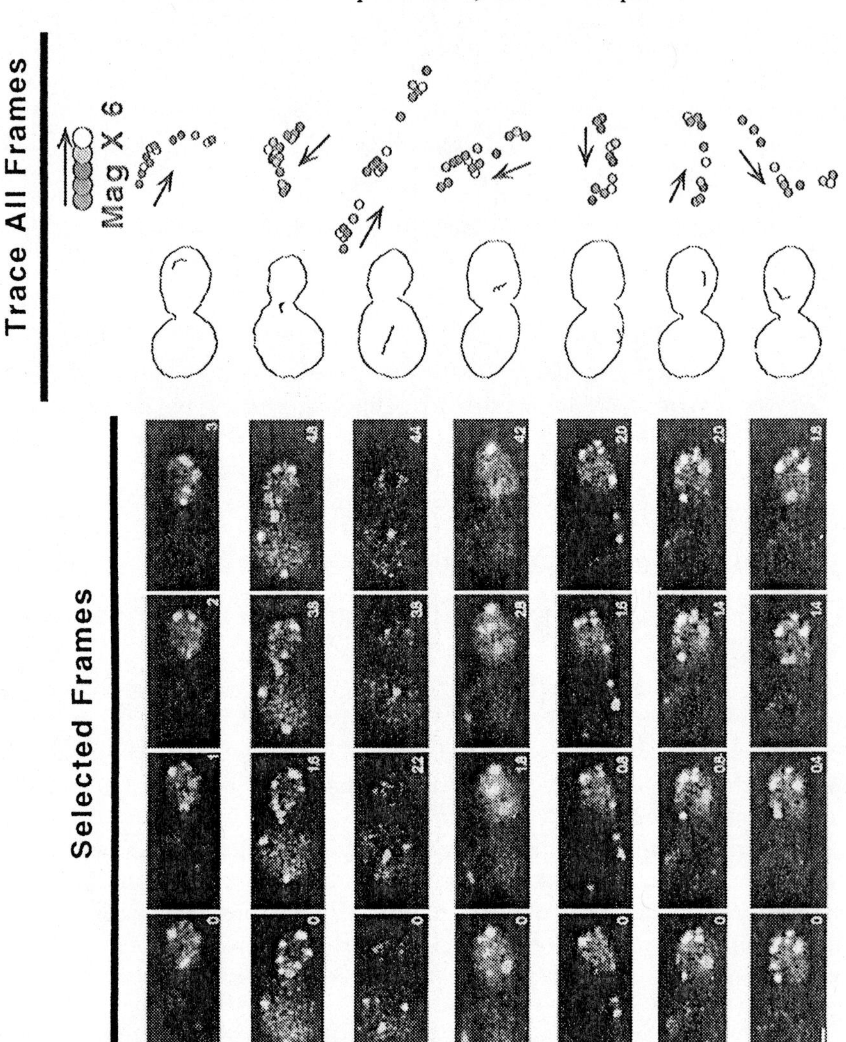

Figure 2. Tracking of cortical actin patches (11). Movies were made at a single plane of focus with frames collected at 0.2 sec intervals. Each row illustrates the movement of one patch. (*Left*) Four selected frames from the movie; the moving patch is coloured yellow. (*Right*) Path of the highlighted patch, tracked in every frame. Under 'Trace All Frames' the path of the highlighted patch is drawn, first as a black line inside an outline of the cell at the same magnification as the micrographs on the left. Under 'Mag X 6' the individual successive positions at each frame are shown, coloured in the sequence red–green–blue–orange–white, at a sixfold increased magnification. Since the time interval between frames is 0.2 sec, the interval between red dots is 1 sec. In this illustration, the coloured dots are placed on top of each other as time proceeds, which means that if a patch does not move, then that dot is covered over and masked by a subsequent dot. Therefore, when colours are missing from the sequence red–green–blue–orange–white, the patch has not moved or has returned to a previous position. Of course, the movies illustrate this movement better. Bar = 3 μm.

fusion of GFP to actin does not rescue the *act1* disruption (10). Both C- and N-terminal fusions of GFP to *CAP2* (β-subunit of capping protein), rescue the disruption well. Individual patches may be tracked for a limited amount of time in 2D and their velocities determined (*Figure 2*).

4.2 Microscopy of GFP-tagged patches

GFP-tagged patches provide a weak signal, are small, and move fast (10, 11). Therefore, to monitor patch movement one needs high magnification and a sensitive video-rate camera.

Our video system consisted of an Olympus Bmax-60F microscope equipped with a 1.35 NA × 100 UPlanApo objective and a U-MWIBA interference filter set (Olympus, Lake Success, NY); a DAGE ISIT-68 camera, an RC68 controller, a DSP-2000 processor, a stage and shutter controller (MAC2000, Ludl Electronic Products Ltd., Hawthorne, NY), and a Power Macintosh with a frame grabber (AG-5, Scion Corporation, Frederick, MD). This system was co-ordinated by custom macros in *NIH Image* (program written by Wayne Rasband at the National Institutes.of Health, available by anonymous ftp at `zippy.nimh.nih.gov`).

Sample preparation was important to minimize background fluorescence, immobilize the cells, and prevent drying. A drop of molten 2% agarose in either distilled water or a simplified (non-fluorescent) synthetic medium (KH_2PO_4 0.9 g/litre, K_2HPO_4 0.23 g/litre, $MgSO_4$ 0.5 g/litre, $(NH_4)_2SO_4$ 3.5 g /litre, glucose 20 g/litre, CSM-Trp media supplement) (Bio 101, La Jolla, CA) was placed in a well of a Teflon-coated glass slide (No. 10-1189-A, Cell-Line Associates, Inc., Newfield, NJ) and immediately flattened by covering with an uncoated glass slide. The upper glass slide was removed, leaving behind a circular, flattened agarose pad. Using a razor blade, the pad was trimmed into a square that matched the dimensions of the well. 1–2 μl of a cell suspension was placed in the centre of the pad and then covered with a No. 2 glass coverslip that had a thin (~ 0.5 mm) line of Vaseline at the outer edges. Slight pressure was applied to the coverslip to disperse the cell droplet over the surface of the agarose pad and make a seal.

Living cells were imaged at room temperature, 22–24 °C. Cells remained viable and growing during 6 h experiments. Cells left in the experimental chamber after an experiment proceeded to divide many times and form microcolonies. Based on these criteria and the morphology of the cells and their actin cytoskeleton, we saw no evidence for cell damage caused by the experiment. Long-term experiments were limited by loss of GFP signal due to photobleaching. Photobleaching was minimized by closing the excitation shutter between image collections and reducing the illumination intensity with neutral density filters.

Two different acquisition strategies were used. First, rapid acquisition at a single focal plane for a short time (minutes) was used to monitor cells at sub-second acquisition rates to determine the rate, direction, and range of patch movements. Second, slow, time-lapse acquisition of 10 focal planes 0.5 μm apart for a long time (hours) with projection of all focal planes onto a single two-dimensional image was used to determine the extent of polarization of patches over the cell cycle.

Calculation of velocity. Instantaneous speeds of patch movement were calculated for two groups of patches. Group 1 contained patches that remained visible for > 2 sec and showed obvious movement. Group 2 contained patches that remained visible for > 4 sec regardless of obvious movement. Group 1 is biased in favour of moving patches and Group 2 in favour of non-moving patches. The (x, y) positions of the selected patches at 0.2 sec (Group 1) or 0.4 sec (Group 2) time intervals were collected using *NIH Image*. Speed was calculated as distance divided by time at 0.4 sec intervals for each. The distance between adjacent pixels was 80 nm; therefore, the smallest possible speed greater than 0 was 0.2 μm/sec. For Group 1 patches the mean was 0.49 μm/sec, the median was 0.28 μm/sec, the standard deviation was 0.30 μm/sec, the range was 0–1.7 μm/sec, and N = 183. For Group 2 patches the mean was 0.31 μm/sec, the median was 0.28 μm/sec, the standard deviation was 0.28 μm/sec, the range was 0–1.9 μm/sec, and N = 392.

4.3 Use of inhibitors to study actin dynamics and functions *in vivo*

Several actin assembly inhibitors have been tested in yeast (17). Neither cytochalasin nor jasplakinolide worked in yeast. Swinholide-A was effective only in certain mutants (*erg6*) with higher permeability for swinholide. Only mycaloide-B and latrunculin were capable of inhibiting growth of yeast. Latrunculin A has been proven to be most efficient and has been widely used for studies of actin disassembly *in vivo* (*Protocol 5*). Latrunculin induces actin disassembly by sequestering actin monomers (18). Latrunculin also inhibits nucleotide exchange in actin (17).

The dynamics of patch and cable disappearance may be monitored by rhodamine phalloidin staining of cells fixed at appropriate time intervals (17, 19), or by observation of live cells with patches tagged by GFP-labelled proteins (Karpova, Cooper, unpublished data). This method may be applied to studies of the role of actin-binding proteins in the stability of the F-actin (19).

Protocol 5. The actin disassembly induced by latrunculin treatment

This protocol is derived from ref. 17 with slight modifications (6). Treatment with ≥ 200 mM of latrunculin A rapidly induces almost complete disassembly of cables and patches. Latrunculin B is less efficient than latrunculin A. A small fraction (< 5%) of cells retain cables or patches or both even in presence of higher (up to 900 nM) concentrations of latrunculin A.

Reagents

- YPD: 1% yeast extract, 2% tryptone, 2% glucose
- 800 nM latrunculin A (2 ×)
- DMSO
- 37% formaldehyde

Method

1. Grow yeast cells to log phase (10^6 cells/ml) in YPD.
2. Prepare YPD with 800 nM latrunculin A from 100 mM DMSO stock. Add DMSO to 2% total to maintain latrunculin A in solution.
3. Mix cell suspension and YPD with 2 × latrunculin A in 1:1 ratio. For live cells, mix directly on a slide, and then observe immediately.
4. Alternatively, at the desired times obtain aliquots of the sample and mix them with 0.1 vol. of 37% formaldehyde by trituration.
5. Stain the cells with rhodamine phalloidin as described in *Protocol 3* and observe the dynamics of disassembly of the actin cables and patches.

5. Binding of proteins to actin filaments

Purified proteins may be tested in a sedimentation assay to determine their ability to bind wild-type or mutant actins (*Protocol 6*, *Figure 3*).

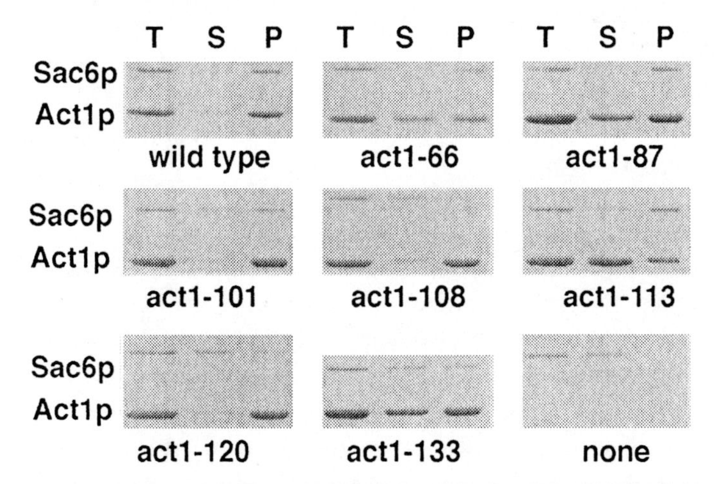

Figure 3. Binding of fimbrin to actin from a wild-type strain and from mutants with a fimbrin-binding defect (3). The fimbrin (Sac6p) and actin bands of SDS gels stained with Coomassie blue are shown for total sample (T) before sedimentation, and the supernatant (S) and pellet (P) after sedimentation.

Protocol 6. Sedimentation assay for binding of proteins to actin

This version of the assay was used to measure fimbrin (Sac6p) binding to actin. The procedure was modified from ref. 20.

Equipment and reagents

- Equipment and reagents for SDS–PAGE gels
- Reagents for Bradford assay
- 1 × MKEI: 2 mM $MgCl_2$, 100 mM KCl, 1 mM EGTA, 20 mM imidazole–HCl pH 7.0

Method

1. Clarify the actin and Sac6p solution immediately before use. Thaw frozen aliquots of protein rapidly. Centrifuge for 30 min at 150 000 *g* at 4°C (70 000 r.p.m. in a rotor TLA 100.1 in a centrifuge TL100) (Beckman Instruments, Inc.). Keep supernatant on ice. Measure protein concentration by Bradford assay.
2. Prepare a 100 μl solution of purified actin at 3 μM and purified fimbrin at 0.75 μM in polymerization buffer 1 × MKEI. The control reaction contains only actin without fimbrin. Mix the samples gently and incubate for 90 min at room temperature.
3. Remove 25 μl of 100 μl for the sample 'Total' and add to 25 μl of 2 × SDS–PAGE sample loading buffer.
4. Spin the remaining 75 μl at 90 000 *g* for 60 min at 22 °C (50 000 r.p.m. in a TLA 100.1 rotor in a TL100 centrifuge) (Beckman Instruments, Inc.).
5. Remove 25 μl from the meniscus, for the sample 'Supernatant', and add to 25 μl of 2 × SDS sample buffer.
6. Remove the remaining liquid from the tube, avoiding the pellet. Dissolve the pellet in 150 μl of 1 × SDS sample buffer.
7. Boil all three samples. Separate the proteins on 10% SDS–polyacrylamide gels.
8. Stain the gels with Coomassie blue.
9. Scan the gels and quantitate the intensities of the relevant bands with *NIH Image 1.61*. Intensities of protein bands may be converted to absolute amounts by comparison with the internal standards.

6. Conclusions: tests of actin assembly and dynamics applied to studies of wild-type and mutant actin cytoskeletal proteins

In this review we presented and discussed a wide choice of tests available for studies of the actin cytoskeleton of yeast. Now we will summarize the available *in vitro* and *in vivo* methods and their applications. The assays discussed above can be used to study the function of new proteins of the actin cytoskeleton. Potential actin cytoskeletal proteins are frequently identified in genetic and biochemical screens or by homology to known cytoskeletal proteins. Further *in vitro* and *in vivo* tests are necessary:

(a) To prove that these proteins are indeed part of the actin cytoskeleton.
(b) To investigate their subcellular localization and their structural and biochemical function.

In vitro assays with purified proteins (Sections 3 and 5) permit one to test biochemical characteristics of these proteins, namely their binding to actin and their effect on actin polymerization. Most of these assays were described previously (21, 22), therefore we did not provide their descriptions in this publication. Studies of purified cofilin and twinfilin may serve as a good example of a general strategy of testing actin-binding properties of proteins. Binding of a wild-type twinfilin to F-actin and to G-actin was demonstrated by co-sedimentation with F-actin (*Protocol 6*) and by analysis of complexes with actin monomers on native gels (21, 23). After the ability of the given protein to bind actin is established, the next step is to study the effects of binding to actin. Function of twinfilin and cofilin was further tested in *in vitro* assays of actin polymerization and disassembly and of inhibition of actin nucleotide exchange (22–24). The same *in vitro* tests may be applied to studies of the role of separate protein domains or selected amino acid residues, as in the study of purified cofilin-like repeats of twinfilin (23), or in studies of the effects of point mutations on cofilin function (24). *In vitro* studies supply the basic characteristics of the protein. These experiments are preliminary to studies of protein functions *in vivo*.

In vivo assays with wild-type and mutant cells (Sections 2, 3, and 4) permit one to test the localization and cellular functions of actin and actin-binding proteins. These methods are new, and we supply protocols for most of them. The subcellular localization of proteins may be determined by tagging of these proteins with GFP, as in case of twinfilin (23) or verprolin (25). The effect of actin-binding proteins on the assembly of the F-actin may be studied in F-actin and G-actin *in vivo* assays (*Protocols 1–3*). For example, F-actin and G-actin *in vivo* assays permitted one to compare the effects of mutations of capping protein and fimbrin on F-actin stabilization (3). The effect of the *act1–159* mutation on F-actin accumulation in individual cells was compared with wild-type (26). Dynamics of latrunculin-induced patch and cable disassembly *in vivo* was studied in individual wild-type cells and in cofilin mutants (*Protocol 5*) (6, 19). Dependence of F-actin polymerization in the cell cortex on Cdc42p, Sla1p, Sla2p, Bee1p, and Pca1p was studied in intracellular polymerization assays (4) (see Section 3).

Obviously, as all of these methods address separate issues, for totally unknown proteins the whole plethora of methods should be used to get the complete characteristics of the phenotype. Also, among these tests readers may select methods most suitable for their specific tasks.

References

1. Botstein, D., Amberg, D., Mulholland, J., Huffaker, T., Adams, A., Drubin, D., *et al.* (1997). In *The molecular and cellular biology of the yeast Saccharomyces. Cell cycle and cell biology* (ed. J. R. Pringle, J. R. Broach, and E. W. Jones), p. 1. Cold Spring Harbor Laboratory Press. Cold Spring Harbor.
2. Pringle, J. R., Preston, R. A., Adams, A. E. M., Stearns, T., Drubin, D. G., Haarer, B. K., *et al.* (1989). *Methods Cell Biol.*, **31**, 357.
3. Karpova, T. S., Tatchell, K., and Cooper, J. A. (1995). *J. Cell Biol.*, **131**, 1483.
4. Li, R., Zheng, Y., and Drubin, D. G. (1995). *J. Cell Biol.*, **128**, 599.
5. Lillie, S. H. and Brown, S. S. (1994). *J. Cell Biol.*, **125**, 825.
6. Karpova, T. S., McNally, J. G., Moltz, S. L., and Cooper, J. A. (1998). *J. Cell Biol.*, **142**, 1501.
7. Kron, S. J., Drubin, D. G., Botstein, D., and Spudich, J. A. (1992). *Proc. Natl. Acad. Sci. USA*, **89**, 4466.
8. Salmon, E. D., Shaw, S. L., Waters, J., Waterman-Storer, C. M., Maddox, P. S., Yeh, E., *et al.* (1998). *Methods Cell Biol.*, **56**, 185.
9. Shaw, S. L., Yeh, E., Bloom, K., and Salmon, E. D. (1997). *Curr. Biol.*, **7**, 701.
10. Doyle, T. and Botstein, D. (1996). *Proc. Natl. Acad. Sci. USA*, **93**, 3886.
11. Waddle, J. A., Karpova, T. S., Waterston, R. H., and Cooper, J. A. (1996). *J. Cell Biol.*, **132**, 861.
12. Niedenthal, R. K., Riles, L., Johnston, M., and Hegemann, J. H. (1996). *Yeast*, **12**, 773.
13. Karpova, T. S., Moltz, S. L., Riles, L. E., Gueldener, U., Hegemann, J. H., Veronneau, S., *et al.* (1998). *J. Cell Sci.*, **111**, 2689.
14. Carminati, J. L. and Stearns, T. (1997). *J. Cell Biol.*, **138**, 629.
15. Heil-Chapdelaine, R. A., Tran, N. K., and Cooper, J. A. (1998). *Curr. Biol.*, **8**, 1281.
16. Kilmartin, J. V. and Adams, A. E. M. (1984). *J. Cell Biol.*, **98**, 922.
17. Ayscough, K. R., Stryker, J., Pokala, N., Sanders, M., Crews, P., and Drubin, D. G. (1997). *J. Cell Biol.*, **137**, 399.
18. Coue, M., Brenner, S. L., Spector, I., and Korn, E. D. (1987). *FEBS Lett.*, **213**, 316.
19. Lappalainen, P. and Drubin, D. G. (1997). *Nature*, **388**, 78.
20. Honts, J. E., Sandrock, T. S., Brower, S. M., Odell, J. L., and Adams, A. E. M. (1994). *J. Cell Biol.*, **126**, 413.
21. Safer, D. (1989). *Anal. Biochem.*, **178**, 32.
22. Cooper, J. A. (1992). In *The cytoskeleton: a practical approach* (ed. K. L. Carraway and C. A. C. Carraway), p. 47. IRL Press, Oxford.
23. Goode, B. L., Drubin, D. G., and Lappalainen, P. (1998). *J. Cell Biol.*, **142**, 723.
24. Lappalainen, P., Fedorov, E. V., Fedorov, A. A., Almo, S. C., and Drubin, D. G. (1997). *EMBO J.*, **16**, 5520.
25. Vaduva, G., Martin, N. C., and Hopper, A. K. (1997). *J. Cell Biol.*, **139**, 1821.
26. Belmont, L. D. and Drubin, D. G. (1998). *J. Cell Biol.*, **142**, 1289.

4

Signalling complexes: association of signalling proteins with the cytoskeleton

CORALIE A. CAROTHERS CARRAWAY and
KERMIT L. CARRAWAY

1. Introduction

A plethora of studies in the areas of membranes, cytoskeleton and, more recently, signal transduction have contributed to both concepts and technologies for the study of interactions of signalling proteins with the cytoskeleton (for overviews, see refs 1 and 2). These studies have implicated association of signalling proteins with the membrane and with the cytoskeleton in targeting these proteins to appropriate cellular locations for their participation in signal transduction events. Appropriate cellular localization is requisite for activation processes (3) and can occur reversibly, to turn off signal. The approaches used historically are diverse, ranging from tissue and cell localization by microscopy, to protein sequence analyses for the identification of domains for structure–function determinations to genetic perturbations. The recent explosion in genomics/proteomics and informatics is beginning to contribute in a major way to design and interpretation of experiments based on sequence similarities, structural and functional domains. Further, these areas are expected to play significant roles in cross-fertilization of concepts among major areas of technology and of biological investigation.

2. Co-localization by microscopy

Methods for localization in tissues and cells have traditionally taken two major approaches, both employing the specificity and sensitivity of antibodies: screening on immunoblots and screening by immunofluorescence microscopy. There are several major aspects of the question of localization which relate to specificity.

(a) Is the protein of interest species-specific? The answer to this question gives preliminary clues to phylogenetic importance of the protein.
(b) Is the protein tissue-specific? Information on tissue specificity provides perspectives on the potential role for the protein in normal differentiated function.
(c) Is the expression of the protein altered in abnormal cells, such as in neoplasia?
(d) Is the protein specific developmentally, i.e. is it expressed at a particular point in development?

Taken together, the answers to these questions can provide a very valuable perspective in the design of further approaches and questions.

In generating antibodies to be used for these studies the major consideration is, of course, specificity. If phylogenetic studies are to be performed, a second consideration is capability of the antibody to recognize the protein in a non-species-dependent manner. The choice of type of antibody, i.e. monoclonal versus polyclonal, entails several considerations. Whereas polyclonal antibodies, because of the greater number of epitopes recognized, provide an increased chance to visualize the protein in any particular species or tissue, there may be a risk of lack of specificity. Although a monoclonal antibody may be more specific, if elicited against a unique epitope, an antibody to a kinase domain may or may not be specific for a particular kinase, depending on the epitope recognized.

When the presence of the protein in a tissue or cell has been determined, the second level of localization, i.e. subcellular localization, can provide further clues as to its cellular role. Immunofluorescence microscopy, the classic first step in these studies, has been described frequently, including protocols in this book and its precursor volume (4). A particularly interesting example is its use in the study of the association of the cytoskeleton with a signalling protein, the nuclear/cytoplasmic tyrosine kinase c-Abl (5, 6). Interestingly, in this case a key clue to potential function was derived from the primary sequence of the protein, which had sequences at its C-terminus which suggested that it might bind F- and/or G-actin (6). Chapter 4 of the previous volume in this series (4) describes more thoroughly methods for and the importance of identification and characterization of binding domains.

The investigator is not limited to immunomicroscopy to perform this type of localization study. If an appropriate antibody is not available, other methods of protein tagging are useful in localization experiments by microscopy. One can use a fluorescently-labelled compound known to bind specifically to an endogenous protein. A classic example is the frequently utilized method for the detection of microfilaments using fluorescently-labelled phallicidin, which binds specifically to microfilaments. This technique is described in Chapter 6 in this volume and Chapter 3 of the previous volume (4). A second approach is use of a fluorescent tag on the protein, which is then introduced into the cell by a mechanism such as microinjection or transfection. This approach is described in detail for green fluorescent protein labelling of exogenous proteins in Chapter 6 of this volume. Chapter 1 of the previous volume (4) describes labelling of exogenous proteins with conventional fluorescent tags, such as rhodamine, and microinjection.

3. Co-localization by co-purification

A second major approach to demonstrating co-localization of proteins is by co-purification. When microscopy co-localization studies have provided clues to the cellular locale of the proteins of interest, an early step in demonstrating an interaction, either directly or indirectly, is a cell disruption and organelle or compartment purification. Cell fractionation studies have historically received considerable attention, and methods for purification of the major organelles from both tissues and cells have been described (*Table 1*).

Table 1. Subcellular fractionation methods

Fraction	Approach	Reference
Endoplasmic reticulum	Homogenization, followed by differential and gradient centrifugation	55
Golgi	Homogenization, followed by differential and gradient centrifugation	55
Nucleus	Hypotonic lysis in nucleus-stabilizing buffer, differential centrifugation	55
Plasma membrane	Hypotonic lysis or N_2 cavitation, differential centrifugation or separation based on specific membrane properties; lysis after stabilization, yielding 'intact' envelopes	7
Membrane skeleton		26, 27
Erythrocyte	Demembranation by non-ionic detergent lysis, centrifugation	56
Platelet	Demembranation by non-ionic detergent lysis, centrifugation	30, 31
Microvilli	Shearing from cell surface by forcing through needle, differential centrifugation	57, 58
Caveolae	Non-ionic detergent lysis, flotation gradient	38
MF-associated signal transduction particle	Non-ionic detergent lysis, differential or velocity sedimentation centrifugation, or gel filtration	
Cytoskeleton	Demembranation by extraction in non-ionic detergent-containing buffer under optimal stabilization conditions, differential centrifugation	14
MF		11
IF		17, 24
MT		19, 20

Table 2. Receptor–cytoskeleton cellular interaction sites[a]

Site	Cell surface components	Membrane skeletal proteins	Cytoskeletal components	Signal transduction proteins
Erythrocyte membrane skeleton	Band 3 (anion channel)	Ankyrin, spectrin, band 4.1, glycophorin	Actin protofilaments adducin	None?
Focal adhesion	Integrins	α-Actinin, vinculin, talin, VASP, ERMs, fimbrin, filamin	Microfilaments	Focal adhesion kinase (FAK), PKC, MARCKS
Platelet membrane skeleton	Integrin GPIIb/IIIa GPIb-IX	ABP, α-actinin	Actin protofilaments?	Src kinases, others
Caveolae	Anion transporter, GPI-linked proteins (e.g. alk. PTase, uPAR, Thy-1)	Caveolin, annexin II	G-actin	G-proteins, Src family kinases, casein kinase II, PKC
Adherens junctions	Cadherin	α-, β-, and γ-catenins; α-actinin	Microfilaments	Src family kinases
Desmosome	Desmoglein, desmocollin (cadherins)	?	Intermediate filaments	?
Tight junction (*occludens* junction)	Occludin	ZO-1, ZO-2, ZO-3	Intermediate filaments	?
Neuromuscular junction	AChE, dystroglycan, ErbB receptors	Dystrophin, 43K, spectrin	Microfilaments	MuSK, ErbBs, PKC
Mitogenic signalling complexes	Growth factor/cytokine receptors	?	Microfilaments	Mitogenic pathway proteins
Microvillar signal transduction particle	Transmembrane complex gp's; ErbB2, ErbB3; E-cadherin	α-Actinin, calpactin, ERM's, fodrin	Microfilaments	Ras to MAPK pathway, Rsk, Src, Abl, PLCγ, PI-3-kinase

[a]Modified from ref. 1, Table 3.2.

3.1 Association with plasma membranes

An emerging paradigm for the transduction of the pleiotypic cellular effects mediated by signalling molecules describes the localization of signalling complexes in plasma membranes at sites of membrane–microfilament interactions (2, 3). The paradigm describes these sites as complexes of membrane and cytoplasmic proteins which are assembled and associated, at least transiently, with cytoskeletal and signalling proteins (3). The sizes and relative stabilities of the complexes depend in part on the receptor and may be transient, as with growth factor receptor stimulation, or more stable, as in cell–cell interaction sites. Such sites include adhesion plaques, cadherin-containing junctions, microvilli, caveolae, and synaptic junctions (1, 2) (*Table 2*).

Isolation and characterization of such signalling assemblies is complicated by their complex structures, size, heterogeneity, and insolubility. One approach is to isolate a plasma membrane fraction containing the complexes, then solubilize and fractionate the membranes for complex purification. Two general methods have been used for plasma membrane isolation (7). The first seeks to maintain the structure of the membrane by obtaining a relatively intact plasma membrane envelope or 'ghost', similar to the ghosts isolated from erythrocytes (8). In this type of procedure mild, reversible stabilization techniques are used to prevent the breakdown of the membrane into vesicles (7). The objective is to maintain membrane skeleton associations with the membrane as well as prevent dissociation of peripheral membrane components involved in signalling (7). This method can be used successfully with many types of cells in culture and with ascites tumour cells, but not with tissues. Fractionation and identification are generally simplified by the large size of the envelopes compared to fragmented intracellular organelles. Furthermore, additional purification is possible by converting the envelopes to vesicles and further fractionation.

Protocol 1. Isolation of plasma membrane envelopes from 13762 rat mammary ascites cells by a $ZnCl_2$ membrane stabilization method (9)

Equipment and reagents

- Desk top centrifuge and ultracentrifuge with Beckman SW27 rotor
- Gradient fractionator
- 13762 rat mammary ascites cells from Fischer 344 female rats (9)
- 20 mM Hepes-buffered saline
- 40 mM Tris–HCl pH 7.4
- 1 mM $ZnCl_2$ solution
- Solutions of 40%, 45%, 50%, and 55% (w/w) sucrose in 20 mM Tris–HCl pH 7.4

Method

1. Maintain ascites cells by weekly passage in Fischer 344 female rats. Isolate the cells from the peritoneal cavity and wash three times in Hepes-buffered saline with centrifugation at 210 *g*-min.

Protocol 1. *Continued*

2. Suspend the washed cells in 10 vol. of ice-cold 40 mM Tris–HCl pH 7.4 and allow to swell for 4 min at 4 °C before centrifugation at 1900 *g*-min. Repeat this step once to yield a second pellet of swollen cells.
3. Suspend the swollen cells in 11 vol. of ice-cold $ZnCl_2$ for 2 min and homogenize with 20 strokes of a Dounce homogenizer fitted with a tight pestle to yield a mixture of envelopes and plasma membrane sheets with about 85% broken cells.
4. Dilute the homogenization mixture with an equal volume of 40 mM Tris–HCl pH 7.4 and wash three times in 40 mM Tris–HCl pH 7.4 with centrifugation at 2.4×10^5 *g*-min.
5. Fractionate the membranes further by layering onto a discontinuous sucrose density gradient composed of 8 ml each of 40%, 45%, 50%, and 55% (w/w) sucrose in 20 mM Tris–HCl pH 7.4 and centrifuging in a Beckman SW27 rotor at 15 000 r.p.m. for 60 min at 4°C.
6. Isolate the membrane envelopes from the 40/45% interface using a bent needle and syringe or gradient fractionator. Wash twice by suspending in Hepes-buffered saline and centrifuging at 1.2×10^5 *g*-min.

The second method entails disruption of cells and tissues directly into plasma membrane vesicles, which are then purified by fractionation techniques (7). Many different disruption methods have been applied, depending on the cell or tissue being studied. Not surprisingly, cells are usually easier to disrupt than tissues. Traditionally, density gradient methods have been used for most membrane fractionations. *Protocol 2* applies a density perturbation method to enrich membranes by a sequential gradient procedure (10). However, when vesicles containing specific signalling components are being sought, application of affinity methods for vesicle enrichment may be preferable. *Protocols 1* and *2* give examples of the two types of procedures. Both use relatively simple methods and commonly available equipment and can be applied, with some modifications, to most cells and tissues. Many variations of these procedures have been presented, some of which are described or tabulated in ref. 7 (copies of this article are available from the authors on request).

Protocol 2. Isolation of plasma membrane vesicles from rat mammary tissue by a digitonin shift protocol on flotation ultracentrifugation (10)

Equipment and reagents

- Dissecting scissors
- Sorval Omni-mixer
- Cheesecloth
- Ultracentrifuge with Beckman SW27 rotor
- Lactating female Fischer rat (14 days post-partum)

- Sucrose solutions: 0.25 M sucrose/20 mM Tris–HCl pH 7.4; 40%, 36%, 32%, and 0.9% sucrose in sucrose/20 mM Tris–HCl pH 7.4
- Digitonin (Sigma)

Method

1. Excise mammary gland from normal lactating female Fischer 344 rats and place in ice-cold buffered 0.25 M sucrose solution.
2. Cut tissue into small pieces with scissors, wash, and mince with scissors.
3. Dilute to 5–7 ml per gram tissue and homogenize in a Sorval Omnimixer in a 50 ml bucket (two bursts of 30 sec each, power setting 5).
4. Filter the homogenate through four layers of cheesecloth and centrifuge at 755 *g* for 10 min. Centrifuge the supernatant at 17 300 *g* for 10 min.
5. Centrifuge the second supernatant at 27 000 r.p.m. for 75 min in a Beckman SW27 rotor to yield the membrane vesicle pellet.
6. Adjust the pellet to 40% buffered sucrose and a final volume of 8.0 ml.
7. Layer 10 ml each of 36%, 32%, and 0.9% buffered sucrose onto the pellet sample and centrifuge at 96 000 *g* for 4 h to equilibrate vesicles to their equilibrium density.
8. Remove the band at the 0.9/32% interface (designated F1) with a bent needle. Resuspend in 38 ml of 10 mM Tris–HCl pH 7.4, and centrifuge at 96 000 *g* for 60 min to remove sucrose.
9. Suspend the washed F1 fraction in buffered sucrose, add digitonin to 0.03%, and incubate at 23°C for 15 min.
10. Make this sample to 40% sucrose as in step 6 and repeat the differential centrifugation of step 7, using buffered sucrose solutions containing 0.03% digitonin.
11. Remove the band at the 32/36% interface and wash with buffered sucrose.

3.2 Cytoskeletal preparations

A first approach to determining whether a regulatory or signalling protein binds to a cytoskeletal element is to generate a crude 'cytoskeletal preparation'. This simple diagnostic test consists of a brief extraction of cells in a non-ionic detergent-containing isotonic buffer to solubilize the phospholipid bilayer of the membrane. The particulate material can then be observed by microscopy or electron microscopy (11–20) or centrifuged and analysed biochemically (13, 14, 17, 18, 21). All cell membrane and intracellular proteins not associated with the cytoskeleton or subcellular organelle remnants are solubilized and remain in the supernatant, while those associated with the 'cytoskeleton' are sedimented. Analysis of SDS–PAGE gels of the pellets with antibodies to specific proteins gives a preliminary suggestion of an association.

Even with this simplistic first approach, however, there are several important considerations in the design of the experiment. Ideally, the extraction is performed in physiological buffers for a brief time (5–15 min). Both a protease inhibitor cocktail and phosphatase inhibitors should be included to stabilize the components during extraction (22). The stabilities of the three cytoskeletal structures are significantly different, and the composition of the extraction buffer should take these differences into account. Further, the different cytoskeletal structures interact via cytoskeleton-associated proteins, and disruption of one type of structure is followed by alterations in the others.

Intermediate filaments are the most stable and generally do not require further alterations in the extraction buffer. However, the subunits of these structures are sensitive to degradation by the Ca^{2+}-activated neutral protease (23), and it is important to include EGTA and/or protease inhibitors in the extraction buffer. Indeed, intracellular Ca^{2+} is frequently a perturbant of both structure and function of intracellular proteins, and free Ca^{2+} concentrations are tightly regulated by cells. In inactivated cells this value is submicromolar, and extraction buffer compositions should reflect this physiological mandate.

Microfilaments, though dynamic structures in many cells, can be isolated relatively intact in physiological buffer if the extraction time is short. Actin-associated proteins dissociate with time, causing alterations in microfilament bundling. The association constants of some proteins for actin filaments are relatively low and do not necessarily reflect the cellular importance of the interaction. An example is the actin cross-linking protein α-actinin, whose K_d for actin is in the micromolar range. Further, the associations of some actin-binding proteins are sensitive to the presence of specific cations, and the buffer composition should reflect this concern. For example, the presence of Ca^{2+} induces dissociation of α-actinin from microfilaments but enhances association of tropomyosin and calpactin with these structures. Further, microfilament assembly and stability require the relatively higher concentrations of K^+ found in the cytoplasm. Actin-stabilizing buffers usually contain 50–100 mM K^+.

Microtubules are considerably more labile than either of the other two cytoskeletal structures. For stabilization additional components such as polyethylene glycol must be included in the extraction buffer (24). Further, microtubules in some types of cells are cold-labile (25), necessitating extraction at room temperature or above.

Both microfilaments and microtubules can be stabilized *in vitro* by specific compounds, such as phalloidin for the former and Taxol for the latter. These reagents can also enhance either annealing or polymerization of these structures and should be used judiciously. One useful application of these stabilizing agents was employed in the development of a perturbation method for the identification of proteins specifically associated with microfilaments. The method for phalloidin shift of microfilaments and microfilament-associated proteins on velocity sedimentation gradient centrifugation (26) was described in the previous volume on the cytoskeleton (22). Basically, the approach entails extraction of the cell or cell membrane fraction in either the absence or presence of phalloidin, followed by ultracentrifugation. The larger microfilaments and their associated proteins in the 'stabilized' extract sediment more rapidly than in the unstabilized extract. The proteins found in the 'shifted' fractions can be more confidently described as microfilament-associated proteins rather than simply proteins in large complexes fortuitously co-sedimenting with the filaments. This same approach can be used for microtubule stabilization, using Taxol as the 'perturbation' agent.

3.3 Membrane skeletons

Membrane–microfilament interaction sites have been isolated for a number of cells and cell structures (*Table 2*). The pioneering classic work in this area was the description of the membrane skeleton of the erythrocyte (27, 28). These early studies on the relatively simple model systems erythrocytes and platelets provided a general approach to isolating sites of association of cytoskeletal elements with plasma membranes, as well as insights into molecular mechanisms of interactions in the complexes.

The general approach to making a membrane skeleton preparation comprises three steps. The first entails the preparation of a plasma membrane fraction under appropriate stabilizing conditions. The optimal conditions for each cell type must be developed, although the Zn^{2+} stabilized preparation serves as a reasonable first approach. The next step is demembranation by lysis of the membranes in a non-ionic detergent-containing stabilizing buffer. The last step is separation by a method which efficiently fractionates components of the lysate by size, e.g. velocity sedimentation gradient or a gel filtration column. Optimization of the gradient conditions and gel permeation pore size should be based on the size of the complexes obtained, which can be estimated by a preliminary differential centrifugation of the lysate. An example is given below of the purification of a large membrane skeleton from a constitutively activated (29) ascites mammary tumour cell.

In this example the membrane preparation which served as the starting material was microvilli isolated from an ascites mammary tumour subline. Microvilli were physically sheared from cells in isotonic buffer using a wide gauge needle and separated from the remaining intact cell bodies by differential centrifugation (30, 31). This preparation affords a stabilized, highly purified plasma membrane retaining its associated microfilaments and uncontaminated by intracellular material, which remains in the intact cell bodies. From these microvilli a plasma membrane preparation was made by homogenization in a high pH, low ionic strength buffer which depolymerizes microfilaments (31, 32).

Early studies had shown that a large, glycoprotein-containing actin-associated complex could be isolated from either a microvillar microfilament core, made by extraction in non-ionic detergent, or a microvillar membrane preparation, prepared under hypotonic/high pH conditions which depolymerize microfilaments (31, 32). The isolation of the large complex, which contained the stably associated tyrosine kinases p185^{neu} (34), c-Src, and c-Abl (29), was accomplished by either gel filtration or velocity sedimentation (33).

Protocol 3. Preparation of a membrane- and microfilament-associated transmembrane complex from the 13762 ascites rat mammary adenocarcinoma (33)

Equipment and reagents

- Gel filtration column: a 0.75 cm × 75 cm column containing either Sepharose 2B or Sephacryl S500 (Pharmacia), pre-equilibrated with extraction buffer
- Microvillar membranes: prepared by homogenization of microvilli in 5 mM glycine, 2 mM EDTA, 2 mM 2-mercaptoethanol and differential centrifugation (31)
- Velocity sedimentation centrifugation: ultracentrifuge and SW40 rotor with buckets
- Extraction buffer: 0.15 M NaCl, 2 mM $MgCl_2$, 0.2 mM ATP, 0.2 mM dithioerythritol, 0.5% Triton X-100, 5 mM Tris pH 7.6; just prior to extracting membranes add a protease inhibitor cocktail (22)[a]

Method

1. Extract the microvillar membrane preparation (≤ 150 μg) by lysis in 1 ml of extraction buffer,[b] 15 min at room temperature.
2. Perform the fractionation by either gel filtration or velocity sedimentation centrifugation.
   (a) Gel filtration. After washing column with ten column volumes of extraction buffer, load the solubilized membrane preparation onto the gel filtration column. Elute with extraction buffer, collecting 1 ml fractions.
   (b) Velocity sedimentation centrifugation.
      (i) Pour 0.5 ml of 60% sucrose into the bottom of a Beckman SW40 rotor tube (total volume approx. 13 ml).
      (ii) Layer a gradient of 7–25% sucrose in extraction buffer (11.5 ml) onto the sucrose cushion.
      (iii) Load the extract, in a volume ≤ 1 ml, onto the gradient.[c]
      (iv) Place the carefully balanced buckets onto the SW40 rotor and centrifuge at 30 000 *g* for 16 h.
      (v) Collect 1 ml fractions by either extrusion from the top, using a peristaltic pump, or insertion of a 20 gauge needle into the bottom of the tube and collection by gravity flow.

3. To each fraction from either the column or the gradient, add SDS to 1.0%,[d] mixing thoroughly. Take 0.1 ml aliquots for scintillation counting, if samples were metabolically labelled.
4. Dialyse the remainder of each sample against 0.1% SDS, lyophilize, and resolubilize in 100–200 μl of electrophoresis buffer for SDS–PAGE or IEF/SDS–PAGE.[e]

[a] Pre-mixed protease inhibitor cocktails specially designed for use with mammalian cells, yeast, or bacteria are commercially available (Sigma).
[b] This ratio of extraction buffer to membrane protein gave optimal extraction.
[c] For tube stability during ultracentrifugation, the tube contents after loading the extract should be within 2–3 mm of the top of the tube.
[d] The addition of SDS prior to dialysis aids in displacement of Triton from proteins and prevents irreversible aggregation of proteins upon lyophilization.
[e] Boil for 3 min after addition of sample buffer to lyophilized fractions. Membrane and other hydrophobic proteins are frequently difficult to solubilize completely.

For simple, rapid quantification of protein across the gradient or column fractions, the cells can be pre-labelled metabolically with either [^{3}H]leucine for protein or [^{14}C]glucosamine (New England Nuclear) for glycosylated protein or both (see ref. 35).

3.3.1 Analyses

At this point a number of informative analyses may be performed (*Table 3*). An important predictor for success in understanding the molecular basis for

Table 3. Analysis of large complexes

Assay	Approach	Reference
Assay for complete protein composition	Protein stains Metabolic labelling Biotinylation	4, 60
Assays for specific proteins	Immunoblots; in-gel or on-blot enzyme assays	
Detection of receptors with lectins	Lectin blots	22
Detection of phosphorylated proteins	Metabolic labelling with $^{32}P_i$ or *in vitro* labelling with [^{32}P]ATP Immunoblot with anti-P-Tyr, anti-P-Ser, anti-P-Thr	57, 65, 66
Detection of activated signal transduction proteins	Immunoblot with antibodies specific for activated form	
Miscellaneous signal transduction protein assays		
PI3K	Hydrolysis of PIP_2	61
PKC	Activation of kinase by phosphatidylserine, DAG	62
PLCγ	Hydrolysis of phospholipid	63
Assays with inhibitors		64

cellular function is the ability of the investigator to recognize Mother Nature's major *caveats*. The prime *caveat* is 'Things are always more complex than one would like them to be'. The first corollary to the prime *caveat* is 'You will not find it unless you look for it' and conversely 'You can make things happen that do not occur in nature'. The literature is fraught with examples of deductions or extrapolations from oversimplified or inadequately controlled experiments. For this reason it is crucial to perform a thorough analysis of the membrane skeletal complex, which is often quite large. If its migration on gel filtration or velocity sedimentation suggests that it is a large complex, both general and specific approaches are available for analysis of the number and kinds of proteins present.

For protein detection the usual protein stains may be used, with a cautionary note concerning the use of silver staining. This procedure tends to cause extraction of certain proteins, particularly membrane glycoproteins (35), from gels. A particularly reliable and sensitive method is biotinylation of proteins in solution or on blots and detection with streptavidin linked to an enzyme such as horseradish peroxidase or alkaline phosphatase, using chemiluminescence detection. Excellent kits are available for this purpose; one which we have used quite successfully is supplied by Pierce.

Samples may be transferred to blots for probing with lectins, which can provide information on the presence of cell surface glycoproteins/receptors in the complexes (described in detail in ref. 22). Particularly useful lectins include concanavalin A (Con A) and wheat germ agglutinin, which bind to carbohydrate moieties found in many membrane proteins. Lectins which detect less common moieties can be useful when relevant to a specific glycoprotein of interest. This type of information may also provide useful approaches to methods for purification of specific glycoproteins. An example is the identification of the microfilament-associated glycoproteins of the TMC by [^{14}C]glucosamine label and Con A (36), which provided the basis for their purification on Con A–agarose columns (37).

More specific detection of signalling proteins may be accomplished by enzymatic assays (*Table 3*). These may be done in solution and, in some cases, on gels or blots. An especially useful application of gel and blot analyses is in the analysis of protein phosphorylation.

3.3.2 Phosphorylation, kinases, and phosphatases

Phosphorylation is the most important type of signalling event in regulating cell functions. A battery of experimental approaches and methods have been developed to analyse phosphorylation and its changes (*Table 4*). A hierarchy of assays can lead the investigator from the simple question 'Does my cell or cell fraction have kinase/phosphatase activity'? to 'What is the role of phosphorylation on a specific residue in the cellular function of the protein'? When assaying a fraction from an activated cell, several questions need to be answered. What proteins are phosphorylated in the cell population? What

Table 4. Approaches to detection of phosphorylation and its functions

Approach	Method
Detection of phosphorylated proteins	Metabolic labelling of cells with phosphate followed by SDS–PAGE and autoradiography [^{32}P]ATP labelling of cell lysates or fractions followed by SDS–PAGE and autoradiography Anti-phosphotyrosine immunoblotting or IP
Identification of phosphorylated amino acids and peptides	Protein hydrolysis under mild conditions, TLC, or TLE Use of Abs against specific phosphorylated forms Proteolytic cleavage of phosphorylated proteins and peptide analysis
Identification of kinases or phosphatases	In-gel or on-blot enzyme assays Inhibition with specific kinase or phosphatase inhibitors Immunoblot or IP with specific antibodies Immune complex assays of IPs

amino acid residues in the proteins are phosphorylated? What are the enzymes involved in the phosphorylation and dephosphorylation reactions? What types of signalling complexes are formed by the phosphorylated proteins?

Identification of proteins phosphorylated *in situ* can be readily accomplished by metabolic labelling with [^{32}P]phosphate followed by SDS–PAGE and autoradiography (57). The kinetics of stimulated phosphorylation can be addressed in this manner, as can the localization of the components by cell fractionation before SDS–PAGE. Alternatively, [γ-^{32}P]ATP can be used for labelling permeabilized cells or cell fractions, though it must be recognized that the results of the two methods will not necessarily be the same. Even the mildest permeabilization method allows for activation of enzymes and relocalization of the enzymes and/or potential substrates.

Amino acid identification encompasses two questions, the type of amino acid phosphorylated, and the site(s) on the protein which is modified. Both require hydrolysis of the protein(s) under conditions which do not cleave phosphate residues. Identification of phospho-Ser, -Thr, or -Tyr can then be accomplished using 1D or 2D thin-layer electrophoresis or chromatography, with appropriate standards, and autoradiography (40). Identification of the site phosphorylated involves protein cleavage, peptide fractionation, and sequencing. Although labelling is usually used for these analyses, advances in peptide fractionation and sequencing technology by mass spectrometry may soon provide a widely available alternative.

The use of anti-phosphotyrosine antibodies has been a huge boon to the signalling field. Not only do they provide simple reagents for identifying tyrosine-phosphorylated proteins, they can also be used for isolating those proteins by immunoprecipitation or other immunoaffinity methods. Moreover, these methods can be used for isolating and characterizing complexes formed by the phosphorylated proteins as one of the first stages of many signalling pathways (38). Comparable antibodies recognizing phospho-Ser and -Thr have not proven as reliable, although the great interest in these reagents is stimulating the generation of new and improved antibodies.

Once phosphorylated proteins have been identified, it is often desirable to identify the enzymes involved in their modification, both kinases and phosphatases. A powerful method which can be used on complex mixtures and does not require immunoprecipitation is the in-gel or on-blot assay, a simple approach to identifying individual kinases and phosphatases in a cell fraction. This method entails separation of the proteins by SDS–PAGE, followed by a complete denaturation–renaturation protocol and assay either in the gel or on a blot of the gel. The former requires incorporation of a visualizable (e.g. stained or radioactive) broad specificity substrate for the class of enzyme to be assayed. For example, if the enzymes of interest are any type of kinase, a protein which can be heavily phosphorylated on both tyrosine and serine/threonine residues will serve adequately as substrate. If the focus is on one of the classes of kinases, one may use a synthetic polypeptide which can be phosphorylated on that particular residue. An example is the synthetic polypeptide E_4Y, a linear random co-polymer of Glu and Tyr in a 4:1 ratio. E_4Y can be used in both solution and in-gel assays. The in-gel procedure for the detection of kinases has been described in several methods references (e.g. 57). The same substrate can also for the detection of tyrosine phosphatases after prior phosphorylation, as described in *Protocol 4* for in-gel detection of phosphatases (39).

Protocol 4. In-gel assay for identification of protein tyrosine phosphatases (modified from ref. 37)

Equipment and reagents

- Tyrosine kinase substrate: synthetic polypeptide E_4Y (Sigma) (M_r 20–50 kDa)
- Phosphorylation buffer: 150 mM NaCl, 2 mM $MnCl_2$, 0.02% TX-100, 5% glycerol, 0.1 mM sodium orthovanadate, 0.2 mM ATP, 50 mM Hepes pH 7.4
- Recombinant c-Src (Upstate Biotechnology, Inc.)
- SDS–PAGE gel (usually 8% or 10%, or a linear gradient of 8–12% acrylamide)
- SDS removal buffer: 20% isopropanol in 50 mM Tris–HCl pH 8.0
- Gel filtration column: 15 × 0.7 cm G-50 Sephadex, equilibrated in 50 mM imidazole pH 7.2
- Gel wash buffer: 50 mM Tris–HCl, 0.3% 2-mercaptoethanol pH 8.0
- Denaturing solution: 6 M guanidine–HCl, 1 mM EDTA
- Renaturation buffer: 1 mM EDTA, 0.3% 2-mercaptoethanol, 0.04% Tween 20, 50 mM Tris–HCl pH 8.0 (plus 4 mM DTT for the last incubation)[a]

Method

1. Incubate for 4–5 h 1 mg of E_4Y in 0.5 ml of phosphorylation buffer containing recombinant Src[b] and 0.25–50 mCi [γ-^{32}P]ATP (3000 Ci/mmol, New England Nuclear).

2. Microcentrifuge for 1 min. Collect supernatant and add an equal volume of cold 20% trichloroacetic acid. Incubate for 30 min on ice.
3. Centrifuge at 12000 *g* for 10 min at 4°C. Dissolve the pellet by adding 100 μl of 2 M Tris base.
4. Fractionate by gel filtration on Sephadex G-50, collecting 1 ml fractions.
5. Perform scintillation counting to identify the large E_4Y peak. Pool the peak fractions, concentrate, and store in aliquots for use in tyrosine phosphatase assays.
6. Thoroughly mix ^{32}P-labelled E_4Y (~ 10^5 c.p.m./ml) with the SDS–PAGE gel components prior to addition of polymerizing agents; polymerize.
7. Subject the cell fraction sample(s) to be analysed to SDS–PAGE electrophoresis in the usual manner.
8. Incubate the gel in SDS removal buffer for 30 min at room temperature with rocking.
9. Wash the gel twice, 30 min each, at room temperature in gel wash buffer.
10. Incubate the gel in denaturing solution for 1.5 h at room temperature.
11. Incubate the gel three times, 1 h each, in renaturation buffer at room temperature, then overnight in renaturation buffer plus 4 mM DTT.
12. Stain, destain, and dry gel; perform autoradiography and quantify by densitometry.

[a] The inclusion of reducing agent in the last step is critical to renaturation of tyrosine phosphatase activity, since the active site of these enzymes contains a conserved Cys residue required for activity.
[b] A crude membrane preparation of A431 cells (67), which contains activated EGF receptor, can also be used for phosphorylation of the substrate.

The in-gel phosphatase assay, like any gel method which entails the consumption of visible substrate, will produce a 'negative' image, as shown in *Figure 1*.

Kinases can be assayed by both in-gel and on-blot methods. For the in-gel kinase assay the non-phosphorylated substrate E_4Y is incorporated into the SDS–PAGE gel. After electrophoresis, kinases are detected by addition of [^{32}P]ATP in an assay buffer containing Mn^{2+}, which activates tyrosine kinases in the absence of ligand (41). After extensive washing of the gel, the kinases are then detected by autoradiography. *In situ* activated kinases may be assayed by a variation of this procedure in which Mn^{2+} is excluded from the renaturation buffer.

A particularly powerful method for the identification of autophosphory-

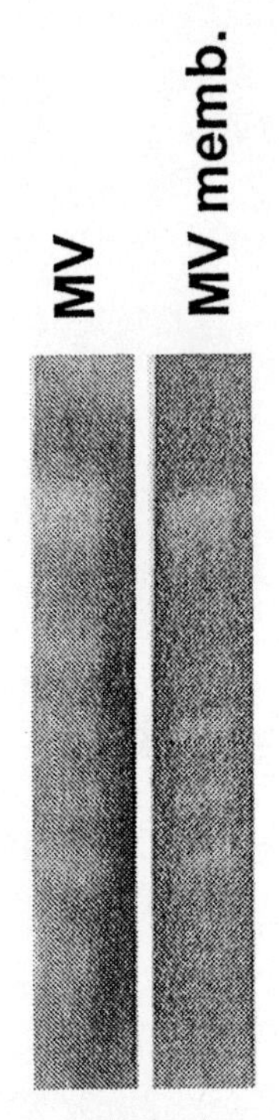

Figure 1. In-gel phosphatase assay of isolated microvilli (30) from 13762 ascites tumour cells and a microvillar membrane preparation made in an actin depolymerizing buffer (31).

lating kinases is the on-blot kinase assay, described in *Protocol 5*. This technique is simpler than the in-gel assay, since no exogenous substrate is required.

Protocol 5. On-blot assay for autophosphorylating kinases

Equipment and reagents

- Immobilon P (Millipore)
- Transfer buffer: 25 mM Tris pH 8.3, 192 mM glycine, 0.01% SDS, 10% methanol
- Denaturing solution: 7 M guanidine–HCl, 50 mM Tris pH 8.3, 2 mM EDTA, 50 mM DTT
- TTBS: 0.1% Tween 20, 150 mM NaCl, 50 mM Tris pH 8.0
- Renaturation buffer: 100 mM NaCl, 2 mM DTT, 2 mM EDTA, 1% BSA, 0.1% NP-40, 50 mM Tris pH 7.5[a]
- TBS: 150 mM NaCl, 50 mM Tris pH 8.0
- Blocking buffer: 5% BSA in TTBS
- Kinase activating buffer: 25 mM Hepes pH 7.5, 10 mM $MgCl_2$, 10 mM $MnCl_2$[b]

Method

1. Subject the cell fraction to electrophoresis on SDS–PAGE. Transfer proteins to Immobilon PVDF paper (Millipore) in transfer buffer at 175 mA for 1 h.[c]

2. Incubate the blot in denaturing solution for 1 h at room temperature with gentle rocking.
3. Wash the membrane briefly in renaturation buffer. Incubate at 4°C in the same buffer for 12–16 h with gentle rocking.
4. Block for 1 h at room temperature with blocking buffer.
5. Detect autophosphorylating kinases by incubating with 50 μCi/ml [γ-^{32}P]ATP in kinase activating buffer for 1 h at room temperature.
6. Wash three times in TBS and twice in TTBS. Treat the blots with 1 M KOH for 20 min at room temperature.
7. Wash three times with water, once in 10% acetic acid, and twice with water. Dry the blot and detect the autophosphorylated kinases by autoradiography.

[a] For the reliable reconstitution of the activities of integral membrane kinases it is important to include a small amount of non-ionic detergent, which replaces phospholipid in binding to the hydrophobic transmembrane helix.
[b] Mn^{2+} may be excluded for the assay of activated kinases.
[c] A longer transfer time is recommended for high M_r proteins and integral membrane kinases such as receptors. We recommend overnight transfer at 60 mA.

For further characterization of specific kinases and phosphatases, antibodies are available for most of the well known signalling enzymes and can be used for both immunoblotting and immunoprecipitation. For analysing the enzyme activities, immune complex assays are often used, in which appropriate substrates, e.g. peptide plus [^{32}P]ATP, are added to the immunoprecipitate (57).

4. Co-localization by co-immunoprecipitation

One of the most direct approaches to demonstrating an interaction between two proteins is co-immunoprecipitation from a lysate of a cell or cell fraction. The basic method involves lysis in a non-ionic detergent-containing buffer and immunoprecipitation with antibody to one component of interest, followed by immunoblot analysis using antibody to the second protein. *When attempting to demonstrate direct interactions, however, this frequently-used approach is over-simplistic and subject to several caveats.* The first, the you-will-not-find-it-unless-you-look-for-it *caveat*, concerns the inference of a direct complex between two components in co-precipitation experiments. This inference is based on the assumption that the co-immunoprecipitated protein being sought is the only one present in the immunoprecipitate. When mild extraction conditions are used for lysis, this is frequently not the case. In fact, components of cells and subcellular structures often exist in multi-component complexes. This observation is particularly true for proteins which interact with filament systems in both inactivated and activated cells, as exemplified by the complexes listed in *Table 2*.

When the investigator recognizes this problem and attempts to circumvent it by performing a preliminary centrifugation step prior to immunoprecipitation, a second *caveat* is encountered, i.e. the throwing-out-the-baby-with-the-bath-water *caveat*. In many cases membrane proteins tightly associated with cytoskeletal elements in these stable complexes can be largely removed during the centrifugation step. To avoid this potential hazard, one should begin by performing a whole cell lysate centrifugation and quantitative analysis of both the supernatant and the pellet for the protein(s) of interest. If a significant fraction of the protein is in the pellet, further fractionation methods should be sought which yield a 'more soluble', or smaller, complex containing the protein.

One such method is a fractionation of the cell lysate which avoids the problem of co-pelleting. Routine biochemical fractionation methods such as velocity sedimentation centrifugation or gel filtration largely avoid 'co-pelleting' of heterogeneous large complexes and frequently provide a more highly purified starting material for further analysis. These fractionation methods can include a perturbant of some type which will impart a degree of specificity of interactions in the complex of interest. An example is the phalloidin shift perturbation velocity sedimentation to analyse for association of a specific protein with microfilaments (4, 36). In this type of analysis distributions on gradients of lysates made in the presence and absence of this microfilament-stabilizing agent provide additional support for specificity of the interaction of the protein of interest with microfilaments. A further modification of this approach which is particularly useful when working with very large complexes is a modification of the lysis buffer which partially dissociates large complexes. Fractionation on a gradient or gel filtration column can yield smaller complexes containing the protein(s) of interest, which can then be subjected to further analysis by the same types of procedures listed in *Table 3*.

The following example typifies an extreme case, the composition of the ErbB2-containing signal transduction particle (STP) in the constitutively activated (29, 42) 13762 ascites rat mammary adenocarcinoma. Gel filtration and velocity sedimentation suggested that its size is $> 2 \times 10^6$ Da (33, 35). The elucidation of its composition is under investigation, using many of the assays listed in *Table 3*. Early attempts to immunoprecipitate phosphotyrosine-containing proteins from lysates of ascites cells as well as their highly stabilized microvilli identified a major concern in lysing cells and cell fractions for immunoprecipitation of monomeric proteins. Lysis buffer conditions short of complete denaturation in SDS resulted in incomplete dissociation of the large complex (37). Immunoprecipitation with anti-ErbB2 after lysis in RIPA (radioimmunoprecipitation assay) buffer (43), designed for dissociation of complexes, brought down a large complex. Further fractionation and immunoprecipitation studies showed that ErbB2 was associated in a large signalling complex with all of the components of the Ras to MAPK mitogenic pathway (34). The procedure for the fractionation and immunoprecipitation of this signal transduction particle is described in *Protocol 6*.

Protocol 6. Co-immunoprecipitation from constitutively activated 13762 ascites cells of p185neu/ErbB2 with the complete Ras to MAPK pathway

Equipment and reagents

- SDS–PAGE minigel (8% acrylamide)
- Nitrocellulose membranes
- Ascites cell microvilli, prepared as previously described (35)
- Lysis buffer: 0.5% Triton X-100, 2 mM EGTA, 5 mM glycine pH 9.5 (a microfilament-depolymerizing buffer) containing a protease inhibitor cocktail (Sigma)
- PBS (phosphate-buffered saline) pH 7.4
- TTBS (see *Protocol 5*)
- TBS (see *Protocol 5*)
- Second antibodies: mouse (for polyclonal) or rabbit (for monoclonal) IgG (1:14 000; Promega)
- Protein A–agarose (Sigma)
- Blocking buffer: TTBS containing 0.2% non-fat dry milk[a]
- Primary antibodies:[b] p185^{neu} mouse monoclonal (1:4000; Calbiochem Ab-3); Shc rabbit polyclonal (1:1000; Transduction Laboratories); Grb2/Sem5 rabbit polyclonal (1:1000; Upstate Biotechnology); Sos (1 + 2) rabbit polyclonal (1:1000; Transduction Laboratories); pan-Ras mouse monoclonal (1:750; Oncogene Science); human Raf-1 polyclonal rabbit (1:2500; Upstate Biotechnology, Inc.); MAPKK (Mek 1 + 2) rabbit polyclonal (1:1500; (Transduction Laboratories); MAPK (Erk 1 + 2) (1:2500; Zymed); and Rsk mouse monoclonal (1:500; Transduction Laboratories)

Method

1. Pre-incubate 25 μl of monoclonal anti-ErbB2 or 25 μl of control mouse non-immune antisera with protein A–agarose (Sigma) overnight at 4°C with rocking. Pellet by centrifugation at 150 g for 5 min and wash with PBS.
2. Incubate 300 μl of isolated microvilli on ice for 30 min in 700 μl of lysis buffer. Centrifuge at 15 000 *g* for 15 min at 4°C.
3. Add approx. 500 μl of the lysate supernatant to each antibody–agarose pellet. Incubate overnight at 4°C with rocking.
4. Collect the pellets by centrifugation at 150 g for 5 min. Wash three times with PBS.
5. Elute the agarose-bound protein with SDS–PAGE sample buffer and analyse by SDS–PAGE.
6. Transfer proteins to nitrocellulose blots as in *Protocol 5*. Block with blocking buffer.
7. Detect mitogenic pathway proteins by incubating the replicas with primary antibodies in TTBS (*Protocol 5*) containing 1% BSA for 1 h.
8. Remove the antibody solutions and incubate the blots for 1 h with horseradish peroxidase-conjugated antibodies to the appropriate rabbit or mouse immunoglobulin.

Protocol 6. *Continued*

9. Wash the blots three times (2 h each) with TTBS and once with TBS as in *Protocol 5.* Detect the reactive bands using a chemiluminescence detection kit (DuPont/NEN Renaissance™) according the manufacturer's instructions.

[a] When assaying blots with anti-phosphotyrosine, block the blot with 5% BSA in TTBS. The milk protein casein is heavily phosphorylated on tyrosine, causing a high background with this antibody.
[b] The dilution for each of the antibodies was optimized for the detection of the signalling proteins in the ascites cells.

5. Reconstitution

The reverse approach to co-purification of complexes is reassembly, to demonstrate the feasibility of direct interactions between the components. Although these are basically binding studies, the approaches are becoming quite diverse and sophisticated. An example of a simple solution binding experiment is given below to demonstrate the association of actin with the EGF receptor, whose sequence contains a motif similar to that of the G-actin-binding domain of profilin (44). As a demonstration of the specificity of binding, a synthetic peptide corresponding to the purported actin binding motif was used in a parallel competition experiment.

Protocol 7. Binding of epidermal growth factor receptor (EGFR) to F-actin (44)

Equipment and reagents

- Ultracentrifuge
- G-actin buffer: 2 mM Tris–HCl pH 7.4, 0.2 mM $CaCl_2$, 0.2 mM dithiothreitol, 0.5 mM ATP
- Densitometer
- Synthetic EGFR competitive peptide (DDVVDADEYLIPQ)

Method

1. Purify EGFR from A431 cells as previously described (44).
2. Purify actin from rabbit muscle as previously described (4).[a]
3. Incubate 2 μg EGFR with 10 μg actin in 200 μl G-actin buffer.
4. As a control, mix the competitive peptide from EGFR (5 mg) with 10 μg actin and 100 ng EGFR.
5. Induce actin polymerization by adding KCl to 75 mM and $MgCl_2$ to 2 mM, respectively, and incubating for 1 h at room temperature.
6. Centrifuge for 1 h at 100 000 *g*. Prepare pellets for SDS–PAGE. Concentrate supernatants and prepare for SDS–PAGE.

7. Stain SDS–PAGE gels with Coomassie and quantitate EGFR bands by densitometric scanning. Calculate actin-bound versus unbound fraction after subtracting control level.

[a] If the procedure is not used routinely, smaller amounts of actin can be obtained commercially (e.g. from Sigma). Commercial actin should be purified by two rounds of polymerization–depolymerization in the appropriate buffers before use.

In addition to the traditional solution studies for binding, a growing number of solid phase assays are available which simplify both presentation of proteins and retrieval of complexes. A solid phase binding assay which has great utility is the overlay of gels or blots of cell fractions containing the prospective binding protein with a solution containing the protein of interest. The actin overlay blot (45) is an excellent approach to the identification of prospective actin-binding signalling or other regulatory proteins. Although this method has been described in several methods treatises (e.g. ref. 4), its simplicity and broad utility justify its reiteration here. A modification of the original procedure is given in the example below, using biotinylation and an avidin-linked enzyme for sensitive detection.

Protocol 8. Analysis of actin-binding proteins by blot overlay with biotinylated actin

Equipment and reagents

- Immobilon PVDF membrane (Millipore)
- Biotinylated actin, prepared using the biotinylation kit and instructions supplied by Pierce; store in aliquots at −80°C
- Actin polymerization buffer: 100 mM KCl, 2 mM $MgCl_2$, 1 mM ATP, 0.1 mM $CaCl_2$, 10 mM Tris–HCl pH 7.5
- Actin depolymerization buffer: 2 mM Tris–HCl pH 7.5, 0.1 mM $CaCl_2$, 0.5 mM ATP, 0.2 mM DTT
- TTBS
- Denaturation buffer: 7 M guanidine–HCl, 50 mM TES pH 8.3, 2 mM EDTA, 50 mM DTT
- Renaturation buffer: 50 mM TES, 100 mM KCl, 0.5 mM DTT, 2 mM EDTA, 1% BSA, 0.1% Nonidet P-40 pH 7.5
- Membrane blocking buffer: 5% BSA in 120 mM KCl, 30 mM TES pH 7.5
- TBS
- Avidin-conjugated to horseradish peroxidase
- 4-Chloro-1-naphthol

Method

1. Fractionate the preparation to be analysed by SDS–PAGE and transfer to an Immobilon PVDF membrane.
2. Incubate the membrane at room temperature in denaturation buffer at 4°C overnight with gentle rocking.
3. Block the membrane with blocking buffer at room temperature for 1 h. Incubate with biotinylated actin (approx. 1 μM) in actin polymerization buffer containing 0.02% Triton at 4°C overnight.
4. Wash three times in TTBS and twice in TBS. Incubate the blot with avidin conjugated to horseradish peroxidase at room temperature for 1 h, and develop with 4-chloro-1-naphthol to identify actin-binding proteins.

The actin overlay assay offers several advantages, prominent among these being that the binding protein in question need not be pure, or even identified, for that matter. Consequently, only a small amount of crude preparation containing the protein need be used. Another is that actin is inexpensive and simple to use. Prevailing dogma that actin is 'sticky' and binds many proteins non-specifically has discouraged the use of the actin overlay technique. Use of the appropriate blot blocking methods, positive and negative controls, and competition with excess unlabelled actin afford quite reliable data using this powerful tool.

The reverse approach is to use the authentic potential actin-binding protein in an overlay procedure. This approach requires either that an abundant source of purified protein is available or that authentic protein can be labelled such that it can be used in very small quantity. If the sequence is known, *in vitro* translation in the presence of a high specific activity ^{35}S-labelled amino acid or mixture provides an excellent source of protein. *Protocol 9* demonstrates the utility of this approach in showing that p58^{gag} (29, 46, 47), a membrane skeletal protein implicated in receptor stabilization and xenotransplantablility (33) in the 13762 ascites rat mammary adenocarcinoma, binds directly to TMC-gp65 and -55 of the large, ErbB2-containing signal transduction particle (37).

Protocol 9. Binding of [^{35}S]methionine labelled, *in vitro* translated p58^{gag} to TMC-glycoproteins (37)

Equipment and reagents

- Immobilon P membranes (Millipore)
- [^{35}S]methionine labelled, *in vitro* translated p58^{gag} (produced from 4 μg DNA) (48)
- TTBS
- Purified TMC-gp complex (prepared by Con A affinity column) (37)
- Blocking buffer: 5% BSA in 30 mM TES pH 7.5, 120 mM KCl

Method

1. Separate the Con A–agarose purified glycoproteins by SDS–PAGE and transfer to Immobilon P membranes (37).
2. Denature and renature the blotted proteins as described in the biotinylated actin overlay protocol (*Protocol 8*).
3. Block with blocking buffer at room temperature for 1 h. Incubate with [^{35}S]methionine labelled, *in vitro* translated p58 in PBS plus 0.05% Triton X-100 at 4°C overnight.
4. For the control perform the incubation in the same manner in the presence of a 10^4-fold excess of unlabelled *in vitro* translated p58. Begin the competition by incubating the blot with cold p58 at room temperature for 1 h. Add ^{35}S-labelled p58, then incubate at 4°C overnight.
5. Wash the blots three times in TTBS and twice in TBS. Perform autoradiography.

An extension of this approach in which the binding studies are performed using truncations of one or both of the proteins of interest can be particularly informative. Frequently, binding domains or motifs can be generated which retain their active conformations under physiological binding conditions. This approach has been very useful in studying the specificity of binding to proteins containing SH2 or SH3 domains, which refold even after solubilization in SDS (49). Basically, the procedure involves the generation of the appropriate domain by transfection into a protein expression system. The transfected cells are lysed and the lysate used for blot overlay or for solid phase binding assay using cell fractions enriched in target protein(s). Some of these signalling domain peptides, for example the SH2 domain of Src and the SH3 domain of PI-3-kinase (Calbiochem), are commercially available. *Protocol 10* illustrates the use of the c-Src SH3 domain peptide to demonstrate the specific binding of the membrane skeletal protein $p58^{gag}$ to c-Src via its SH3 domain, the consequent activation of Src, and tyrosine phosphorylation of $p58^{gag}$ (47).

Protocol 10. Binding of $p58^{gag}$ to the SH3 domain of c-Src (47)

Equipment and reagents

- Nitrocellulose blots
- Biotinylated recombinant GST-Src SH3 fusion proteins, prepared as previously described (50, 51)
- Agarose-bound GST-Src SH3 fusion proteins (52)[a]
- Microvilli and microvillar fractions, prepared as described previously (35)
- SDS–PAGE gel (8% acrylamide)
- Streptavidin-conjugated alkaline phosphatase in TTBS (*Protocol 5*)

A. *Blot*

1. Separate the proteins in the cell fractions by SDS–PAGE.
2. Block the filters in TTBS containing 0.2% non-fat dry milk and incubate with biotinylated probes (1 μg/ml) in TTBS for 2 h.
3. Wash the filters four times with TTBS and detect the bound biotinylated proteins using streptavidin-conjugated alkaline phosphatase in TTBS.

B. *Affinity purification*

1. Incubate 0.5 ml of flow-through from the Con A–agarose column with Src SH3 domain-agarose or GST-agarose overnight at 4°C.
2. Wash the agarose conjugates four times with 0.05% Triton in PBS and once with PBS.
3. Elute the proteins with electrophoresis sample buffer and analyse by SDS–PAGE.
4. Analyse the proteins in the SH3 domain-purified preparation with anti-$p58^{gag}$ (46, 48).

As a control for specificity of binding, the GST fusion protein can be incubated with recombinant Src SH3 domain in the presence of authentic c-Src, which can be obtained from Src-infected insect cells. Recombinant c-Src is also available commercially (Upstate Biotechnology, Inc.). A competition curve can be generated by using several ratios of c-Src to binding protein and analysing by anti-Src immunoblotting and densitometric comparison of bands on the blot to a standard curve of recombinant c-Src. Since $p58^{gag}$ binding activated c-Src and was phosphorylated by it, the level of phosphorylation of $p58^{gag}$ at each ratio of intact Src to peptide could be assayed by stripping the blot and probing with anti-phosphotyrosine antibody (Transduction Laboratories).

6. Perspectives (for a more comprehensive view, see Chapter 8, ref. 1)

The cytoskeleton serves a multifunctional role in regulating the cellular signalling processes which determine cell behaviours. Many of the important signalling pathways are associated with membranes, thus membrane transport may provide the mode of information transfer. Cytoskeletal structures provide the highway and motors for transport of information-carrying molecules within the cell, e.g. from the plasma membrane to the nucleus. Both microtubules and microfilaments have been implicated in vesicle transport and in signalling. The connections between information transfer and cytoskeletal behaviour remain elusive, but an increasing number of molecules appear to contribute to both, including the phosphoinositide kinases and Rho family of small G-proteins.

The cytoskeleton also acts as a scaffolding on which signalling elements are concentrated and organized. This mechanism may serve to enhance the kinetics of the flow of information through individual signalling pathways. Considerable circumstantial evidence supports such a mechanism. One example may be the proposal that the translational apparatus is organized on cytoskeletal elements to facilitate 'channelling' of substrates and protein synthesis. By the same token, cytoskeletal structures may serve to sequester signalling proteins until an activation event triggers their release. Many signalling proteins appear to shuttle between the cytoplasm and the nucleus upon activation. Association of transcription-modulating proteins with cytoskeletal structures is an appealing general mechanism for retention of the signalling proteins in the cytoplasm in inactivated cells.

An equally important function is the integration of multiple signalling pathways for 'cross-talk' after a signalling event which leads to pleiotypic cellular effects. It is at these sites that one expects to obtain insights into the transduction of signals which elicit morphological and major functional changes in cells. Scaffolding proteins which are key to linkage of pathways will be expected to associate with cytoskeletal machinery. The study of the complexes which comprise these integration sites will be a primary focus of studies on cross-talk between pathways. A major effort is now underway to identify these complexes and determine their compositions and interactions, especially those which interact with cytoskeletal proteins.

It seems reasonable to extrapolate that application of increasingly sophisticated microscopic, biophysical, biochemical, immunological, and molecular biological tools will unravel most of the complex interactions determining the role of the cytoskeleton in cell signalling. A battery of diverse approaches toward the analysis of potential protein binding partners is currently available, including phage display, yeast two-hybrid systems, and a growing number of solid phase assay techniques. The last approach has initiated a revolution in our capabilities to screen peptide combinatorial libraries (68), and is already making substantial contributions to our understanding of the specificities of signalling domains such as SH2, SH3, PDZ, and Lim domains. The oriented peptide library has the capability of identifying protein–protein interaction sites for almost any interacting system involving a linear peptide sequence. These techniques allow insights not only into potential physiological binding proteins (53), but also into specific substrates for enzymes (54, 69).

A major focus in the next decade will be the elucidation of the structure and organization of multimolecular complexes. Contributions to our understanding of the organization and reorganization of signalling complexes are expected with advances in the capabilities of high resolution structural analyses. Innovations in electron microscopy will likely make the most immediate contributions to the study of static structures of large complexes, though EM lacks the necessary resolution capability for molecular characterization. At the opposite extreme X-ray crystallography offers excellent resolution and capability for the study of binding sequences and conformation (e.g. ref. 70) in static systems, while lacking the capacity to study large complexes. Other tools, such as NMR spectroscopy, are quite useful for studying interactions, conformations, and dynamic processes within ligand–protein complexes (71); the limitation of NMR is the size of the complex which can be analysed. Mass spectrometry, using electrospray ionization (ESI)- and matrix-assisted laser desorption/ionization (MALDI)-based approaches, has great potential in characterizing non-covalent protein complexes and assessing the contribution of individual amino acid residues to a protein's function (72). In addition, it can be used to study relatively weak interactions and to assess stoichiometries of components (73). Delineating the molecular details of the interactions involved in such supramolecular complexes remains one of the challenges for the future of both signalling and cytoskeleton research. It is clear that a multifaceted approach will be required for their characterization.

References

1. Carraway, K. L., Carraway, C. A. C., and Carraway, K. L. III. (1998). *Signaling and the cytoskeleton*. Springer, Berlin.

2. Carraway, C. A. C. and Carraway, K. L. (1996). In *Treatise on the cytoskeleton* (ed. J. E. Hesketh and I. F. Pryme), Vol. 2, p. 207. JAI Press, Inc., Greenwich, CT.
3. Carraway, K. L. and Carraway, C. A. C. (1995). *BioEssays*, **17**, 171.
4. Carraway, K. L. and Carraway, C. A. C. (ed.) (1992). *The cytoskeleton: a practical approach*. IRL/Oxford University Press, Oxford.
5. Van Etten, R. A., Jackson, P., and Baltimore, D. (1989). *Cell*, **58**, 669.
6. Van Etten, R. A., Jackson, P. K., Baltimore, D., Sanders, M. C., Matsudaira, P. T., and Janmey, P. A. (1994). *J. Cell Biol.*, **124**, 325.
7. Carraway, K. L. and Carraway, C. A. C. (1982). In *Antibody as a tool: the application of immunochemistry* (ed. J. J. Marchalonis and G. W. Warr), p. 509. John Wiley, Chichester, England.
8. Carraway, K. L. (1975). *Biochim. Biophys. Acta*, **415**, 379.
9. Carraway, K. L., Fogle, D. D., Chesnut, R. W., Huggins, J. W., and Carraway, C. A. C. (1976). *J. Biol. Chem.*, **251**, 6173.
10. Huggins, J. W. and Carraway, K. L. (1976). *J. Supramol. Struct.*, **5**, 59.
11. Mooseker, M. S. and Tilney, L. G. (1975). *J. Cell Biol.*, **67**, 725.
12. Mazia, D., Schatten, G., and Sale, W. (1975). *J. Cell Biol.*, **66**, 198.
13. Tilney, L. G. (1976). *J. Cell Biol.*, **69**, 73.
14. Brown, S., Levinson, W., and Spudich, J. A. (1976). *J. Supramol. Struct.*, **5**, 119.
15. Hainfeld, J. F. and Steck, T. L. (1977). *J. Supramol. Struct.*, **6**, 301.
16. Small, J. V. (1977). *J. Cell Sci.*, **24**, 327.
17. Starger, J. M., Brown, W. E., Goldman, A. E., and Goldman, R. D. (1978). *J. Cell Biol.*, **78**, 93.
18. Trotter, J. A., Foerder, B. A., and Keller, J. M. (1978). *J. Cell Sci.*, **31**, 369.
19. Osborn, M., Webster, R. E., and Weber, K. (1978). *J. Cell Biol.*, **77**, R27.
20. Webster, R. E., Henderson, D., Osborn, M., and Weber, K. (1978). *Proc. Natl. Acad. Sci. USA*, **75**, 5511.
21. Weihing, R. R. (1976). *J. Cell Biol.*, **71**, 303.
22. Carraway, C. A. C. (1992). In *The cytoskeleton: a practical approach* (ed. K. L. Carraway and C. A. C. Carraway), p. 123. IRL Press, Oxford, UK.
23. Walker, J. H., Boustead, C. M., Witzemann, V., Shaw, G., Weber, K., and Osborn, M. (1985). *Eur. J. Cell Biol.*, **38**, 123.
24. Osborn, M. and Weber, K. (1977). *Cell*, **12**, 561.
25. Wallin, M. and Stromberg, E. (1995). *Int. Rev. Cytol.*, **157**, 1.
26. Carraway, C. A. C. and Weiss, M. (1985). *Exp. Cell Res.*, **161**, 150.
27. Palek, J. and Liu, S. C. (1980). *Prog. Clin. Biol. Res.*, **43**, 21.
28. Bennett, V. (1982). *J. Cell. Biochem.*, **18**, 49.
29. Juang, S.-H., Carvajal, M. E., Whitney, M., Liu, Y., and Carraway, C. A. C. (1996). *Oncogene*, **12**, 1033.
30. Carraway, K. L., Huggins, J. W., Cerra, R. F., Yeltman, D. R., and Carraway, C. A. C. (1980). *Nature*, **285**, 508.
31. Carraway, C. A. C., Cerra, R. F., Bell, P. B., and Carraway, K. L. (1982). *Biochim. Biophys. Acta*, **719**, 126.
32. Carraway, K. L., Cerra, R. F., Jung, G., and Carraway, C. A. C. (1982). *J. Cell Biol.*, **94**, 624.
33. Carraway, C. A. C., Jung, G., and Carraway, K. L. (1983). *Proc. Natl. Acad. Sci. USA*, **80**, 430.

34. Carraway, C. A. C., Carvajal, M. E., Li, Y., and Carraway, K. L. (1993). *J. Biol. Chem.*, **268**, 5582.
35. Carraway, C. A. C., Fang, H., Ye, X., Juang, S.-H., Liu, Y., Carvajal, M., *et al.* (1991). *J. Biol. Chem.*, **266**, 16238.
36. Carraway, C. A. C. and Weiss, M. (1985). *Exp. Cell Res.*, **161**, 150.
37. Li, Y., Hua, F., Carraway, K. L., and Carraway, C. A. C. (1999). *J. Biol. Chem.*, **274**, 25651.
38. Carraway, K. L., Carvajal, M. E., and Carraway, C. A. C. (1999). *J. Biol. Chem.*, **274**, 25659.
39. Burridge, K. and Nelson, A. (1995). *Anal. Biochem.*, **232**, 56.
40. Boyle, W. J., van der Geer, P., and Hunter, T. (1991). In *Methods in enzymology*, Vol. 201, p. 110.
41. Koland, J. G. and Cerione, R. A. (1990). *Biochim. Biophys. Acta*, **1052**, 489.
42. Carraway, K. L. III, Rossi, E. A., Komatsu, M., Price-Schiavi, S. A., Huang, D., Guy, P. M., *et al.* (1999). *J. Biol. Chem.*, **274**, 5263.
43. Burr, J. G., Dreyfuss, G., Penman, S., and Buchanan, J. M. (1980). *Proc. Natl. Acad. Sci. USA*, **77**, 3484.
44. den Hartigh, J. C., van Bergen en Henegouwen, P. M. P., Verkleij, A. J., and Boonstra, J. (1992). *J. Cell Biol.*, **119**, 349.
45. Snabes, M. C., Boyd, A. E. 3d, and Bryan, J. (1983). *Exp. Cell Res.*, **146**, 63.
46. Liu, Y., Carraway, K. L., and Carraway, C. A. C. (1989). *J. Biol. Chem.*, **264**, 1208.
47. Huang, J., Zhang, B.-T., Li, Y., Juang, S.-H., Mayer, B. J., Carraway, K. L., *et al.* (1999). *Oncogene*, **18**, 4099.
48. Juang, S.-H., Huang, J., Li, Y., Salas, P., Fregien, N., Carraway, C. A. C., *et al.* (1994). *J. Biol. Chem.*, **269**, 15067.
49. Cantley, L. C., Auger, K. R., Carpenter, C., Duckworth, B., Graziani, A., Kapeller, R., *et al.* (1991). *Cell*, **64**, 281.
50. Mayer, B. J., Jackson, P. K., and Baltimore, D. (1991). *Proc. Natl. Acad. Sci. USA*, **88**, 627.
51. Cicchetti, P., Mayer, B. J., Thiel, G., and Baltimore, D. (1992). *Science*, **257**, 803.
52. Weng, Z., Taylor, J. A., Turner, C. E., Brugge, J. S., and Seidel-Dugan, C. (1993). *J. Biol. Chem.*, **268**, 14956.
53. Zhou, S. and Cantley, L. C. (1995). In *Methods in enzymology*, Vol. 254, p. 523.
54. Zhou, S. and Cantley, L. C. (1998). *Methods Mol. Biol.*, **87**, 87.
55. Graham, J. M. and Rickwood, D. (ed.) (1997). *Subcellular fractionation: a practical approach*. IRL Press, Oxford.
56. Fox, J. E., Reynolds, C. C., and Boyles, J. K. (1992). In *Methods in enzymology*, Vol. 215, p. 42.
57. Chang, W. J., Ying, Y. S., Rothberg, K. G., Hooper, N. M., Turner, A. J., Gambliel, H. A., *et al.* (1994). *J. Cell Biol.*, **126**, 127.
58. Sargiacomo, M., Sudol, M., Tang, Z., and Lisanti, M. P. (1993). *J. Cell Biol.*, **122**, 789.
59. Franklin, C. C., Adler, V., and Kraft, A. S. (1995). In *Methods in enzymology*, Vol. 254, p. 551.
60. Pierce Catalog (1999) Rockford, IL, USA.
61. Auger, K. R., Carpenter, C. L., Cantley, L. C., and Varticovski, L. (1989). *J. Biol Chem.*, **264**, 20181.
62. Epand, R. M. (1994). *Anal. Biochem.*, **218**, 241.

63. Ulitzur, S. and Heller, M. (1978). *Anal. Biochem.*, **91**, 421.
64. Biomol Catalog (1999) Plymouth Meeting, PA, USA.
65. Cooper, J. A., Sefton, B. M., and Hunter, T. (1983). In *Methods in enzymology*, Vol. 99, p. 387.
66. Sahal, D., Li, S. L., and Fujita-Yamaguchi, Y. (1991). In *Methods in enzymology*, Vol. 200, p. 90.
67. Lin, P. H., Selinfreund, R. H., Wakshull, E., and Wharton, W. (1988). *Anal. Biochem.*, **168**, 300.
68. Lam, K. S. (1997). *Anticancer Drug Des.*, **12**, 145.
69. Songyang, Z., Blechner, S., Hoagland, N., Hoekstra, M. F., Piwnica-Worms, H., and Cantley, L. C. (1994). *Curr. Biol.*, **4**, 973.
70. Edmundson, A. B., Harris, D. L., Tribbick, G., and Geysen, H. M. (1991). *Ciba Found. Symp.*, **158**, 213.
71. Clore, G. M. and Gronenborn, A. M. (1998). *Trends Biotechnol.*, **16**, 22.
72. Winston, R. L. and Fitzgerald, M. C. (1997). *Mass Spectrom. Rev.*, **16**, 165.
73. Loo, J. A. (1997). *Mass Spectrom. Rev.*, **16**, 1.

5

Ras-related GTPases and the cytoskeleton

ANNE J. RIDLEY and HELEN M. FLINN

1. Introduction

The Ras superfamily of GTPases all bind GTP, and can hydrolyse GTP to yield GDP. They act as molecular switches in cells and are active when bound to GTP, assuming a conformation which allows them to interact with their downstream target proteins. In the GDP-bound conformation they no longer bind to most target proteins (1, 2). Under physiological conditions, exchange of guanine nucleotides on the GTPases is very slow and is catalysed by exchange factors, while GTP hydrolysis is catalysed by GTPase activating proteins (GAPs) (*Figure 1*).

The Ras superfamily can be divided into several subfamilies, based on sequence homology (1, 3). The Rho family of GTPases consists of at least 12 mammalian proteins, of which Rho, Rac, and Cdc42 have been best characterized (4, 5). Initially, they were identified as key regulators of actin cytoskeletal reorganization induced by extracellular signals. This led to research showing that they regulate many aspects of cell behaviour that are dependent on the actin cytoskeleton, including cell migration, phagocytosis, smooth muscle cell contraction, and intracellular vesicle trafficking. Genetic analysis in *Saccharomyces cerevisiae*, *Schizosaccharomyces pombe*, *Drosophila*, and *Caenorhabditis elegans* has also been important in elucidating the function of Rho GTPases and interacting proteins in regulating actin organization (6). More recently, Rho GTPases have been shown to be involved in mitogenic signalling and regulation of transcription factor activity, and at least in part these responses are independent of their effects on the cytoskeleton (4, 7).

A number of approaches have been taken to define the functions of Ras GTPases in regulating cell morphology and cytoskeletal organization. Initially, microinjection studies with purified recombinant proteins established their roles in mediating the formation of distinct types of actin cytoskeletal structures. Rho was shown to regulate actin stress fibre formation, while Rac regulates membrane ruffling and the formation of lamellipodia, and Cdc42 regulates filopodia (8–11). Injection of recombinant Ras protein stimulates

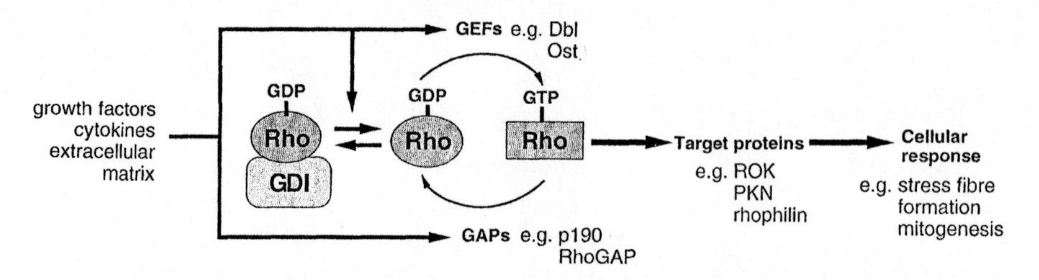

Figure 1. Regulation of Rho GTPases. Rho GTPases form an inactive cytosolic complex with GDIs (guanine nucleotide dissociation inhibitors), in which the GDI masks the carboxy-terminal isoprenyl group and thereby prevents membrane association of the GTPase. This complex is in equilibrium with the membrane-bound form of the GTPase. Activation of cells by extracellular signals such as growth factors, cytokines, and extracellular matrix proteins stimulates an increase in the membrane-bound form of the GTPase, presumably by stimulating dissociation of the Rho-GDI complex. At the membrane exchange of nucleotide is stimulated by exchange factors (GEFs), leading to an increase in the level of Rho-GTP. The level of Rho-GTP can also be regulated by altering the activity of GTPase activating proteins (GAPs), which stimulate GTP hydrolysis on Rho proteins, converting them from the GTP- to GDP-bound form. Rho-GTP is the active form of the protein that interacts with downstream target proteins, in turn activating them and ultimately inducing cellular responses.

membrane ruffling in fibroblasts and epithelial cells via Rac (8, 12, 13). Microinjection or transfection of DNA expression vectors has also been used to analyse Rho GTPase function and intracellular localization (e.g. 14–16). More recently, introduction of recombinant proteins into permeabilized cells or *in vitro* assay systems and biochemical analysis of intracellular proteins required for specific responses have allowed further analysis of the effects of these proteins on actin organization. For example, in permeabilized mast cells it is possible to follow the changes in actin organization induced by Rac and Rho (17). In permeabilized fibroblasts biochemical fractionation of cell extracts has led to the identification of moesin as a protein involved in Rho-induced stress fibre formation (18). In addition, Cdc42 can stimulate actin polymerization in cell extracts. This *in vitro* system should greatly facilitate the biochemical identification of the components linking Cdc42 activation to actin reorganization (19–21).

Once functions for Rho, Rac, and Cdc42 had been identified, a major goal of subsequent research has been to determine how they induce specific responses, both by mapping the regions of each Rho GTPase required for that response and by identifying downstream interacting partners. A variety of techniques have been used in these experiments. Mutant proteins have been created either by making point mutations or small deletions, or by making chimeric proteins between closely related family members (22–26). These mutant proteins are introduced into cells either as recombinant proteins by microinjection, or in expression vectors by transient transfection or micro-injection. Downstream partners have been identified by biochemical purification of proteins binding to GTP-loaded Rho, Rac, or Cdc42 or by yeast two-hybrid screening. Expression of these downstream targets has also been important in defining the signalling pathways leading to cytoskeletal reorganization.

Here, we describe the technique of microinjection as applied to Rho GTPases and their interaction partners, as this has been widely and successfully used to dissect these signalling pathways. In addition, we describe scrape loading as a technique for introducing proteins into cells which allows rapid biochemical analysis of pathways activated by these proteins. Finally, we describe the methods used to analyse cytoskeletal changes induced by Rho GTPases and other signalling proteins.

2. Introduction of recombinant proteins into cells

A major advantage of introducing recombinant proteins into cells over transfection of DNA is that it is possible to analyse very early responses to proteins. Responses to proteins can be detected within minutes, whereas DNA requires hours to express, and expression levels vary greatly between cells. Microinjection has been widely used as a technique to introduce proteins into mammalian cells. Most cells, including primary cells, are microinjectable, whereas many cell types are not readily transfectable. Analysis of responses in microinjected cells is usually based on immunocytochemical approaches as, in general, it is not possible to inject sufficient numbers of cells to carry out biochemical studies. In some cases, however, microinjection has been used to analyse changes in protein phosphorylation, for example, following injection of fibroblasts with cAMP-dependent protein kinase (27).

In addition to microinjection, other methods have been used to introduce recombinant proteins directly into cells. Scrape loading is a technique whereby proteins enter cells through holes transiently generated by detachment of adherent cells from tissue culture plates (28, 29). Its advantage over microinjection is that proteins can be introduced into large numbers of cells, allowing analysis at a biochemical level of signalling pathways activated acutely in response to an active protein. It has been used to analyse signalling pathways activated by Ras and Rho (30–33). One disadvantage compared to microinjection is that the cells are detached from the substratum, so that signals provided by the extracellular matrix are disrupted. This has to be taken into account when analysing the results. A variety of other techniques have been devised to introduce proteins into cells, including glass bead loading (34), osmotic lysis of pinosomes (35, 36), and syringe loading (37), although none of these have been reported to be used with Rho GTPases. Macromolecules, including proteins and antibodies, can also be introduced into cells by electroporation (38). This approach is particularly suitable for non-adherent cells, including haematopoietic cells, as microinjection can only be carried out on adherent cells. Finally, Rho GTPases have been successfully introduced into permeabilized cell systems, including mast cells and fibroblasts (17, 18, 39), where it is then possible to carry out further analysis of signalling proteins involved on a biochemical level.

2.1 Purification of recombinant proteins

Recombinant Rho GTPases are typically purified from *E. coli* as glutathione *S*-transferase (GST) fusion proteins, as this allows a straightforward one-step purification on glutathione beads (40). In addition, the GST can be enzymatically cleaved from the GTPase subsequent to purification, so that unfused GTPase can be introduced into cells. A disadvantage to the GST-fusion proteins is that generally only approximately 10% of the purified protein is in an active conformation and can bind to guanine nucleotides (40). Despite this problem, these proteins have been used successfully for structural analyses (41, 42). Other types of fusion proteins can also be used, but for introduction of proteins into cells, it is preferable that the fusion partner or tag be removable, as it may alter the activity of the protein and/or its localization within the cell. Any fusion partner or tag must be attached to the amino-terminus of the GTPases, as the carboxy-terminus is post-translationally modified by isoprenylation, removal of the last three amino acids, and carboxymethylation. The last four amino acids, the CAAX box, are essential for these modifications to occur (43).

E. coli-derived recombinant proteins are not post-translationally modified, and it is presumed that at least a proportion of the recombinant protein becomes modified following introduction into cells. In some *in vitro* assays with Rho GTPases, however, it is necessary to use carboxy-terminally modified recombinant proteins (19, 21). For this, the major approach has been to express the proteins using baculovirus vectors in insect cells such as *Spodoptera frugiperda* (Sf9) cells. In these cells the proteins are correctly post-translationally modified and can be purified in isoprenylated forms (44).

Protocol 1. Purification of recombinant Rho GTPases from *E. coli*

Reagents

- L broth, reconstituted from powder (Gibco BRL) with double-distilled H_2O; ampicillin is added from a 100 mg/ml stock (stored at –20°C) immediately prior to incubation
- Lysis buffer: 50 mM Tris–HCl pH 7.5, 50 mM NaCl, 5 mM $MgCl_2$; add 1 mM DTT, 1 mM PMSF immediately before use
- Thrombin buffer: 50 mM Tris–HCl pH 7.5, 2.5 mM $CaCl_2$, 100 mM NaCl, 5 mM $MgCl_2$; add 1 mM DTT immediately before use
- IPTG: stock solution 1 M in H_2O stored at –20°C
- Dialysis buffer: 50 mM Tris–HCl pH 7.5, 100 mM NaCl, 5 mM $MgCl_2$; add 1 mM DTT immediately before use
- Thrombin: dissolve bovine thrombin (Sigma, Cat. No. T7513) at 0.5 U/μl in H_2O, and store in 10 μl aliquots at –20°C; use approx. 3 U for each litre of *E. coli*
- Glutathione–Sepharose beads (Pharmacia)

Method

1. Inoculate *E. coli* from an agar plate or glycerol stock into 100 ml of L broth containing 100 μg/ml ampicillin and shake the culture at 37 °C overnight. Dilute the culture 1:10 into fresh, pre-warmed L broth containing 50 μg/ml ampicillin and grow for 1 h at 37 °C.
2. Induce protein expression with IPTG (0.2 mM) for 3 h at 37 °C.[a] Centrifuge cells (10 min, 5000 g, 4 °C) and resuspend the pelleted *E. coli* in 6 ml of cold lysis buffer. Keep on ice.
3. Lyse the bacteria by sonication[b] for 3 × 45 sec with 1 min cooling between each burst. Keep the tubes on ice throughout sonication. Centrifuge at 100 000 g for 10 min at 4 °C. Transfer the supernatant into a 15 ml tube.[c]
4. Prepare the glutathione–Sepharose beads by centrifuging 500 μl of the bead suspension briefly in a microcentrifuge (< 10 sec), taking off the supernatant and resuspending the beads in 1 ml of ice-cold lysis buffer. Repeat this three times to wash the beads, then resuspend the beads in 500 μl lysis buffer
5. Add the beads to the supernatant obtained in step 3.
6. Invert the bead/supernatant mixture for 30–60 min at 4 °C. Centrifuge at 1000 g for 1 min at room temperature, remove the supernatant carefully, and wash the beads three times with 5 ml of cold lysis buffer without PMSF[d] to remove unbound proteins.
7. Resuspend beads in the same volume (approx. 500 μl) of thrombin buffer and transfer to a microcentrifuge tube. Remove 1 μl for subsequent SDS–PAGE analysis (store at –20 °C). Add bovine thrombin to a concentration of 5 U/ml of beads and incubate with inversion overnight at 4 °C (or at room temperature for 1 h).[e]
8. Centrifuge the beads briefly in a microcentrifuge and transfer the supernatant to a fresh microcentrifuge tube. Wash the beads with 500 μl of lysis buffer, spin briefly, and add this supernatant to the first supernatant. Remove 1 μl each of supernatant and beads for subsequent SDS–PAGE analysis.
9. Add 10 μl of *p*-aminobenzamidine–agarose bead suspension (Sigma) (to remove thrombin), and incubate with inversion for 30 min at 4 °C. Centrifuge briefly (< 10 sec) in a microcentrifuge to pellet the beads, and transfer the supernatant to a fresh microcentrifuge tube.
10. Dialyse the supernatant twice for at least 3 h against 1 litre of dialysis buffer at 4 °C. Remove 1 μl for subsequent gel analysis.
11. Concentrate the dialysate in Centricon 10 (Amicon) tubes by centrifugation at 5000 g (4 °C) until the volume in the upper chamber is approx. 100 μl (this takes 20–60 min). Collect the concentrate by inverting the tubes and centrifuging for 1 min at 1000 g.

Protocol 1. *Continued*

12. Estimate the concentration and purity by running 1 μl of the final concentrate on a 12.5% polyacrylamide gel containing SDS, as well as analysing efficiency of the process by running the 1 μl aliquots saved from each step. Flash-freeze the remaining protein in 10 μl aliquots in liquid N_2.
13. For long-term storage, Rho GTPases are more stable in liquid N_2, but −70 °C is adequate for short-term storage.
14. For GTP-binding proteins, determine the activity of the protein by performing a GTP-binding assay (40). In addition, determine the protein concentration using a standard assay kit, such as that manufactured by Bio-Rad.

[a] Some proteins, such as RhoE (45) are unstable or not soluble in *E. coli* incubated at 37 °C. If this is the case, the induction conditions in the presence of IPTG are altered either by reducing the temperature of incubation, reducing the concentration of IPTG, and/or altering the length of induction time.
[b] We use a small probe on an MSE Soniprep 150 sonicator at an amplitude of 14 μm.
[c] Up to 50% of the GST-fusion protein is found in the pellet at this stage, presumably denatured.
[d] PMSF inhibits thrombin activity, so it is important to wash out residual PMSF prior to thrombin addition.
[e] Rac is susceptible to cleavage by thrombin when incubated at room temperature, or after prolonged incubation at 4 °C, so it is important to minimize the time that Rac protein is incubated with thrombin.

2.2 Microinjection of cells

The microinjection technique was initially described in detail by Graessmann and Graessmann (46). Protein, RNA, or DNA can be microinjected into cells, and the choice of which of these to inject is dependent on the individual experimental situation, although in practice the majority of experimenters use protein or DNA. In the microinjection process, protein or DNA solution is delivered into cells via glass pipettes which have been pulled to a fine point (approx. 0.5–1 μm diameter) at one end. A micromanipulator is used to position the point of the glass pipette very close to the cells to be injected. The other end of the pipette is attached via tubing to a pressure regulator. Air pressure applied to this end of the pipette forces the protein–DNA solution out of the pointed end of the pipette at a constant rate. The pipette is manipulated so that it transiently pierces the plasma membrane of a cell, allowing the solution in the pipette to enter the cell. The pipette remains within the cell for only a very short period (< 0.5 sec) and then is removed, allowing the membrane to reseal. The volume of solution introduced into cells is between 5% and 10% of their total volume, or approx. 10^{-14} litres (46).

In our microinjection experiments, we aim to inject 100–150 cells on one coverslip within 15 minutes. This minimizes the time that cells are exposed to light from the microscope, which may affect their responses. The cells are subsequently incubated for varying lengths of time with or without addition of growth factors, then fixed, permeabilized, and stained to identify injected cells. In addition, cells are co-stained to show other cellular structures of interest, for example staining with phalloidin will visualize actin filaments, while specific antibodies can be used to detect focal adhesion proteins.

Protocol 2. Preparation of cells for microinjection

Equipment

- 13 mm diameter glass coverslips (Chance Propper, No. ½), cleaned by washing first with nitric acid, then rinsing extensively with distilled water, and finally with ethanol; bake prior to use
- Tissue culture dishes with four wells each with a diameter of approx. 15 mm
- Tissue culture dishes with a diameter of 35 mm or 60 mm

Method

1. Prepare coverslips by drawing a cross on each one with a diamond-tipped marker pen, which facilitates localization of injected cells. Sterilize by dipping in 100% ethanol and flaming. Place in 15 mm diameter wells in 4-well dishes.
2. Seed cells at an appropriate density on the coverslips in at least 500 μl of their normal growth medium. Cells are normally microinjected at least 24 h after seeding, but can be microinjected as soon as they have spread sufficiently.
3. If cells are to be starved, remove the growth medium after an appropriate number of days, replace with medium containing low or no serum, and incubate the cells in this medium overnight.
4. Approx. 1 h before microinjection, transfer the coverslips to 35 mm or 60 mm diameter dishes containing the appropriate medium. Keep cells in an incubator close to the microinjector to minimize changes in temperature and medium pH during transfer to and from the microinjector.[a]

[a] Responses similar to those induced by Rho GTPases, including membrane ruffling and stress fibre formation, can be induced in cells by a drop in pH or by shaking the dishes. It is therefore very important that cells are handled gently when analysing morphological changes.

Protocol 3. Microinjection of recombinant proteins into cells

Equipment and reagents

- Microinjection buffer 1: 10 mM Tris–HCl pH 7.5, 150 mM NaCl, 5 mM $MgCl_2$
- Microinjection buffer 2: 20 mM Hepes–HCl pH 7.2, 100 mM KCl, 5 mM $MgCl_2$[a]
- Programmable pipette puller (Campden Instruments; Model No. 773)
- Glass pipettes: 1.2 mm bore with inner threads (Clark Electroinstruments)

Protocol 2. *Continued*

- Microinjection apparatus: our apparatus consists of an inverted phase-contrast microscope (Zeiss) fitted with a heated stage and an enclosed Perspex chamber. The temperature and CO_2 concentration in the chamber are maintained by a temperature regulator TRZ3700 and CTI controller 3700, respectively, obtained from Zeiss. Humidity is provided by placing a Perspex dish containing sterile distilled water in the chamber. Cells are injected using an Eppendorf transinjector and micromanipulator.
- Marker for injected cells: either immunoglobulin (IgG) (Pierce) or fluorescently-labelled dextran (M_r 10000) (Molecular Probes). The species chosen for IgG depends on which other antibodies are being used to stain cells after microinjection. We use goat, rat, or rabbit IgG, obtained as a 10 mg/ml solution (Pierce) and stored at 4°C. Dextran is stored at 100 mg/ml in microinjection buffer in small aliquots at –20°C.
- Microloader tips for loading micropipettes (Eppendorf)

Method

1. Thaw protein aliquots on ice.[b]
2. Turn on the temperature regulator and CO_2 controller at least 20 min prior to beginning microinjection to allow the temperature to reach 37°C and CO_2 levels to reach 10%.
3. Pull pipettes on a pipette puller, according to the manufacturer's instructions. Pipettes can be pulled in advance and stored by pressing the middle of each pipette onto a strip of Blu-Tak adhesive, in a 150 mm diameter plastic dish with a lid.
4. Dilute IgG to 1 mg/ml or dextran to 10 mg/ml[c] in protein injection buffer. Centrifuge the proteins, protein injection buffer, and diluted IgG/dextran for 5 min at 4°C at 13000 *g* to pellet small particles that will block the microinjection pipettes.
5. Mix proteins, buffer, and IgG/dextran in sterile 600 μl microcentrifuge tubes to give the required concentrations of proteins[d] and final concentration of 0.5 mg/ml IgG or 5 mg/ml dextran. Store proteins on ice until adding to the microinjection needle.
6. Take a dish containing cells on a coverslip from the incubator. Gently press down the coverslip at the edge onto the dish with a pipette tip, to exclude air bubbles and prevent the coverslip moving during microinjection. Place the dish on the microscope stage and localize the etched cross (*Protocol 2*) using a low power objective.
7. Load approx. 1 μl of protein solution into a microloader tip, then load this into a micropipette. Care should be taken to ensure that bubbles are not present in the solution in the pipette.
8. Insert the pipette into the holder, then using the joystick, move it to the centre of the coverslip, initially looking from above the stage. Then, looking down the microscope, bring the pipette down so that it is nearly in-focus above the cells. A bright spot, representing the meniscus, should appear first. On higher power, again bring the pipette to be nearly in-focus but just above the surface of the cells.

9. Inject cells in manual mode using a × 32/0.4 objective lens and × 10/18 eyepieces. Clear the pipette at high pressure (3000–6000 hPa) briefly (< 5 sec) before injecting cells at a working pressure of between 100 and 1000 hPa. Adjust the pressure as necessary during the course of injection. Between 100–150 cells are normally injected over 10 min,[e] then the dish is returned to the incubator.
10. To determine the effects of injected proteins on growth factor responses, add growth factors to the medium 15–30 min after finishing injections, and mix gently.

[a] Rho GTPases are normally prepared in buffer 1 (*Protocol 1*), as this is compatible with biochemical assays carried out with the proteins. For most studies microinjecting Rho GTPases, this buffer has proved adequate. Buffer 2 is closer in composition to the cytoplasm of cells, and therefore is preferable where cells or responses monitored may be highly sensitive to changes in, for example, levels of sodium. A variety of different microinjection buffer compositions have been used by others and are documented in the literature.
[b] After thawing, Rho GTPases can be stored for several days at 4°C. They should not be re-frozen, as this results in loss of activity.
[c] We have found that dextran is toxic to some cells, particularly in longer-term analyses, for example where cells are analysed after 16 h. It can be detected if injected at concentrations down to 1 mg/ml. We have found no toxic effect of injecting immunoglobulin.
[d] Proteins can be injected at any concentration up to 5 mg/ml (total protein). For each new protein the minimum concentration required to give a response is determined by titration. In practice, the active protein concentration (determined by GTP-binding assay) of Rho required to give a reproducible response in Swiss 3T3 fibroblasts is approx. 20 μg/ml, whereas with Rac it is 100 μg/ml (47).
[e] To minimize the possibility of altered cellular responses through exposure to light from the microscope, we do not inject cells on any one coverslip for any longer than 15 min. To increase the number of cells injected, inject multiple coverslips sequentially.

2.3 Scrape loading of recombinant proteins into cells

Scrape loading has been used to dissect signalling pathways downstream of Ras, including activation of protein kinase C (30), elevation of c-myc expression (31), and activation of mitogen-activated protein kinases (32). It has also been used to analyse signalling pathways activated by RhoA and to investigate the involvement of RhoA in integrin-mediated signalling (33, 48). Basically, cells are scraped gently off dishes in the presence of a concentrated solution of protein in physiological buffer. They can then either be incubated in suspension, or replated on extracellular matrix proteins to investigate the effects of a particular protein on signalling via integrins and cell spreading. The design of scraper is extremely important in order not to damage cells physically. Commercially available scrapers or rubber policemen are not ideal, and we make our own scrapers. Using these, over 90% of cells survive scraping, and over 80% of cells routinely take up protein, as determined using fluorescently-labelled dextran of the same approximate size as the protein (33). However, it is important to note that the efficiency of protein uptake decreases with the size of the protein (28).

Protocol 4. Scrape loading with recombinant proteins

Equipment and reagents

- Scrapers are constructed using rubber Pasteur pipette teats, which are cut and wrapped around the end of a stainless steel spatula (180 mm × 10 mm). The last 20 mm of the metal spatula is bent at an angle of 45°. The rubber is held in place with copper wire (1.5 mm) wound around it (*Figure 2*). A different scraper is used for each protein preparation. Scrapers are sterilized in ethanol prior to use, and then allowed to dry in the tissue culture hood. They are extensively washed in water then ethanol after use.[a]
- Phosphate-buffered saline (PBS), tissue culture grade (Gibco BRL)
- Scrape loading (SL) buffer: 10 mM Tris–HCl pH 7.0, 114 mM KCl, 15 mM NaCl, 5.5 mM $MgCl_2$
- 10 μl aliquots of recombinant proteins (*Protocol 1*), thawed immediately prior to use and kept on ice
- Fibronectin-coated coverslips, prepared by incubating coverslips with 50 μg/ml human fibronectin (Sigma) for at least 90 min at 37 °C; wash the coverslips twice with PBS immediately prior to use

Method

1. Seed cells onto 100 mm tissue culture dishes, grow to confluence, and starve, if relevant, in serum-free medium.
2. Mix protein with SL buffer to give a final concentration of approx. 20 μg/ml of active protein. Do not dilute protein less than 1:10 in SL buffer, as this will significantly alter the composition of the buffer.
3. Wash cells twice with PBS.
4. Add 160 μl of protein in SL buffer to a dish of cells and leave for 1 min to allow the solution to disperse over the surface of the dish.[b]
5. Wet the scraper in SL buffer, then glide it gently back and forth over the surface of the dish to detach the cells. The weight of the scraper in itself is enough to remove the cells; therefore, no force need be applied.
6. After scraping, leave cells for 1 min (to allow them to recover), then wash them off the dish with serum-free medium (equilibrated to 37 °C, 10% CO_2). Transfer to 20 ml Universals (for biochemical analysis)[c] or seed onto fibronectin-coated coverslips (for immunocytochemical analysis).[d]
7. For biochemical studies incubate cells in suspension at 37 °C, 10% CO_2, for the required lengths of time. Transfer them to 15 ml centrifuge tubes, centrifuge at 1500 r.p.m. for 5 min, remove the medium, and lyse the cells in an appropriate lysis buffer for subsequent analysis.

[a] The rubber on the scrapers will disintegrate if left in ethanol for long periods of time.
[b] It is not possible to process more than one or two dishes simultaneously; therefore, in a large experiment, time points have to be staggered. For biochemical analysis two dishes of cells are required per point.
[c] Detached cells from two 100 mm dishes can be combined in 4.5 ml serum-free medium.
[d] Cells from one 100 mm dish can be seeded onto eight to ten coverslips (13 mm diameter) on 15 mm wells.

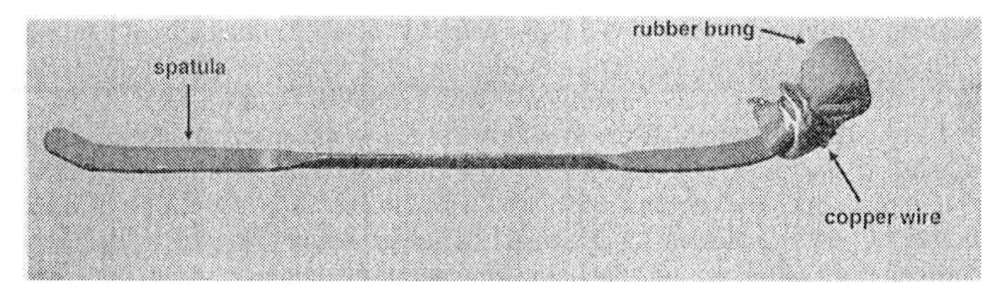

Figure 2. A scrapeloader. The construction of a scrapeloader is described in *Protocol 4*.

3. Microinjection of DNA into cells

Although recombinant proteins provide an ideal approach for investigating cytoskeletal reorganization induced by Rho GTPases, there are many situations in which this is not a realistic proposition. Many signalling proteins are too large to be purified in adequate quantities from *E. coli*, and alternative approaches such as purification from insect cells are very long and labour-intensive. Many investigators have instead introduced expression vectors encoding the protein of interest into cells. DNA can be introduced into cells by many different methods, including calcium phosphate-mediated transfection, lipofection, DEAE–dextran-mediated transfection, and electroporation. Here, we will concentrate on microinjection as an approach which allows rapid analysis of changes to the cytoskeleton without the accumulation of long-term changes in gene expression which may indirectly affect cell morphology. Microinjection of DNA also provides a rapid means of assessing the localization of proteins in cells. By tagging a protein with an epitope, it is possible to follow its localization independently of endogenous proteins (14).

For microinjection we have found that it is important to have highly purified DNA, preferably purified by CsCl density gradient centrifugation (49). This step minimizes the levels of toxic contaminants from *E. coli* that affect expression of plasmid-encoded genes, interfering with effective expression of the protein of interest in injected cells. In addition, when working with cells sensitive to lipopolysaccharide, such as monocytes or endothelial cells, we have found that higher percentages of microinjected cells express from plasmids prepared with endotoxin-free reagents. In this case, DNA purified with a commercially available plasmid purification kit (for example, those supplied by Qiagen), using endotoxin-free reagents provides high quality DNA. With highly purified DNA, the percentage of microinjected cells that express protein from the microinjected plasmid can approach 100%. Optimal expression in cells also depends on the enhancer/promoter in the expression vector; different cell types prefer different vectors. In addition, quiescent cells express only some expression vectors efficiently. Generally, SV40- or CMV-based promoter/enhancer systems work well in most, although not all, cell types. In quiescent cells or primary cells, it is worth experimenting with different expression vectors to find one that expresses efficiently in the cells of interest, and then subclone all cDNAs into this vector.

The concentration of DNA required for microinjection can vary greatly depending on the vector, insert, cell type, and conditions. To minimize toxic side-effects due to overexpression of a protein, the DNA should be titrated down to find the lowest level which still gives detectable, high efficiency protein expression. In addition, the time after injection at which cells are analysed should be kept to a minimum, as increased accumulation of protein over time can be toxic. With Rho GTPases protein expression and cellular responses can be detected within two hours of microinjecting expression vectors. If left for 16 hours, however, most injected cells die. With some larger proteins it may be necessary to incubate cells for longer than two hours, but in our experience four hours is always sufficient.

Protocol 5. Microinjection of DNA into cells

Equipment and reagents

- Microinjection equipment and pipettes as in *Protocol 3*
- DNA microinjection buffer: PBS or one of the buffers in *Protocol 3*, without $MgCl_2$[a]
- Expression plasmids: DNA is stored at a concentration of at least 1 mg/ml in 10 mM Tris–HCl pH 7.5, 1 mM EDTA, at −20 °C

Method

1. Prepare cells for microinjection as described in *Protocol 2*.
2. Immediately prior to injection, dilute DNA to an appropriate concentration (1–100 ng/μl) in DNA microinjection buffer.
3. Centrifuge the diluted DNA at 13 000 *g* for 5 min at 4 °C, then carefully transfer approx. 5 μl to another microcentrifuge tube.
4. Load DNA into a micropipette and inject (*Protocol 3*) into the nucleus of cells.
5. Fix cells 1–24 h after microinjection.

[a] Mg^{2+} ions are required for DNase activity, so to minimize degradation of DNA, buffers should not include Mg^{2+}.

4. Immunolocalization analyses of injected cells and analysis of cytoskeletal changes

Cells injected with recombinant protein can be identified either by staining directly for the injected protein, or by staining for an inert co-injected marker protein, such as IgG, or by co-injecting fluorescently-labelled dextran (see *Protocol 3*).

Cells injected with DNA should preferably be detected by staining directly for the protein encoded by the expression plasmid. Often, detection of this exogenous protein is facilitated by tagging it at the amino- or carboxy-terminus with a short amino acid stretch that is recognized by a specific monoclonal antibody. Epitope tags that have been used successfully with Rho GTPases include the 9E10 (myc) epitope tag (8) and the haemagglutinin (HA) tag (50). With GTPases, tags are placed at the amino-terminus, because a carboxy-terminus tag would prevent the post-translational processing required for membrane localization (43). Expression vectors carrying the sequence of these tags, with restriction enzyme sites downstream of the tag, have been generated to facilitate cloning of the gene of interest directly and are available commercially (e.g. vectors with the myc tag are available from Invitrogen).

In general, when staining with primary antibodies, we invert coverslips onto a drop of diluted antibody solution on Parafilm, to limit the amount of antibody required. For incubation with secondary antibodies and/or fluorescently-labelled phalloidin, we incubate coverslips in 15 mm diameter wells, as these reagents are not usually limiting. This approach reduces the amount of manipulation of coverslips and facilitates washing and processing.

Protocol 6. Fixation and permeabilization of cells

Reagents

- Formaldehyde: obtained as a 40% solution containing 9–11% methanol. It is toxic by inhalation; therefore, to minimize exposure, fixing cells are placed in a fume-cupboard and formaldehyde is disposed in the fume-cupboard outlet. Dilute fresh 1:10 (v/v) in PBS(CM) immediately before use.
- PBS(CM): PBS containing 0.9 mM $CaCl_2$ and 0.5 mM $MgCl_2$; this can be purchased from a number of companies (e.g. Gibco BRL)
- 0.2% Triton X-100/PBS: a stock solution of 10% Triton X-100 is used

Method

1. Fix cells at time points after injection ranging from 5 min to 24 h. At the appropriate time point, wash the cells with PBS(CM), then fix in 4% formaldehyde for 15 min.
2. Following fixation, transfer coverslips to 18 mm diameter wells containing PBS(CM), then wash six times with PBS(CM). An optional incubation step with 50 mM ammonium chloride in PBS(CM) (10 min) can be included to quench residual formaldehyde.
3. Permeabilize for 5 min with 0.2% Triton X-100 in PBS(CM),[a] then wash twice with PBS(CM).

[a] Antibodies that recognize extracellular epitopes of cell surface proteins often work better on non-permeabilized cells.

Protocol 7. Staining of cells for F-actin

Reagents

- PBS(CM) (see *Protocol 6*)
- Fluorescently-labelled phalloidin: dissolve TRITC-phalloidin (Sigma) in sterile distilled water at a concentration of 50 μg/ml and store in small aliquots at –20°C in a light-sealed container. Dilute immediately prior to use to 0.1 μg/ml in PBS(CM). Other fluorescently-labelled phalloidins (e.g. FITC, Bodipy) are supplied by Molecular Probes in methanol and stored at –20°C. Immediately prior to use, place the required volume of stock solution in a microcentrifuge tube, allow the methanol to evaporate, and then redissolve the phalloidin in PBS(CM) with vortexing, to a concentration of 50 nM.
- Fluorescently-conjugated anti-species-specific IgG: if TRITC-phalloidin is used to locate actin filaments, then FITC-conjugated anti-species-specific (rat/rabbit etc.) IgG is used to detect microinjected IgG (see *Protocol 8* for sources of these 'secondary' antibodies).
- Mountant: 0.1% *p*-phenylenediamine (antiquench), 10% (w/v) Mowiol (Calbiochem), 25% (w/v) glycerol, 100 mM Tris–HCl pH 8.5. Mowiol solution with glycerol and Tris can be stored for several months at 4°C without *p*-phenylenediamine; once the *p*-phenylenediamine is added, it is stored in 100 μl aliquots at –70°C. Once thawed, these aliquots can be stored wrapped in aluminium foil at –20°C for approx. one week. However the *p*-phenylenediamine is light- and temperature-sensitive, and gradually turns from a pale yellow to a deep brown colour over time at 4°C or –20°C, but not at –70°C.

Method

1. Incubate each coverslip of fixed and permeabilized cells in a 15–18 mm diameter well with 200 μl of diluted fluorescently-labelled phalloidin in PBS(CM) for 30–60 min. If cells have been injected with IgG to mark injected cells, include fluorescently-conjugated anti-IgG in this incubation.[a] During this incubation, place dishes on a rocking table at low speed and keep the dishes in the dark by covering with aluminium foil.
2. Wash coverslips in multiwell dishes six times with PBS(CM), and place on a rocking table for a final wash in PBS(CM) for 5 min.
3. Mount coverslips on slides with Mowiol solution containing *p*-phenylenediamine as antiquench. This mountant takes approx. 1 h to set permanently at room temperature.[b] Before this time, coverslips cannot be viewed using oil immersion objectives.
4. Store slides at 4°C in a light-tight slide container.
5. View and photograph cells using a conventional epifluorescence microscope or a confocal microscope.[c] Locate the etched cross under phase-contrast microscopy at low power, and subsequently locate microinjected cells using epifluorescence.

[a] The concentration of fluorescently-labelled anti-IgG should be determined by performing a titration. However, we find that a dilution of 1:400 of Jackson Immunoresearch antibodies gives good staining with little background. It may be necessary to include a blocking reagent such as 1% BSA (*Protocol 8*), depending on the source of antibody.
[b] While the mountant is setting, slides should be stored on a completely flat surface to prevent the mountant from setting with a greater depth at one side of the coverslip. If this happens, the coverslip will not be flat with respect to the slide, which causes problems when generating images.
[c] It is advisable to photograph cells within one week of staining, as non-specific background fluorescence increases gradually over time.

In addition to localizing actin filaments, the localization of a number of other cytoskeletally-associated proteins has been studied in cells micro-injected with Rho GTPases or downstream target proteins, particularly those associated with cell–extracellular matrix or cell–cell adhesion sites. By stimulating stress fibre formation, Rho concomitantly induces the assembly of focal adhesions, or sites of close contact between cells and the substratum, where stress fibres are indirectly linked to the plasma membrane. Focal adhesions contain transmembrane integrins, which bind extracellularly to extracellular matrix proteins such as fibronectin and vitronectin, and are associated intracellularly with a complex of cytoplasmic proteins. These include vinculin, talin, the tyrosine kinase FAK, and paxillin, among many others (51). To detect focal adhesions, the protein used most frequently in immunolocalization studies is vinculin (9). Anti-phosphotyrosine antibodies are also used to localize focal adhesions, as focal adhesions contain many tyrosine phosphorylated proteins (52).

Although Rho is classically associated with the formation of focal adhesions, Rac and Cdc42 induce the assembly of smaller structures, termed 'focal complexes' which have been shown to contain many of the same proteins found in classic focal adhesions (10). In addition, the effects of Rho and Rac on sites of intracellular adhesion, and particularly adherens junctions containing cadherins, has been investigated (53, 54).

Generally, the effect of a Rho GTPase on the localization of a protein is studied by dual labelling with antibodies to detect injected cells and antibodies to detect the protein of interest. Triple labelling with phalloidin is useful to determine actin organization, particularly when there is access to a confocal laser scanning microscope with the option of three different laser beam wavelengths.

Protocol 8. Immunolocalization of adhesion-associated proteins

Reagents

- PBS(CM) (as in *Protocol 6*)
- PBS(CM) with 1% bovine serum albumin (PBS/BSA) as a blocking agent for non-specific protein binding sites; this can be stored for many weeks at 4°C provided 0.02% azide is added to prevent growth of micro-organisms
- Primary antibodies: antibodies are stored at 4°C or −20°C, as recommended by the supplier; dilute immediately prior to use in PBS/BSA
- Fluorescently-labelled secondary antibodies: good sources are Jackson Immunochemicals and Southern Biotechnology; dilute immediately prior to use in PBS/BSA

Method

1. Remove coverslips of fixed[a] and permeabilized cells (*Protocol 6*) from wells or dishes using fine forceps and a 21 gauge syringe needle bent at the end to hook under the coverslip.
2. Immediately invert the coverslip, cell-side down, onto a 15 μl drop of

Protocol 8. *Continued*

antibody solution on Parafilm. Place the Parafilm on top of a dish in a plastic box containing a small amount of distilled water, to maintain humidity.

3. After 45 min,[b] transfer coverslips back to 15 mm wells and wash five times with PBS.
4. Incubate coverslips with 200 μl of secondary antibody, diluted in PBS/BSA, for 30–45 min. If secondary antibody is limiting, this stage can be performed on Parafilm.
5. Wash extensively with PBS(CM), including a final 5 min wash.
6. Mount coverslips and view as described (*Protocol 7*).

[a] With some primary antibodies, optimal staining is achieved by fixing in methanol, rather than in paraformaldehyde. Most relevant to cytoskeletal organization, microtubules are only preserved satisfactorily using methanol fixation. Conversely, methanol fixation is not suitable for subsequent staining with phalloidin.
[b] The optimal length of incubation with primary antibody varies depending on the antibody, but normally incubation times are between 30 min and 150 min. With some antibodies, lower background is achieved by incubating overnight at 4°C.

5. Use of Rho GTPase mutants to analyse functional domains

5.1 Point mutations

In order to analyse how Rho GTPases activate different signalling pathways, one approach has been to map the regions and critical amino acids of each GTPase required to induce specific responses in cells. Following the identification of downstream targets for the GTPases, this has been combined with an analysis of the ability of the mutant proteins to interact with these targets, allowing correlations to be drawn between an observed response and downstream targets (23–25). *Table 1* lists the point mutations that have been made in Rho, Rac, and Cdc42, and the effects of these mutations on their ability to induce actin cytoskeletal changes. One problem sometimes encountered when purifying recombinant mutant proteins is that mutations can grossly affect their structure and/or their solubility. For example, RhoA T19N is insoluble in *E. coli*, although wild-type RhoA is soluble (40). For this reason, and also because purifying multiple recombinant proteins is time-consuming compared to purifying DNA, most workers have microinjected or transfected expression vectors encoding mutant Rho GTPases rather than recombinant proteins (24, 25). It has to been borne in mind, however, that some mutant proteins may be unstable when expressed in mammalian cells, and it is therefore important to analyse protein integrity by Western blotting of transfected cells.

Table 1. Effects of point mutations in Rho, Rac, and Cdc42 on actin cytoskeletal responses

GTPase	Mutation[a]	Activity (actin)	Reference
Rho	D28N[b]	Active	60
	V33E	Active	60
	T37S	Active	60
	F39A/V	Inactive	25
	E40L/W	Inactive	25
	E40N/T	Active	25
	Y42C	Active	25
	A44K	Active	60
	D45Q	Active	60
	E47V	Active	60
	T19N	Dom negative[c]	61
Rac	F37L	Inactive	24
	F37A	Dom negative[c]	62
	Y40C/H/K	Active	24
	N43D	Active	24
	T17N	Dom negative[c]	9
Cdc42	F37A	Active	23
	Y40C	Active	23
	T17N	Dom negative[c]	11

[a] All mutations are present in an activated, GTPase-defective background (L61 for Rac and Cdc42, L63 for Rho).
[b] Rho amino acid numbering differs by +2 from Rac/Cdc42, thus RhoY42C is equivalent to RacY40C.
[c] Dom = dominant.

5.2 Chimeric proteins

By making chimeric proteins between Ras and Rho (55), or Rac and Rho (22, 26), it has been possible to map regions of the two proteins required for specific responses regulated by each protein, such as stress fibre formation. As with point mutants, a problem encountered with purifying recombinant chimeric proteins is solubility and correct folding. A number of chimeras are insoluble in *E. coli* or poorly expressed (22, 55). Transfection or microinjection of expression vectors has therefore been used as a more rapid and simpler approach to analysing the responses induced by chimeric proteins (26).

6. Analysis of the function of interacting partners of Rho GTPases

Proteins interacting with the active, GTP-bound conformation of Rho GTPases have been isolated either through biochemical purification or through yeast

Table 2. Morphological effects of Rho GTPase effectors

Target	GTPase	Actin morphology	Reference
ROCK	Rho	Stress fibres	63–65
Citron-K	Rho	Cytokinesis	66
P140mDia	Rho	Disorganized actin filaments	67
PAK	Cdc42/Rac	Lamellipodia	68
WASP	Cdc42	Punctate F-actin	69
N-WASP	Cdc42	Filopodia/punctate F-actin	70

two-hybrid screens (56–58). The function of these proteins in regulating cytoskeletal organization has been determined primarily by transfecting or microinjecting expression constructs into cells, followed by staining with phalloidin and/or immunostaining with antibodies to proteins as described in *Protocols 6–8*. Those targets that have been shown to regulate actin organization are listed in *Table 2*.

7. Conclusions and prospects

Microinjection has been a very useful tool for dissecting the functions of Rho GTPases in regulating cell morphology and cytoskeletal organization, but interpretation of microinjection results has to be treated with caution. In general, the quantity of protein introduced into a cell is much higher than the endogenous level of that protein, and therefore the introduced protein may interact with proteins that the endogenous protein does not normally encounter, and may aberrantly activate and/or titrate out proteins. In addition, the microinjection process itself alters the volume of cells and the cytosolic salt composition, and therefore can stress cells and alter their behaviour. For example, in macrophages microinjection can reduce the rate of cell migration (59). It is therefore important that the microinjection approach is complemented by other techniques for analysing the function of these GTPases, and transfection studies have reinforced the basic conclusions of the original microinjection studies concerning cytoskeletal changes induced by each Rho GTPase. A major future goal will be to determine in detail where the proteins and their downstream targets are in cells, and this will be achieved both through light microscopical and electron microscopical techniques.

References

1. Boguski, M. S. and McCormick, F. (1993). *Nature*, **366**, 643.
2. Wittinghofer, F. (1998). *Nature*, **394**, 317.
3. Hall, A. (1994). *Annu. Rev. Cell Biol.*, **10**, 31.
4. Van Aelst, L. and D'Souza-Schorey, C. (1997). *Genes Dev.*, **11**, 2295.

5. Nobes, C. D., Lauritzen, I., Mattei, M. G., Paris, S., Hall, A., and Chardin, P. (1998). *J. Cell Biol.*, **141**, 187.
6. Hall, A. (1998). *Science*, **279**, 509.
7. Mackay, D. J. and Hall, A. (1998). *J. Biol. Chem.*, **273**, 20685.
8. Ridley, A. J., Paterson, H. F., Johnston, C. L., Diekmann, D., and Hall, A. (1992). *Cell*, **70**, 401.
9. Ridley, A. J. and Hall, A. (1992). *Cell*, **70**, 389.
10. Nobes, C. D. and Hall, A. (1995). *Cell*, **81**, 53.
11. Kozma, R., Ahmed, S., Best, A., and Lim, L. (1995). *Mol. Cell. Biol.*, **15**, 1942.
12. Bar-Sagi, D. and Feramisco, J. R. (1986). *Cell*, **88**, 521.
13. Ridley, A. J., Comoglio, P. M., and Hall, A. (1995). *Mol. Cell. Biol.*, **15**, 1110.
14. Paterson, H., Adamson, P., and Robertson, D. (1995). In *Methods in enzymology*, Editors: W. E. Balch, C. J. Der, and A. Hall. Academic Press. Vol. 256, p. 162.
15. Stam, J. C., Michiels, F., van der Kammen, R. A., Moolenaar, W. H., and Collard, J. G. (1998). *EMBO J.*, **17**, 4066.
16. Jou, T. S. and Nelson, W. J. (1998). *J. Cell Biol.*, **142**, 85.
17. Norman, J. C., Price, L. S., Ridley, A. J., Hall, A., and Koffer, A. (1994). *J. Cell Biol.*, **126**, 1005.
18. Mackay, D. J., Esch, F., Furthmayr, H., and Hall, A. (1997). *J. Cell Biol.*, **138**, 927.
19. Zigmond, S. H., Joyce, M., Borleis, J., Bokoch, G. M., and Devreotes, P. N. (1997). *J. Cell Biol.*, **138**, 363.
20. Zigmond, S. H., Joyce, M., Yang, C., Brown, K., Huang, M., and Pring, M. (1998). *J. Cell Biol.*, **142**, 1001.
21. Ma, L., Cantley, L. C., Janmey, P. A., and Kirschner, M. W. (1998). *J. Cell Biol.*, **140**, 1125.
22. Diekmann, D., Nobes, C. D., Burbelo, P. D., Abo, A., and Hall, A. (1995). *EMBO J.*, **14**, 5297.
23. Lamarche, N., Tapon, N., Stowers, L., Burbelo, P. D., Aspenstrom, P., Bridges, T., *et al.* (1996). *Cell*, **87**, 519.
24. Westwick, J. K., Lambert, Q. T., Clark, G. J., Symons, M., Van Aelst, L., Pestell, R. G., *et al.* (1997). *Mol. Cell. Biol.*, **17**, 1324.
25. Sahai, E., Alberts, A. S., and Treisman, R. (1998). *EMBO J.*, **17**, 1350.
26. Fujisawa, K., Madaule, P., Ishizaki, T., Watanabe, G., Bito, H., Saito, Y., *et al.* (1998). *J. Biol. Chem.*, **273**, 18943.
27. Lamb, N. J., Fernandez, A., Conti, M. A., Adelstein, R., Glass, D. B., Welch, W. J., *et al.* (1988). *J. Cell Biol.*, **106**, 1955.
28. McNeil, P. L., Murphy, R. F., Lanni, F., and Taylor, D. L. (1984). *J. Cell Biol.*, **98**, 1556.
29. McNeil, P. L. (1989). *Methods Cell Biol.*, **29**, 153.
30. Morris, J. D., Price, B., Lloyd, A. C., Self, A. J., Marshall, C. J., and Hall, A. (1989). *Oncogene*, **4**, 27.
31. Lloyd, A. C., Paterson, H. F., Morris, J. D., Hall, A., and Marshall, C. J. (1989). *EMBO J.*, **8**, 1099.
32. Leevers, S. J. and Marshall, C. J. (1992). *EMBO J.*, **11**, 569.
33. Flinn, H. M. and Ridley, A. J. (1996). *J. Cell Sci.*, **109**, 1131.
34. McNeil, P. L. and Warder, E. (1987). *J. Cell Sci.*, **88**, 669.
35. Okada, C. Y. and Rechsteiner, M. (1982). *Cell*, **29**, 33.

36. Lee, G., Delohery, T. M., Ronai, Z., Brandt-Rauf, P. W., Pincus, M. R., Murphy, R. B., *et al.* (1993). *Cytometry*, **14**, 265.
37. Clarke, M. S. and McNeil, P. L. (1992). *J. Cell Sci.*, **102**, 533.
38. Baum, C., Forster, P., Hegewisch-Becker, S., and Harbers, K. (1994). *BioTechniques*, **17**, 1058.
39. Norman, J. C., Price, L. S., Ridley, A. J., and Koffer, A. (1996). *Mol. Biol. Cell*, **7**, 1429.
40. Self, A. J. and Hall, A. (1995). In *Methods in enzymology*, editors: W. E. Balch, C. J. Der, and A. Hall. Academic Press. Vol. 256, p. 3.
41. Wei, Y., Zhang, Y., Derewenda, U., Liu, X., Minor, W., Nakamoto, R. K., *et al.* (1997). *Nature Struct. Biol.*, **4**, 699.
42. Hirshberg, M., Stockley, R. W., Dodson, G., and Webb, M. R. (1997). *Nature Struct. Biol.*, **4**, 147.
43. Marshall, C. J. (1993). *Science*, **259**, 1865.
44. Cerione, R. A., Leonard, D., and Zheng, Y. (1995). In *Methods in enzymology*, editors: W. E. Balch, C. J. Der, and A. Hall. Academic Press. Vol. 256, p. 11.
45. Guasch, R. M., Scambler, P., Jones, G. E., and Ridley, A. J. (1998). *Mol. Cell. Biol.*, **18**, 4761.
46. Graessmann, M. and Graessmann, A. (1983). In *Methods in enzymology*, Vol. 101, Academic Press, eds R. Wu, L. Grossman, K. Moldave, p. 482.
47. Ridley, A. J. (1995). In *Methods in enzymology*, editors: W. E. Balch, C. J. Der, and A. Hall. Academic Press. Vol. 256, p. 313.
48. Barry, S. T., Flinn, H. M., Humphries, M. J., Critchley, D. R., and Ridley, A. J. (1997). *Cell Adhes. Commun.*, **4**, 387.
49. Sambrook, J., Fritsch, E. F., and Maniatis, T. (ed.) (1989). *Molecular cloning: a laboratory manual.* Cold Spring Harbor Laboratory Press, New York.
50. Neudauer, C. L., Joberty, G., Tatsis, N., and Macara, I. G. (1998). *Curr. Biol.*, **8**, 1151.
51. Yamada, K. M. and Geiger, B. (1997). *Curr. Opin. Cell Biol.*, **9**, 76.
52. Ridley, A. J. and Hall, A. (1994). *EMBO J.*, **13**, 2600.
53. Braga, V. M., Machesky, L. M., Hall, A., and Hotchin, N. A. (1997). *J. Cell Biol.*, **137**, 1421.
54. Hordijk, P. L., ten Klooster, J. P., van der Kammen, R. A., Michiels, F., Oomen, L. C., and Collard, J. G. (1997). *Science*, **278**, 1464.
55. Self, A. J., Paterson, H. F., and Hall, A. (1993). *Oncogene*, **8**, 655.
56. Manser, E., Leung, T., and Lim, L. (1995). In *Methods in enzymology*, editors: W. E. Balch, C. J. Der, and A. Hall. Academic Press. Vol. 256, p. 130.
57. Manser, E., Leung, T., and Lim, L. (1995). In *Methods in enzymology*, editors: W. E. Balch, C. J. Der, and A. Hall. Academic Press. Vol. 256, p. 215.
58. Aspenstrom, P. and Olson, M. F. (1995). In *Methods in enzymology*, editors: W. E. Balch, C. J. Der, and A. Hall. Academic Press. Vol. 256, p. 228.
59. Allen, W. E., Zicha, D., Ridley, A. J., and Jones, G. E. (1998). *J. Cell Biol.*, **141**, 1147.
60. Zohar, M., Teramoto, H., Katz, B. Z., Yamada, K. M., and Gutkind, J. S. (1998). *Oncogene*, **17**, 991.
61. Clark, E. A., King, W. G., Brugge, J. S., Symons, M., and Hynes, R. O. (1998). *J. Cell Biol.*, **142**, 573.
62. Schwartz, M. A., Meredith, J. E., and Kiosses, W. B. (1998). *Oncogene*, **17**, 625.

63. Leung, T., Chen, X. Q., Manser, E., and Lim, L. (1996). *Mol. Cell. Biol.*, **16**, 5313.
64. Amano, M., Chihara, K., Kimura, K., Fukata, Y., Nakamura, N., Matsuura, Y., *et al.* (1997). *Science*, **275**, 1308.
65. Ishizaki, T., Naito, M., Fujisawa, K., Maekawa, M., Watanabe, N., Saito, Y., *et al.* (1997). *FEBS Lett.*, **404**, 118.
66. Madaule, P., Eda, M., Watanabe, N., Fujisawa, K., Matsuoka, T., Bito, H., *et al.* (1998). *Nature*, **394**, 491.
67. Watanabe, N., Madaule, P., Reid, T., Ishizaki, T., Watanabe, G., Kakizuka, A., *et al.* (1997). *EMBO J.*, **16**, 3044.
68. Sells, M. A., Knaus, U. G., Bagrodia, S., Ambrose, D. M., Bokoch, G. M., and Chernoff, J. (1997). *Curr. Biol.*, **7**, 202.
69. Symons, M., Derry, J. M., Karlak, B., Jiang, S., Lemahieu, V., Mccormick, F., *et al.* (1996). *Cell*, **84**, 723.
70. Miki, H., Sasaki, T., Takai, Y., and Takenawa, T. (1998). *Nature*, **391**, 93.

6

Applications for analysis of contractile functions in dividing and migrating tissue culture cells

DOUGLAS J. FISHKIND, IRINA ROMENSKAIA, YUN BAO, and CHRISTINE M. MANUBAY

1. Introduction

The cytoskeleton is composed of a dynamic network of filament systems that form the scaffolding and machinery responsible for spatially and temporally integrating a variety of cell functions. In addition to maintaining cell shape and polarity, cytoskeletal polymers orchestrate a variety of cell movements, including intracellular transport, chromosome segregation, cell locomotion, and cell division. Such activities result from the direct interaction of polymer systems such as actin filaments and microtubules that combine with a vast array of complementary binding proteins and molecular motors to transduce mechanochemical energy into active force production. The exquisite sensitivity and timing of such events is controlled in turn by a complex network of signalling molecules that co-operatively function to stimulate and regulate productive interactions leading to force generation and mechanical transduction (1).

Motile and dividing tissue culture cells have long been used as model systems for elucidating protein dynamics, mechanical force production, and cell signalling processes (2, 3). By combining advances in low-light level fluorescence imaging, computational-based image processing, single-cell microinjection, and transfection approaches, together with a continually expanding list of readily available molecular probes, laboratories now have a variety of new tools at their disposal for dissecting motile, signalling, and contractile functions (4). In many cases, procedures for analysing protein function are carried out on living cells, with the main advantage being the acquisition of precise temporal and spatial information on intracellular events. However, at other times, especially when large numbers of cells are required for adequate quantitation, investigators rely on preserved or fixed fluorescent samples which can enhance the acquisition speed, reliability, and spatial resolution of data. Most comprehensive studies typically incorporate a combination of approaches utilizing both live, *in vivo* approaches and data obtained from fixed cell samples.

The purpose of this chapter is to provide a series of protocols we have used to study the complex interactions and interrelationships actin, myosin, and microtubule systems demonstrate during signal-dependent, force transducing events. Protocols in this section are focused on both the temporal detection and analysis of motile processes such as cortical flow during cleavage furrow contraction, the application and use of GFP-fusion protein vectors in motile tissue culture cells, and high-resolution fluorescent mapping of actin, myosin, and microtubule organization and orientation within dividing cells. For additional information on related topics and procedures the reader is directed to a variety of excellent references (2, 4–7).

2. Single-particle tracking assays for analysis of cortical actin flow in dividing cells

One of the key events accompanying cell division is the reorganization and rapid redistribution of the cortical actin cytoskeleton during cytokinesis (8, 9). This process leads to both an increased concentration of actin and myosin in the cleavage furrow and a lateral movement of cell surface receptors to the equatorial mid-zone (10–14). Gaining an understanding of the mechanism controlling these events has important implications for:

(a) Determining when signal-dependent processes are activated.

(b) Dissecting which molecules or signalling pathways function to stimulate this critical process.

(c) Screening the movement and concentration of additional cell surface receptors that are essential in cleavage furrow formation and function.

An important advancement in the study of this problem came about with the development of single-particle tracking assays and their application to dividing tissue culture cells (13, 14). Such studies have been used both to determine the precise timing and vectorial movement for equatorial cortical flow and elucidate the role spindle microtubules play in facilitating bi-directional cortical movements and cytoskeletal reorganization within the cleavage furrow. Application of the protocols described below are of great use in:

(a) Analysing the movement of specific cell surface receptors directed toward the cleavage furrow.

(b) Identifying and characterizing factors involved in activating and regulating cortical flow and contractile activities during cytokinesis.

(c) Determining the interdependence of cortical flow components on actin filament and microtubule-based systems.

It is important to note that the cell types and culturing conditions used for these assays are usually selected to facilitate vectorial analysis of cortical flow, and thus are based on clones such as the normal rat kidney (NRK) cell, termed NRK-2, that undergoes cell division in an attached, well-spread mode (15). However, other cell lines have been used for such analysis and can be substituted when needed (13).

Protocol 1. Culturing NRK-2 cells for single-particle tracking assays and cell division studies

Equipment and reagents

- 100 mm plastic tissue culture grade Petri dishes (Becton Dickinson, Falcon)
- Custom-made 70 × 50 × 6 mm Plexiglass culture chambers (16)
- 45 × 50 mm No. 1.5 or No. 2 glass coverslips (Fisher), alcohol dipped and flamed just prior to mounting
- High-vacuum silicone grease (Dow Corning) loaded into a 25 ml syringe
- STE: 150 mM NaCl, 50 mM Tris, 1 mM Na_2EDTA pH 7.2
- NRK-2 or equivalent cell line
- F12K complete media: Kaighn's modified F12K media (Sigma) supplemented with 10% FCS (BioWhittaker), 2 mM L-glutamine (Life Technologies), 50 μg/ml antibiotic/antimycotic solution (Sigma), 30 mM $NaHCO_3$ pH 6.95
- 0.25% trypsin (Worthington) in TBS: 140 mM NaCl, 5 mM KCl, 30 mM Tris–HCl, 20 mM Na_2HPO_4 pH 7.5

Note: all solutions for cell culture are pre-warmed to 37 °C prior to use.

Method[a]

1. Prior to passage, prepare sterilized culture chambers by:
   (a) Dipping the Plexiglass chamber in 90% alcohol.
   (b) Allowing the chamber to dry for 5–10 min within the laminar flow-hood.
   (c) Mounting with a large, flamed coverslip, using a circular bead of silicone grease applied to the bottom of the chamber.
2. Maintain stock cell cultures by following a strict passaging schedule involving trypsinization of a 100 mm stock dish once every third or fourth day.
3. Passage cells by sequentially:
   (a) Aspirating complete media off the 100 mm stock dish.
   (b) Briefly rinsing with 5 ml of STE to remove residual media.
   (c) Then rapid exchange and aspiration with 4 ml of trypsin, followed by a brief, ~ 2–3 min incubation period where cells are carefully monitored for initial signs of rounding and detachment (to limit exposure to trypsin).
   (d) Gentle resuspension of cells by repeated pipetting with ~ 7.5 ml of complete media.

Protocol 1. *Continued*

4. Plate cells into sterilized, assembled chambers at a density of 1–5 × 10^4 cells and culture for 36–72 h (~ 50–70% confluence) prior to performing experiments and analysis.

[a] At the time of experimentation, cultures should be populated by large numbers of circular epithelial colonies (~ 35–100 cells/colony) that contain cells at various stages of the mitotic cell cycle.

Protocol 2. Preparation and video imaging of fluorescent microspheres

Equipment and reagents

- Inverted fluorescence microscope and image processing workstation for live cell and microinjection work (see *Table 1* for details of configured workstation)
- Airfuge (Beckman Instruments) or small volume ultracentrifuge
- 2% suspension of 100 nm carboxylate-modified latex beads (Polysciences)
- PBS: 137 mM NaCl, 2.68 mM KCl, 8 mM Na_2HPO_4, 1.47 mM KH_2PO_4 pH 7.2
- PBS containing 10 mg/ml bovine serum albumin (BSA; Sigma)
- F12K base media (*Protocol 1*): F12K media lacking FCS and antibiotic/antimycotic or Hank's buffered salt solution (Sigma)
- NRK-2 cells plated in culture chambers (*Protocol 1*)
- 2 × stock solutions of 1–10 μM nocodazole (Sigma), 1–10 μM Taxol (NCI), or 1–5 μM cytochalasin D (Sigma) diluted in complete media immediately prior to treatment
- 1% DMSO (Sigma) in complete media as a carrier control

Method

1. Sonicate bead suspension and place a 100 μl aliquot into a microcentrifuge tube.
2. Centrifuge beads at 100 000 *g* for 4 min in airfuge and remove supernatant.
3. Rinse beads two additional times by repeating steps 1 and 2.
4. Resuspend beads in 0.1–1 ml of PBS/10% BSA and store on ice until further use. The final dilution can be varied depending on the bead density required for analysis.
5. Place chamber on pre-warmed microscope stage and use × 40 phase lens to identify a region of the plate containing one or several cells at early stages in mitosis (prometaphase to metaphase).
6. Secure the chamber to the stage with tape and reposition the mitotic cell to the centre of the field.
7. Remove most of the media and gently rinse the cells with two exchanges of 2 ml of warm Hank's salt or incomplete media.
8. Carefully apply a 50–100 μl aliquot of warm suspension of sonicated beads over cells in the centre of the field.

9. Following a 2–3 min incubation, rinse the chamber with several changes of complete media to remove unbound beads. Aliquot 1 ml of fresh media into the chamber.
10. Switch to a × 100 objective lens, and under low-light level conditions continuously monitor metaphase cell with phase optics.
11. Upon initiation of anaphase onset, begin acquiring time-lapse sequences at 15, 30, or 60 sec intervals until the completion of cytokinesis, using dual phase and fluorescence illumination to allow direct correlation of bead movements with the mitotic stage of the cell cycle.
12. To test the effect various cytoskeletal inhibitors have on transport, make a short, 1–2 min time-lapse leader sequence of the initial early phase of cortical flow, prior to addition and rapid mixing of 1 ml of media containing a two-fold concentration of cytoskeletal inhibitor.
13. Measure the rate and direction of particle flow and compute from time-lapse recordings using *Metamorph's* single-particle tracking application.

Table 1. Configuration for live cell image analysis workstation

Zeiss Axiovert TV-135 microscope with a high-light pass bottom port
Mercury-arc and quartz-halogen epifluorescent light sources
Phase and DIC condenser turret for bright-field illumination
Rhodamine, fluorescein, and Hoechst fluorescent filter sets
Range of high numerical aperture phase and fluorescent objectives
- × 10 (0.2), × 25 (0.8), × 40 (1.3), × 63 (1.3), and × 100 (1.3)

Custom designed CO_2/temperature-controlled stage incubator[a]
Mechanical and rotating specimen stage
Princeton Instruments ST-138 cooled CCD detector
- SITe back-illuminated 512 × 512 chip

P-90 PC running *Metamorph* imaging software[b]
Electronic shutter, filter wheel, and z-drive focus motor
Electronic controller for shutters, filter wheels, and z-motor drive[c]
Neutral density filters and polarizers mounted in filter wheel
DAG-MTI, low-light level video rate camera (CCD-100)
Microinjection equipment:
- microneedle puller (Kopf)
- microneedle holders (Leitz)
- micromanipulator (Leitz)
- air pressure regulator (Adams and List)

[a] Adapted for Zeiss TV-135 according to specifications outlined in ref. 16, and operated at ~ 33–37 °C/5% CO_2 for experiments on dividing and migrating tissue culture cells.
[b] *Metamorph* imaging software (Universal Imaging Corp.) functions to automate and integrate a variety of image acquisition routines, including time-lapse acquisition, motion analysis playback, and automated computational image processing and filtering applications.
[c] *Metamorph* communicates with the controller (Ludl Electronic Products) through serial communication to automatically control electronic triggering and switching of peripheral devices.

Protocol 3. Preparation and injection of rhodamine phalloidin for tracking cortical actin movement in dividing cells (modified from ref. 11)

Equipment and reagents

- Ultracentrifuge and rotor for microtubes (e.g. Beckman Table-Top and TL-100.1 or Type 42.2)
- Microscope imaging workstation with micro-injection accessories described in *Table 1*
- Mineral oil (Sigma)
- N_2 gas
- NRK-2 cells plated in culture chambers (*Protocol 1*)
- Rhodamine phalloidin: 300 units (R415; Molecular Probes)
- Microinjection buffer: 5 mM Tris–acetate in H_2O pH 7.0

Method

1. Freshly prepare rhodamine phalloidin (rhph) for injection into cells as outlined:
   (a) Solubilize 300 U of rhph in 0.75 ml of methanol.
   (b) Pipette 30 μl (12 U) into small Eppendorf tube.
   (c) Using a gentle stream of N_2 gas, slowly evaporate methanol until a small microdroplet of rhph in methanol remains (< 0.4 μl final volume). During the evaporation process, the tube should be intermittently centrifuged to coalesce the probe back to a single microdroplet. This step insures that rhph remains suspended in the organic phase and does not dry to the wall of the centrifuge tube.
   (d) Gently mix 10 μl of microinjection buffer with rhph.
   (e) Centrifuge at 100 000 *g* for 20 min prior to using for microinjection.

   Note: following preparation, rhph should be used within three to four days and be re-centrifuged prior to use.
2. Transfer cells in culture chambers to pre-warmed microscope stage.
3. Survey culture with × 40 objective and identify regions occupied by well-spread dividing cells.
4. Place × 10 objective lens in position for locating microneedle.
5. Prepare microneedle and position for injection of dividing cells by the following procedure:
   (a) Back load pulled microneedle with 0.5–1 μl of rhph.
   (b) Assemble microneedle and microneedle holder assembly.
   (c) Position and align microneedle holder on micromanipulator and connect air pressure line. Position, align, and centre needle over the centre of the field.

(d) Adjust air pressure to initiate a slow continuous flow of rhph from the needle as it is simultaneously being lowered into the media.

(e) Switch to a × 40 objective, locate the needle, and determine that a positive flow of rhph has been established. Perform by illuminating the tip of the needle with rhodamine excitation and observing a stream of fluorescence billowing from the needle.

(f) Carefully approach the cell with the needle, being certain to avoid the spindle region.

(g) With the needle tip located at the cell membrane, advance the needle into the cell using one smooth motion to penetrate the bilayer, and then back the needle out once loading has occurred.

Note: if the cell resists entry, a gentle tap of the finger on the fine adjustment knob of the z-translator control knob can be used to facilitate entry into the cytoplasm. If tapping method is used, recoil of the manipulator will often automatically cause the needle to retract from the cell.

6. Remove the needle holder from the scope, reposition × 63 (1.3) objective lens, focus, and monitor mitotic progression with phase illumination. Often a thin film of mineral oil is layered over the media to help maintain CO_2 conditions within the culture and avoid dramatic shifts in pH.

7. Just prior to anaphase onset, begin acquiring time-lapse images at 60 sec intervals until the completion of cytokinesis. Standard precautions for low-light level imaging should be followed (Section 3.3, *Protocol 7*).

Protocol 4. Preparation of rhodamine phalloidin labelled actin filaments for tracking cortical flow in dividing cells (modified from refs 12 and 17)

Equipment and reagents

- Bath sonicator
- NRK-2 cells plated in culture chambers (*Protocol 1*)
- 1 mg lyophilized G-actin (gel filtered)
- Buffer G: 2 mM Tris, 0.2 mM $CaCl_2$, 0.25 mM DTT, 0.02% NaN_3 pH 8.0
- Buffer A: 75 mM KCl, 5 mM Tris–acetate, 2 mM $MgCl_2$ pH 7.0
- Rhodamine phalloidin: 300 U diluted in 1.5 ml of methanol
- Microinjection buffer: 5 mM Tris–acetate in H_2O pH 7.0

Method

1. Preparation of rhph labelled actin filaments:

(a) Resuspend actin in 1 ml of buffer G and dialyse overnight at 4°C.

(b) Centrifuge G-actin at 100 000 *g* for 30 min at 4°C and retain supernatant.

Protocol 4. *Continued*

(c) Mix 20 μl of G-actin with 20 μl of 2 × buffer A and polymerize for 1 h at RT.

(d) Reduce 50 μl of rhph to 1 μl by evaporation of methanol with N_2 gas (see *Protocol 3*).

(e) Add actin filament solution (step 1c) to microdroplet of rhph and incubate for 30 min.

(f) dilute rhph labelled filaments twofold with 1 × buffer A and centrifuge at 100 000 *g* for 30 min at RT.

(g) Remove supernatant and resuspend pellet in microinjection buffer.

2. Prior to microinjection, sonicate actin filaments with three to five short, high frequency pulses (3–5 sec/pulse) to reduce filament length. Note: this step should be repeated prior to each injection to maintain short filament size, since annealing can occur over time resulting in restricted flow from the microneedle.
3. Perform microinjection, image acquisition, and analysis as described in *Protocol 3*, steps 5–7.

3. Green fluorescent protein (GFP)-tagged fusion proteins as reporters for *in vivo* contractile functions in tissue culture cells

A variety of past studies have successfully utilized conventional fluorescent analogue cytochemistry to track the behaviour and mobility of proteins in living cells (10, 18, 19). While an extremely valuable approach, the process can be quite labour-intensive, often requiring complex procedures involving purification of milligram quantities of proteins from a host tissue, chemical modification and labelling of the purified protein, re-isolation of the fluorescently conjugated probe, and introduction into the cell by microinjection or other cell loading techniques (18, 20). Moreover, data from such studies hinge on the capacity of the probe to faithfully recapitulate endogenous protein localization and function. Such issues are typically addressed by comparing the distribution of the analogue relative to endogenous protein, and characterizing its *in vitro* biochemical properties using conventional assays (18). Since the functional properties of the analogue are dependent on the type of isoform expressed and purified from the host tissue, concerns are raised over whether derivatized analogues properly share functions with endogenous protein isoforms present in cells under study. Finally, such analogues often have relatively short shelf-lives (i.e. one to three days) following derivatization and reconstitution from storage conditions, as reflected by progressive decreases in biochemical and *in vivo* cellular activities.

One alternative to conventional analogue cytochemistry has arisen with the development, advances, and application of GFP technology (6, 21). Such approaches take advantage of the ability to fuse the GFP gene in-frame to a specific gene of interest, allowing *de novo* expression of proteins tagged with the GFP chromophore (22, 23). This method is particularly advantageous in terms of studying protein dynamics in living cells, since it eliminates the need to biochemically produce fluorescent analogues, allows direct study of 'context'-specific, endogenous isoforms, eliminates concerns of storage and shelf-life activity, and allows observation of protein function without the need for demanding loading procedures.

While a wide variety of different cellular and molecular techniques are utilized to perform GFP analogue-based experiments, we describe here a set of standard protocols used in our laboratory to examine contractile protein function in several lines of motile and dividing tissue culture cells. The protocols are arranged sequentially to guide the investigator through basic methods of transient transfection of mammalian tissue culture cells, cellular protein analysis, and image acquisition procedures. Due to the variability of transfection efficiencies among differing cell lines, the protocols provided here are modified and adapted from a variety of previous studies. Readers are encouraged to consult original references for issues that may pertain to specific questions or particular concerns of cell lines under study.

3.1 Calcium phosphate transfections

We have found calcium phosphate-mediated transfections to be a good general method for introducing DNA into mammalian tissue culture cells for studies on cell motility and division. Depending on the cell type and method, transfection efficiencies of ~ 1–10% can be obtained. However, consistent results are contingent upon the quality, concentration, and specific DNA construct being used, as well as the state of cells in terms of cell passage number, timing of application of DNA precipitate relative to the time following cell passage, and the recovery phase response time following transfection. The following protocol is a modified calcium phosphate method that utilizes a glycerol shock step to facilitate transfection of NRK-2 epithelial cells (24).

Protocol 5. Calcium phosphate transfection of NRK-2 cells (adapted and modified from ref. 24)

Equipment and reagents

- 15 ml polystyrene centrifuge tubes
- NRK-2 cells seeded in culture chambers as described in *Protocol 1* (alternatively, cells can be seeded onto No. 1.5 22 × 22 mm coverslips that are prepared by alcohol dipping, flaming, and placing into sterile 35 mm plastic Petri dishes)
- F12K complete media (*Protocol 1*)
- F12K base media (*Protocol 2*)
- 2 M $CaCl_2$ (sterilized by filtration with 0.22 μm filter)
- 31.25% glycerol (Mallinckrodt) in base media (sterilized by filtration)
- Nanopure H_2O (autoclaved)

Protocol 5. *Continued*

- CsCl prepped GFP-fusion protein in mammalian expression vector (CsCl preparation of DNA was adapted and modified from ref. 25)
- Pipes-buffered saline (PipesBS): 280 mM NaCl, 1.5 mM sodium phosphate, 40 mM Pipes pH 6.95, sterilized by filtration through a 0.22 μm filter

Method

1. Passage and plate NRK-2 cells in culture chambers according to *Protocol 1* and culture for 12–16 h (~ 25–40% confluence).
2. 1 h prior to transfection, remove complete media, rinse cultures with two exchanges of warm base media, and incubate cells in 1 ml of base media for the remaining period of time.
3. Preparation of $CaPO_4$–DNA precipitate (for final incubation in a 2 ml volume). Immediately prior to application to the cells, prepare the precipitate by sequentially mixing the following components in a sterile 15 ml polystyrene tube:
   (a) Pipette 10–15 μg DNA into sterile H_2O to a final volume of 87.5 μl.
   (b) Add 12.5 μl of 2 M $CaCl_2$, and rapidly vortex for ~ 1 sec.
   (c) Add 100 μl of PipesBS, and rapidly vortex for ~ 1 sec.
   (d) Incubate the mixture for 1 min at RT.
   (e) Add 800 μl of 31.25% glycerol solution and vortex for ~ 1 sec.
   (f) Incubate mixture for 1 min at RT.
4. Add $CaPO_4$–DNA mixture dropwise into the culture, swirling the chambers gently to mix and distribute the precipitate evenly.
5. Return the culture chamber to the 37 °C CO_2 incubator for 3.5 h.
6. Remove the culture media by aspiration, rinse cells three times with complete media, add 2 ml of complete media, and return culture to the incubator.
7. 1–4 h prior to examination, rinse and exchange culture with fresh complete media.
8. Examine cells for GFP fluorescence 18–24 h post-transfection (*Protocol 7*).

3.2 Lipofectin–transferrin-mediated transfections

As an alternative to transfection with calcium phosphate, we have found liposome-mediated delivery approaches to be especially useful in introducing DNA into cell lines, such as rat embryo fibroblasts (REF-52) cells that can be relatively resistant to conventional transfection methods such as those described above (*Protocol 5*). The following protocol represents a modified lipofectin–transferrin-mediated transfection method (26) that we have found particularly useful in transiently transfecting REF-52 fibroblasts used for motility studies, examining non-muscle myosin II dynamics in cell locomotion and spreading (27, 28). Transferrin is thought to enhance gene transfer and expression in the presence of Lipofectin by facilitating the entry of DNA into cells through receptor-mediated uptake of a lipofectin–DNA–transferrin complex (26).

Protocol 6. Lipofectin–transferrin-mediated transfection of mammalian tissue culture cells (modified from ref. 26)

Equipment and reagents

- 15 ml polystyrene centrifuge tubes
- REF-52 cells seeded in culture chambers or on coverslips as outlined in *Protocol 1* for NRK-2 cells
- F12K complete media (*Protocol 1*)
- F12K base media (*Protocol 2*)
- 10 × Hepes-buffered saline (HBS): 20 mM Hepes, 100 mM NaCl pH 7.4
- 2 mg/ml human transferrin (iron-saturated, heat inactivated; Collaborative Biomedical, Becton Dickinson) diluted in sterilized H_2O, and stored as 25 μl frozen aliquots
- 1 mg/ml lipofectin (Life Technologies, Gibco BRL)
- CsCl purified full-length GFP-myosin IIB DNA vector (Dr Robert Adelstein, NIH)

Method

1. Passage, plate, and culture REF-52 cells for 18–24 h until the cells are 50–60% confluent.
2. Prepare lipofectin–transferrin–DNA complex from a two-part solution.
   - *Solution A*: lipofectin–transferrin complex. In a sterile polystyrene tube sequentially add:
     (a) 10 μl of 10 × HBS.
     (b) 25 μl of 2 mg/ml human transferrin.
     (c) 3 μl of 1 mg/ml lipofectin.
     (d) Nanopure grade deionized water to a final volume of 100 μl.
     (e) Gently mix and incubate at RT for 30 min.
   - *Solution B*: DNA solution. In a sterile polystyrene tube, sequentially add:
     (a) 10 μl of 10 × HBS.
     (b) 4–10 μg of DNA.
     (c) Nanopure grade deionized water to a final volume of 100 μl.
3. Combine the two solutions, mix gently, and incubate at RT for 10–15 min.
4. 1 h prior to transfection, remove complete media from cells, rinse cultures with two exchanges of incomplete media, then place 500 μl of incomplete media on cells.

Protocol 6. *Continued*

5. Using a glass Pasteur pipette, transfer the lipofectin–DNA–transferrin mixture to cells, swirling the chambers gently to mix and distribute the solution evenly.
6. Return cells to incubator and culture for an additional 6 h. Incubations can range from 6–24 h.
7. Following the incubation, remove transfection mixture, rinse the cells with two exchanges of complete media, add 2 ml of complete media, and incubate cells for an additional 18–24 h.
8. 1–4 h prior to examination, rinse and exchange culture with fresh complete media.
9. Assay cells for expression of fluorescent protein 18–24 h post-transfection (*Protocol 7*).

3.3 Analysis of GFP in live tissue culture cells

Due to the need to acquire high-resolution images of GFP-expressing cells over varied periods of time, it is important both to control culture conditions and to utilize a sensitive, low-light image acquisition system. The latter avoids problems associated with photobleaching of GFP-tagged probes and 'over-sampling', which can induce phototoxicity. Such issues can be addressed by applying a variety of standard instrumentation and technology for low-light level analysis that can dramatically reduce sample illumination and over-exposure of cells to excitation light. In our laboratory we have combined the use of a highly sensitive, back-illuminated CCD chip camera, together with a Zeiss TV-135 Axiovert microscope engineered with a bottom camera port for maximizing the amount of light directed to the detector (*Table 1*). In addition, we utilize a stable, variably controlled quartz-halogen epifluorescence lamp source, together with a set of selectable neutral density filters positioned in an automated filter wheel. By combining this equipment with a set of sequentially timed acquisition routines, we can routinely acquire 20–60 high-resolution images per series without significant image degradation or cellular exposure. The following protocol details our standard method for analysis of protein dynamics using GFP-tagged fusion proteins in tissue culture cells. Similar precautions should be used in all experiments where live imaging of fluorescent compounds are conducted.

Protocol 7. Fluorescence imaging of GFP in living cells

Equipment and reagents

- Wide-field fluorescence imaging work-station (*Table 1*)
- Mineral oil (Sigma)
- GFP-expressing NRK-2 or REF-52 cells (*Protocols 5, 6*) plated in culture chambers

Method

1. Pre-warm and equilibrate CO_2 controlled microscope stage incubator chamber for 10–20 min.
2. Obtain culture chamber with transfected cells from incubator.
3. Using a large bore transfer pipette, carefully overlay media with a thin film of mineral oil. Note: if cells are to be further manipulated by the application of beads (*Protocol 2*) or dye loading by microinjection (*Protocols 3, 4*), oil should be applied following such procedures.
4. Place culture chamber on microscope stage and, using a $\times$ 25 or $\times$ 40 oil objective, illuminate the sample using the least amount of light needed to survey the dish for GFP-expressing cells. Note: this is performed either by viewing the cells through the oculars or, preferably, by displaying a live image collected with a video rate, low-light level camera mounted to a second camera port and displayed to a video monitor.
5. After locating and centering the cell of interest, close the epifluorescence shutter, attenuate light source further with neutral density filters (or by lowering the lamp current), open shutter, and focus image using low-light level video camera. To reduce bleaching and cellular exposure, focusing activity should be carried out as quickly as possible.
6. Prior to acquiring an image with the cooled CCD camera, acquire a background image of the electronic noise using the identical exposure settings to be used for sample acquisition.
7. Sample acquisition. To help reduce sample exposure, automate the acquisition process by performing the following events in rapid succession:
   (a) Open illumination shutter.
   (b) Open camera shutter for set exposure time.
   (c) Close camera shutter.
   (d) Close illumination shutter.
   (e) Transfer image data from the camera chip to the computer.
   (f) Subtract the background image from the sample image.
   (g) Display the image.
8. Following acquisition of the first 'test' image, assess the overall pixel intensity values quickly to determine the minimal amount of light and exposure time needed to obtain a well-'saturated' image. Note: pixel values of fluorescent structures $> 2 \times$ above that of non-fluorescent background values are generally found to provide acceptable image quality for acquisition and data analysis. However, longer exposures

Protocol 7. *Continued*

can be acquired depending on the sensitivity and type of acquisition being performed.

9. Adjust exposure times and illumination settings accordingly, and record time-lapse series automatically by acquiring images at 30, 60, or 120 sec intervals, depending on the cellular activity under study. Note: data quality and duration of sample imaging is highly dependent on the amount of GFP expressed, the intensity of illumination required, and sensitivity of instrumentation used for fluorescence detection of probes in living cells.

3.4 Analysis of GFP-fusion proteins in fixed cells

In some experiments it is often important to analyse the localization of GFP-fusion proteins relative to other cellular components. This process requires preserving cells by chemical fixation, followed by additional labelling procedures employing indirect immunocytochemistry or fluorescent phalloidin staining methods. Although GFP and GFP-tagged proteins can be preserved using conventional PBS–formaldehyde fixation (29), we have determined that there can be up to a five-fold loss in fluorescence intensity of GFP within expressing cells, and intracellular structures can often be disrupted or distorted immediately following application of formaldehyde. To optimize preservation of GFP fluorescence while maintaining structural integrity, we have established a formaldehyde-based fixation protocol that retains *in vivo* fluorescence levels of GFP emission with little to no disruption of cytoskeletal structures. This method was experimentally determined by observing live GFP labelled cells during the fixation process. Note that other conventional fixation methods employing either immersion in –20 °C acetone or methanol have not worked particularly well for fixation of our GFP spectral variants.

Protocol 8. Fixation and preservation of GFP fluorescence in cultured cells

Equipment and reagents

- GFP expressing NRK-2 or REF-52 cells plated in culture chambers or coverslips (*Protocols 5, 6*)
- Fixation boxes (recycled 45 × 50 mm plastic coverslip containers)
- PBS, 37 °C (*Protocol 2*)
- Triton X-100 (Sigma)
- 16% formaldehyde (EM grade) stored in sealed ampules (Polyscience)
- Fixative I (freshly prepared): 0.3% Triton X-100, 4% formaldehyde in PBS at 37 °C
- Fixative II (freshly prepared): 4% formaldehyde in PBS at RT

Method

1. Remove media and briefly rinse cells with a 5 sec exchange of warm PBS.

2. Fix cells by rapidly immersing coverslip (cell-side up) into fixative I placed in fixation box.
3. Transfer box to a gently rocking shaker table and allow reaction to continue for 7.5 min at RT.
4. Remove fixative I, rinse coverslips with two rapid exchanges of PBS, and react with fixative II for an additional 7.5 min by agitation on shaker table at RT.
5. Wash coverslips with three exchanges of PBS for 5 min each at RT.
6. Proceed with established procedures for immunofluorescent staining or labelling with rhodamine phalloidin (4, 14, 15, 30).

3.5 Analysis of GFP expressed fusion products

Due to the relatively low transfection efficiencies (i.e. 1–10%) of many optically useful cell lines, the amount of GFP-fusion protein expressed in these cultures can be relatively small. While this does not create a major problem for fluorescent analysis of protein dynamics in single cells, it is quite difficult to assess fully the biochemical characteristics (e.g. molecular size, integrity, and degradation patterns) of expressed GFP-fusion constructs. To address this issue, we perform parallel experiments with either COS-7 or CHO cell lines, since they display both high transfection efficiencies (~ 40–60%) and protein expression levels (*Protocol 9*). In combination with anti-GFP immunoblotting procedures, this approach allows detection and analysis of construct size, stability, and integrity of fusion proteins expressed within tissue culture cells. For direct analysis of protein products within specific cell lines, methods such as fluorescent cell sorting may be required to obtain enriched populations of GFP expressing cells (31).

Protocol 9. Lipofectin-mediated transfection of COS-7 cell line for analysis of protein expression (modified from Gibco BRL's protocol for lipofectin-mediated transfection)

Equipment and reagents

- Microprobe sonicator
- 60 mm sterile glass Petri plates
- COS-7 (African green monkey kidney) fibroblast cells
- Complete DMEM: Dulbecco's modified Eagle's media (DMEM) containing 4.5 g/litre glucose (Sigma) supplemented with 10% FCS (BioWhittaker), 2 mM L-glutamine (Gibco BRL), 0.1 mM non-essential amino acids (Sigma), 50 μg/ml antibiotic/antimycotic solution (Sigma), and 42 mM $NaHCO_3$ pH 7.8 at 37 °C, 5% CO_2
- 15 ml polystyrene centrifuge tubes
- Incomplete DMEM: DMEM base media lacking FCS and antibiotic/antimycotic solution
- 1 mg/ml lipofectin (Gibco BRL)
- CsCl prepared GFP-fusion protein in mammalian expression vector
- 0.05% trypsin, STE at 37 °C (*Protocol 1*), PBS at 4 °C (*Protocol 2*)
- 2 × Laemmli SDS–PAGE sample buffer: 200 mM DTT, 4% SDS, 125 mM Tris, 20% glycerol, 0.006% bromophenol blue pH 6.8

Protocol 7. *Continued*

Method

1. Plate cells in glass Petri plates at a density of ~ 1×10^5 cells per dish with 5 ml complete DMEM.
2. Culture cells at 37 °C in a CO_2 incubator for 2 h. Confirm plating efficiency of 40–50% confluence.
3. Prepare the following solutions in sterile polystyrene tubes (recipe for transfection of one dish):
   (a) *Solution A*: dilute 10 μl of lipofectin into 100 μl of nanopure grade H_2O and allow to stand at RT for ~ 45 min.
   (b) *Solution B*: dilute 2–5 μg of DNA into 100 μl of nanopure grade H_2O.
   Combine solution A and B. Mix by flicking gently for 30 sec, and incubate at RT for 10–15 min.
4. After 3 h following plating, rinse cells with two exchanges of 2 ml incomplete media.
5. For each transfection add 0.8 ml incomplete media to each tube containing the lipofectin–DNA complex. Mix gently and overlay the complex onto cells using glass Pasteur pipette.
6. Incubate the cells for 8–18 h at 37 °C in a CO_2 incubator.
7. Replace transfection media with 2 ml complete DMEM and incubate for an additional 18–48 h.
8. To prepare protein extracts rinse cells with a rapid exchange of warm STE with trypsin, and then recover cells by gently pipetting 1.5 ml cold PBS across the detached cell monolayer.
9. Place cell suspension in microcentrifuge tube (1.5 ml) and spin at 500 *g* for 1–2 min to pellet cells.
10. Carefully remove supernatant, gently resuspend cell pellet in residual PBS by repeated flicking, and add 25 μl of hot 2 × sample buffer.
11. Mix the extract by repeated pipetting, spin briefly to collect sample at the bottom of the tube, and then sonicate with microprobe sonicator for 2–5 sec using cyclic, high frequency pulses.
12. Assay GFP-fusion protein products by transferring SDS–PAGE separated cellular proteins to nitrocellulose and reacting immunoblots with anti-GFP antibodies according to conventional protocols (32).

4. Methods for analysing actin filament and myosin II organization in motile and dividing cells

One of the key features contractile filaments display in non-muscle systems is a relatively high degree of disorder. Actin filaments form extensive isotropic networks (9, 15), while myosin II structures are distributed in a complex pattern of punctate dots, spots, fibres, and ribbon structures (10, 27, 28, 33–36). Sorting through these complex filament arrays can be a tortuous process, especially when one is trying to understand the underlying principles of how form determines function. By utilizing various optical approaches in combination with digital imaging methods, we have been able to apply several unique tools to help elucidate the organization and orientation of contractile filaments in dividing and motile cells (14, 15). Studies employing fluorescence-detected linear dichroism (FDLD), digital optical sectioning microscopy, and three-dimensional reconstructions of the cytoskeleton have allowed the development of new models for understanding the complex interactions spindle microtubules, actin filaments, and myosin II display in regulating contractile functions associated with cytokinesis. The following protocols describe the basic approaches we utilize for applying FDLD, digital optical sectioning microscopy, and three-dimensional reconstructions to the study of actin filaments in dividing cells. Such approaches are applicable to a variety of biological problems centered on defining the organization and orientation of polymer systems. Detailed descriptions and further discussions of these methods can be found in several recent publications (37–40).

Protocol 10. Fixation and labelling of cells for FDLD microscopy

Equipment and reagents

- NRK-2 cells seeded in culture chambers (*Protocol 1*) or on coverslips (*Protocol 5*)
- PBS at 37 °C (*Protocol 2*)
- PBS containing 1% BSA (Fraction V, Sigma)
- Cytoskeleton buffer: 137 mM NaCl, 5 mM KCl, 1.1 mM Na_2HPO_4, 0.4 mM KH_2PO_4, 2 mM $MgCl_2$, 2 mM EGTA, 5 mM Pipes, 5.5 mM glucose pH 6.1 (41)
- 0.1% $NaBH_4$ in PBS (freshly prepared)
- Fixative A: 0.5% glutaraldehyde, 0.2% Triton X-100 in cytoskeleton buffer (42)
- Fixative B: 1% glutaraldehyde in cytoskeleton buffer
- 200 nM rhodamine phalloidin (Molecular Probes) in PBS (freshly prepared) (note: different lots of phalloidin are pre-screened since the extent of FDLD can vary among lots)

Method

1. Prior to fixation, rinse coverslips briefly (< 5 sec) with warm PBS, then rapidly immerse into fixative A for 1 min at RT.
2. Rinse coverslips briefly with two exchanges of cytoskeleton buffer and fix for an additional 10 min in fixative B.
3. Following fixation, rinse coverslips in cytoskeleton buffer for 5 min, and PBS for an additional 10 min.
4. Quench autofluorescence with a 5 min incubation in 0.1% $NaBH_4$, and incubate for 1 h in PBS-BSA to block non-specific binding.
5. Stain cells by incubating coverslips with 200 μl rhodamine phalloidin for 1 h. Rinse in PBS for an additional 30 min before examination by FDLD microscopy.

Protocol 11. Acquisition and generation of images using FDLD microscopy

Equipment and reagents

- Wide-field fluorescence imaging workstation described in *Table 1*, outfitted with rotating stage, × 40 (NA 1.0) apochromatic objective lens with iris diaphragm, and filter wheel containing polarizers oriented parallel and perpendicular to the optical axis of the microscope
- Rhodamine phalloidin stained NRK-2 cells (*Protocol 10*)

Method

1. Adjust the aperture of the objective lens to ~ 0.75. Note: to increase depth of field, the objective is typically set to a numerical aperture of ~ 0.75. This reduces the fluorescence intensity of the image to ~ 31.3% of normal, assuming fluorescence intensity is proportional to the fourth power of the NA of the objective (43).
2. Position the culture chamber or coverslip on the stage.
3. While illuminating with rhodamine excitation, survey the sample, locate and centre the cell of interest, and orient the cell with respect to filament structures to be measured. Note: for dividing cells, the sample is usually oriented on the specimen stage so that the spindle axis, defined by the position of chromosomes, is aligned parallel to the horizontal axis of the microscope.
4. After acquiring a background image (*Protocol 7*, step 5), illuminate the sample with 550 nm light, and acquire images with plane-polarized light oriented parallel (||) and perpendicular ($\perp$) to the optical axis using appropriately oriented polarizers placed within the filter wheel. Note: images are acquired using these orientations to minimize any depolarizing effects of the dichoric mirror.
5. Archive the || and $\perp$ images to storage for further mathematical operations.
6. Use pairs of || and $\perp$ images to generate a ratio image by dividing the difference of the pair by the sum of the pair ($\| - \perp / \| + \perp$), following correction for the difference in excitation intensity at the two orientations. Note: the correction factor must be experimentally derived for individual optical systems by imaging a mounted drop of 0.5 mg/ml rhodamine dye in glycerol at the two defined orientations and taking the ratio of the two intensities. In our system, this ratio is 0.9 and is used to adjust the brighter image.
7. Derive the angular distribution of filaments relative to the cleavage plane from differential FDLD measurements using the mathematical relationship outlined below. It should be noted that even with the

precaution taken, the values represent only a relative measure of the linear dichroism, since the purity of polarization can be significantly affected following passage through the objective lens (44), and structures on different planes are likely to make variable contributions to the measured value depending on the thickness of the sample relative to the depth of field.

8. Generate FDLD images by properly scaling the results such that FDLD values in the range of –0.25 to 0.25 are mapped as grey values in the range of 0 to 255, with zero FDLD giving a display value of 128.
9. Display the grey value image using a 'spectrum' pseudocolour look-up table to allow discrimination of filament structures aligned along a given axis.
10. Quantify temporal and spatial analysis of FDLD images further by averaging pixel values within a defined region (e.g. 5 μm^2 spot) positioned at places of interest within the cell.

4.1 Mathematical relationship between FDLD values and the angular distribution of filaments

Assuming a two-dimensional structure, the mathematical relationship between differential FDLD measurements and angular distribution of actin filaments can be calculated as follows, if:

$F_{\parallel}$ = fluorescence intensity with excitation light $\parallel$ to the equator
$F_{\perp}$ = fluorescence intensity with excitation light $\perp$ to the equator
$S_{\parallel}$ = cross-section for absorption along the axis of each filament
$S_{\perp}$ = cross-section for absorption perpendicular to the axis of each filament
$n(\theta)\, d\theta$ = number of filaments oriented at angle θ to $\theta + d\theta$

Then:

$$F_{\parallel} \propto \int_0^{2\pi} n(\theta) \quad S_{\parallel} \cos^2\theta \, d\theta + \int_0^{2\pi} n(\theta) \quad S_{\perp} \sin^2\theta \, d\theta$$
$$F_{\perp} \propto \int_0^{2\pi} n(\theta) \quad S_{\parallel} \sin^2\theta \, d\theta + \int_0^{2\pi} n(\theta) \quad S_{\perp} \cos^2\theta \, d\theta$$
$$F_{\parallel} + F_{\perp} \propto \int_0^{2\pi} n(\theta)\, d\theta \, (S_{\parallel} + S_{\perp}) = N + (S_{\parallel} + S_{\perp})$$
$$F_{\parallel} - F_{\perp} \propto \left(\int_0^{2\pi} n(\theta)\, (\cos^2\theta - \sin^2\theta)\, d\theta\right) (S_{\parallel} - S_{\perp})$$
$$\frac{F_{\parallel} - F_{\perp}}{F_{\parallel} + F_{\perp}} \propto \left[\frac{S_{\parallel} + S_{\perp}}{S_{\parallel} - S_{\perp}}\right] \int_0^{2\pi} \frac{n(\theta)}{N} (\cos^2\theta - \sin^2\theta)\, d\theta$$

where,

$$\int_0^{2\pi} \frac{n(\theta)}{N} (\cos^2\theta - \sin^2\theta)\, d\theta$$

gives a weighted average angular filament distribution, with a weight equalling +1 if oriented along the parallel axis, 0 if oriented 45° to this axis, and –1 if oriented perpendicular to the parallel axis. Thus, structures preferentially oriented parallel and perpendicular to a given axis give opposite signs for FDLD values and are reflective of a given filament orientation within the cell.

Protocol 12. Digital optical sectioning microscopy and three-dimensional analysis

Equipment and reagents

- Wide-field fluorescence imaging workstation described in *Table 1* and × 100 (1.3) objective
- Rhodamine phalloidin stained NRK-2 cells (*Protocol 10*) mounted in anti-photobleach reagent

Method

1. Place the sample on the stage, locate and centre the cell of interest.
2. Assess the thickness of the sample by manually focusing through the sample and monitoring the distance the z-focus drive travels. Note: this can be assessed manually or digitally via the z-focus motor interface.
3. Acquire a complete through-focus series of optical sections by sequentially advancing the z-motor drive in 0.3 μm increments and collecting images at each optical plane. Be certain to obtain the out-of-focus focal plane below and above the first and last planes of focus when acquiring the image stack. Note: to ensure precise steps we always collect sections by advancing the objective 'upward' away from gravitational forces on an inverted microscope stand. In addition, we utilize an electronically stable power supply for illumination since computational steps involving processing of digital images are highly sensitive to changes or fluctuations in lamp intensities.
4. Collect image planes of a sample from a small region in the centre of the field to limit spherical aberration and acquire with a cooled CCD camera using identical acquisition settings. Note: it is important to collect well-saturated images for nearest-neighbour deconvolution processing, so we typically acquire several images from the brightest focal plane to determine the integration or exposure time to be used for collecting an optical stack.
5. Archive optical stacks to the hard disk and then render out-of-focus fluorescence and restoration of optical slices using a nearest-neighbour deconvolution algorithm routine (39, 45–48). Note: we presently utilize *Metamorph's* haze removal filtering routine with a low pass kernel size of 11, scaling factor of 0.95, and scaled results value at a level appropriate for the image set. For more mathematically rigorous approaches, algorithms integrating the point spread function

acquired from the optical system are highly recommended and can be determined from standard procedures (48, 49).

6. Utilize individual optical planes to generate three-dimensional reconstructions of x-y stereo pairs by stacking a set of deconvolved sections and projecting them at angles +10° and −10° from the optical axis using *Metamorph* 3D reconstruction software.
7. Perform direct examination and analysis of complex three-dimensional data sets by displaying red-green stereo pairs on the monitor or rotating reconstructed stacks at different angles using movie loop routines.

5. Summary and prospects

Recent advances in the technology for studying cytoskeletal dynamics have allowed the development of new models for understanding the complex interactions and processes involved in contractile filament reorganization and cell signalling events regulating cytokinesis and cell locomotion. In this chapter, we have provided a number of cell-based methods and optical approaches that facilitate the study of mitotically active and motile cells involved in stimulus-driven, mechano-sensitive, force production and signal transduction pathways. As these and other approaches continue to grow, experiments designed to couple and integrate such technology will expand investigations focused on gaining a comprehensive understanding of the cytoskeleton's role in signalling and force production.

Acknowledgements

The authors wish to thank Drs Yu-li Wang, Joseph O'Tousa, and Kirian Setty for contributions in the development of these protocols, and Drs Robert Adelstein, Quize Wei, and Tom Hughes for providing reagents used to test some of the approaches. Research for this work is supported by a grant from the NSF, MCB-9808240.

References

1. Janmey, P. A. (1998). *Physiol. Rev.*, **78**, 763.
2. Wang, Y. -L. and Taylor, D. L. (ed.) (1989). *Methods in cell biology*, Vol. 29, p. 503. Academic Press, London.
3. Nuccitelli, R. (ed.) (1994). *Methods in cell biology*, Vol. 40, p. 368. Academic Press, London.
4. Carraway, K. L. and Carraway, C. A. C. (ed.) (1992). In *Cytoskeleton: a practical approach*, p. 268. IRL Press, Oxford.

5. Mason, W. T. (ed.) (1993). *Fluorescent and luminescent probes for biological activity*, p. 433. Academic Press, London.
6. Wilson, L. (ed.) (1982). *Methods in cell biology*, Vol. 24, p. 445. Academic Press, London.
7. Sullivan, K. F. and Kay, S. A. (ed.) (1998). *Methods in cell biology*, Vol. 58, p. 386. Academic Press, London.
8. Satterwhite, L. L. and Pollard, T. D. (1992). *Curr. Opin. Cell Biol.*, **4**, 43.
9. Fishkind, D. J. and Wang, Y.-L. (1995). *Curr. Opin. Cell Biol.*, **7**, 23.
10. Sanger, J. M., Mittal, B., Dome, J. S., and Sanger, J. W. (1989). *Cell Motil. Cytoskel.*, **14**, 201.
11. Cao, L.-G. and Wang, Y.-L. (1990). *J. Cell Biol.*, **110**, 1089.
12. Cao, L.-G. and Wang, Y.-L. (1990). *J. Cell Biol.*, **111**, 1905.
13. Wang, Y.-L., Silverman, J. D., and Cao, L.-G. (1994). *J. Cell Biol.*, **127**, 963.
14. Fishkind, D. J., Silverman, J. D., and Wang, Y.-L. (1996). *J. Cell Sci.*, **109**, 2041.
15. Fishkind, D. J. and Wang, Y.-L. (1993). *J. Cell Biol.*, **123**, 837.
16. McKenna, N. M. and Wang, Y.-L. (1989). In *Methods in cell biology* (ed. Y.-L. Wang and D. L. Taylor), Vol. 29, p. 195. Academic Press, London.
17. Sanders, M. C. and Wang, Y.-L. (1990). *J. Cell Biol.*, **110**, 359.
18. Wang, Y.-L. (1989). In *Methods in cell biology* (ed. Y.-L. Wang and D. L. Taylor), Vol. 29, p. 1. Academic Press, London.
19. Farkas, D. L., Baxter, G., DeBiasio, R. L., Gough, A., Nederlof, M. A., Pane, D., *et al.* (1993). *Annu. Rev. Physiol.*, **55**, 785.
20. McNeil, P. L. (1989). In *Methods in cell biology* (ed. Y.-L. Wang and D. L. Taylor), Vol. 29, p. 153. Academic Press, London.
21. Chalfie, M., Tu, Y., Euskirchen, G., Ward, W. W., and Prasher, D. C. (1994). *Science*, **263**, 802.
22. Gerdes, H. H. and Kaether, C. (1996). *FEBS Lett.*, **389**, 44.
23. Ludin, B. and Matus, A. (1998). *Trends Cell Biol.*, **8**, 72.
24. Wilson, S. P., Liu, F., Wilson, R. E., and Housley, P. R. (1995). *Anal. Biochem.*, **226**, 212.
25. Sambrook, J. (1989). In *Molecular cloning: a laboratory manual* (2nd edn) (ed. I. Sambrook, E. F. Fritsch, and T. Maniatis), Vol. 1, p. 21. Cold Spring Harbor Laboratory Press, NY.
26. Cheng, P.-W. (1996). *Human Gene Therapy*, **7**, 275.
27. Verkhovsky, A. B. and Borisy, G. G. (1993). *J. Cell Biol.*, **123**, 637.
28. Verkhovsky, A. B., Svitkina, T. M., and Borisy, G. G. (1995). *J. Cell Biol.*, **131**, 989.
29. Girotti, M. and Banting, G. (1996). *J. Cell Sci.*, **109**, 2915.
30. Wang, Y.-L. (1992). In *Cytoskeleton: a practical approach* (ed. K. L. Carraway and C. A. C. Carraway), p. 1. IRL Press, Oxford.
31. Mackay, A. M., Ainsztein, A. M., Eckley, D. M., and Earnshaw, W. C. (1998). *J. Cell Biol.*, **140**, 991.
32. Kain, S. R., Mai, K., and Sinai, P. (1994). *BioTechniques*, **17**, 982.
33. Fishkind, D. J., Cao, L.-G., and Wang, Y.-L. (1991). *J. Cell Biol.*, **114**, 967.
34. Maupin, P., Phillips, C. L., Adelstein, R. S., and Pollard, T. D. (1994). *J. Cell Sci.*, **107**, 3077.
35. Sanger, J. M., Dome, J. S., Hock, R. S., Mittal, B., and Sanger, J. W. (1994). *Cell Motil. Cytoskel.*, **27**, 26.

36. DeBiasio, R. L., LaRocca, G. M., Post, P. L., and Taylor, D. L. (1996). *Mol. Biol. Cell*, **7**, 1259.
37. Borejdo, J. and Burlacu, S. (1993). *Biophys. J.*, **65**, 300.
38. Zhukarev, V., Sanger, J. M., Sanger, J. W., Goldman, Y. E., and Shuman, H. (1997). *Cell Motil. Cytoskel.*, **37**, 363.
39. Agard, D. A. (1984). *Annu. Rev. Biophys. Bioeng.*, **13**, 191.
40. Shaw, P. J. (1993). In *Electronic light microscopy* (ed. D. Shotton), p. 211. Wiley-Liss, NY.
41. Small, J. V. (1981). *J. Cell Biol.*, **91**, 695.
42. Small, J. V., Rinnerthaler, G., and Hinssen, H. (1982). *Cold Spring Harb. Symp. Quant. Biol.*, **46**, 599.
43. Inoué, S. (1986). *Video microscopy*, p. 584. Plenum Press, NY.
44. Axelrod, D. (1989). In *Methods in cell biology* (ed. Y.-L. Wang and D. L. Taylor), Vol. 30, p. 333. Academic Press, London.
45. Castleman, K. R. (1979). *Digital image processing*, p. 429. Prentice-Hall, NJ.
46. Agard, D. A., Hiraoka, Y., Shaw, P., and Sedat, J. W. (1989). In *Methods in cell biology* (ed. Y.-L. Wang and D. L. Taylor), Vol. 30, p. 353. Academic Press, London.
47. Shaw, P. J. and Rawlins, D. J. (1991). *Prog. Biophys. Mol. Biol.*, **56**, 187.
48. Wang, Y.-L. (1998). In *Methods in cell biology* (ed. G. Sluder and D. E. Wolfe), Vol. 56, p. 305. Academic Press, London.
49. Hiraoka, Y., Sedat, J. W., and Agard, D. A. (1990). *Biophys. J.*, **57**, 325.

Video microscopy: protocols for examining the actin-based motility of *Listeria*, *Shigella*, vaccinia, and lanthanum-induced endosomes

W. L. ZEILE, D. L. PURICH, and F. S. SOUTHWICK

1. Introduction

Cellular locomotion takes on many forms—from macrophage tracking and capturing a bacterium during infection to platelet spreading during blood clotting to tumour cell crawling during metastasis to cells dividing during mitosis or meiosis, but the common element is exquisitely choreographed directed movement. There are literally hundreds of proteins and enzymes involved in cell motility, and many of these proteins have multiple and often redundant roles in promoting specific types of motility (e.g. lamellipodial versus filipodial extension). The challenge then for cell biologists is to unravel the protein–protein interactions and other regulatory signals that constitute each distinct mechanism of vectorial movement. To meet this challenge requires both intuition and invention in the development and systematic application of new experimental techniques. Although this chapter describes the specifics of applying video microscopy, microinjection, and immunolocalization to investigate actin-based motility of intracellular pathogens, it is also true that the strategies presented here will be of value to those interested in other forms of cell motility and studies in cell biology in general. Significantly, video microscopy, microinjection, and immunofluorescence have been used with great success in the study of signal transduction pathways, specifically the role of monomeric GTPases in cytoskeletal remodelling (1).

Intracellular parasites have been exploited in the study of actin cytoskeletal dynamics and structure as these organisms usurp a relatively small and tractable component of an otherwise large and multifaceted structure for their motility. Complex cellular structures, such as a lamellipodia with multiple interacting components, are extremely difficult to study. Opposed to the study of complex cellular structures, pathogen movement and location within the cell is readily assayed as described by procedures in this chapter. With careful experimental design, perturbations of movement and location from normal levels or positions can be used as assays to narrow the range of possible mechanisms or interactions under consideration to a few. Investigations based on these methods have yielded significant insights into processes of actin polymerization/depolymerization, cytoskeletal protein binding interactions, and biochemical pathway characterization (2–5).

This chapter will explore the use of the pathogen/host cell interaction as a way of dissecting complex interactions of the cytoskeleton. Although a number of organisms have been identified that use actin-based motility as a part of their life cycle, the examples given here are the Gram-positive bacteria *Listeria monocytogenes*, the Gram-negative bacteria *Shigella flexneri*, and the double-stranded DNA virus vaccinia virus. We also will look at a host cell-derived vesicle, the lanthanum-induced endosome that moves by actin assembly.

The discussion begins with the preparation of the host cells and the pathogens for productive and consistent study by video microscopy and covers the process of infection by *Listeria*, *Shigella*, and vaccinia virus and the conditions that we have found to be optimal for experimental culture. The next section describes the preparation of bone marrow macrophages for the study of endosomal rocketing and the procedure for stimulating formation of endosomes and their actin-based motility. The third and final section of this chapter covers the analysis of motile processes by video microscopy. We first look at the use of time-lapse imaging in recording movement and analysing speed. Microinjection is discussed next as an important method for delivering accurately metered amounts of exogenous treatment substances, specifically for activation and inhibition studies. Immunolocalization using epifluorescence microscopy is also covered as it applies to elucidating actin-based mechanisms. Finally, methods of image data analysis and presentation are discussed with specific examples.

2. Biological preparations for intracellular motility experiments

2.1 Mammalian host cells and cell culture conditions

A broad range of eukaryotic host cells has been effectively used in studies of motility by intracellular pathogens. This list includes, but is not limited to HeLa (human cervical epithelial), PtK2 (kangaroo rat, *Potorous tridactylis*, kidney epithelial), Cos-7 (African green monkey kidney fibroblast), J774 (mouse monocyte-macrophage), Henle 407(human intestinal epithelial), human 293 (human primary embryonal kidney), Caco-2 (human adenocarcinoma colonic epithelial), chicken embryo fibroblasts (CEF), F9 (mouse embryonal carcinoma), and Vero (African green monkey kidney fibroblast). Although selection of cell type will be dictated by the experimental design, if it is possible, the use of a cell with a spreading morphology and with strong adherence to glass is ideal for many types of cytoskeletal studies. For example PtK2 cells grown on glass with space to spread exhibit large lamellipodia or regions of thin, extended cytoplasm. When these cells are infected with *Listeria,* the bacteria and its polymerized actin tail structure is clearly visible. This clarity of structure is not often found in cells such as the J774 macrophage cell. Because these cells have a rounded morphology and lack large thin lamellipoda, imaging of the bacteria and its attendant structures is much more difficult.

Preparation and maintenance of healthy cells in tissue culture are large topics, and detailed descriptions will not be presented here. On the other hand, it is vital for the investigator to have healthy, well-adhered, semi-confluent cells for unambiguous and consistent experimental results. We have used a broad range of host cells in our experiments, but have found PtK2 and HeLa cells to be the best all round hosts for *Listeria*, *Shigella*, and vaccinia. Therefore, a brief protocol on preparation and maintenance of these mammalian cells in culture is given below.

An important aspect is the type of serum used in culture of these cells. Fetal bovine serum (FBS) that has been heat inactivated must be used for any cells that will be experimentally infected with bacteria. Serum from birthed animals such as calf serum and serum that has not been heat inactivated contains complement and antibodies which are extremely effective at killing bacteria.

Protocol 1. Preparation and maintenance of PtK2 and HeLa cell monolayers in culture

Equipment and reagents

- Temperature controlled, humidified CO_2 incubator set at 37 °C
- 25 cm^2 flasks
- Centrifuge with swinging bucket rotor
- PtK2 cell maintenance media: minimum essential media (MEM) with L-glutamine, without $NaHCO_3$ (Gibco, 11435–039), 10% fetal bovine serum (FBS), certified heat inactivated (Gibco, 10082–139), 1% penicillin (50 U/ml)–streptomycin (50 μg/ml) (Gibco, 15070–030), 1% non-essential amino acids (Gibco, 11140–050), 0.02 M $NaHCO_3$, filter sterilized and pH adjusted to 7.3–7.5
- 35 mm glass-bottom microwell dishes (MatTeK Corp., P35G-1.5-14-C)
- Ampule of frozen cells (PtK2; ATCC, No. CCL 56) (HeLa; ATCC, No. CCL2.2)
- HeLa cell maintenance media: MEM, 10 mM Hepes–NaOH, 10% FBS, 1% penicillin–streptomycin, 1% non-essential amino acids, 0.02 M $NaHCO_3$, filter sterilized and pH adjusted to 7.3–7.5
- Trypsin-EDTA: 0.25% trypsin, 1 mM EDTA-Na_4 (Gibco, 25200–056)
- Cell freezing media: 90% FBS, certified heat inactivated, 10% dimethyl sulfoxide (DMSO)

Method

1. Thaw the frozen cells at 37 °C and resuspend in 10 ml of maintenance media in a 15 ml conical tube. Centrifuge the washed cells at 500 *g* at 4 °C for 5 min. Remove the supernatant and resuspend the cell pellet in

Protocol 1. *Continued*

10 ml of maintenance media and transfer to a 25 cm^2 flask. Place the flask in the 37 °C CO_2 incubator overnight.

2. The next day aspirate the media to remove unattached cells and replace the media with 10 ml of maintenance media. Place the flasks in the 37 °C CO_2 incubator and grow cells to confluence.
3. When cells are confluent in the 25 cm^2 flask (approx. 3×10^6 cells), aspirate the media and add 1 ml of 37 °C trypsin-EDTA to detach the cells. Incubate for several minutes and observe microscopically for complete detachment of the cell layer. Add 9 ml of maintenance media and gently break up cell clumps by pipetting up and down. The cell culture is maintained by inoculating 10 ml of maintenance media with 0.5 ml of the trypsinized cell solution for an ~ 1:20 split in a 25 cm^2 flask. At a ~ 1:20 split the cells will reach confluence and require passage within three to four days. The best experimental results are obtained with cells that have been passed less than ten times.
4. Prepare experimental cell cultures in 35 mm glass-bottom microwell dishes by inoculating 2 ml of maintenance media with varying amounts of the trypsinized cells. A cell density of between 50–75% confluence is usually optimum for a maximum number of fully spread cells. It is desired that this stage of confluence be reached within two days for consistent experimental results. Therefore, one to three drops (50–150 μl) per dish of the trypsinized cells is usually a good level of inoculum to achieve this level of growth in two days.
5. Prepare frozen cell stocks by detaching confluent cell layers as above. Resuspend the cells in maintenance media (10 ml per 25 cm^2 flask) to wash. Place the cell solution in 15 ml conical tube and centrifuge at 150 *g* for 5 min. Remove the media and resuspend the cells in cell culture freezing media to a concentration of 10^6–10^7 cells/ml.
6. Aliquot the cells into cryogenic storage vials, cool on ice, freeze at –70 °C, and then store in liquid nitrogen.

2.2 Isolation and cell culture of mouse bone marrow macrophages

Membrane vesicles or endosomes, generated by chemical treatment of mouse bone marrow macrophages in culture, move by actin assembly through the cytoplasm (6). Study of endosomes has revealed important insights into actin-based motility, as the rocketing endosome represents a mechanism originating solely from the host cell. This is in contrast to the bacterial model system in which extracellular components make up some of the mechanism of movement. Motile endosomes have plucked from the host cell a patch of membrane that must contain all of the components necessary and sufficient for directing and maintaining actin assembly. Studies have been undertaken to record the speed of rocketing endosomes and the identity of critical components contained within the membrane of these endosomes for direct comparison with other actin-based mechanisms (6). These studies confirmed that the mechanism of endosomal rocketing shares common features with bacterial and viral actin-based motors. Even more importantly, the observations and studies of rocketing endosomes indicated directly the minimal components that must comprise any actin-based motor.

To study endosomal rocketing in culture it has been found that macrophages from mouse bone marrow can be induced to form endosomes capable of actin assembly. In *Protocol 2* the preparation of bone marrow macrophages for induction of endosomal rocketing and microscopic study will be described. *Protocol 9* describes the chemical induction of endosomal rocketing in these cultured cells.

Protocol 2. Isolation and cell culture of mouse bone marrow macrophages

Equipment and reagents

- Temperature controlled, humidified CO_2 incubator set at 37 °C
- Centrifuge with swinging bucket rotor
- Haemacytometer
- 6-well culture dishes (Falcon, 3046)
- 25 mm circular cover glass
- 25 mm circular cover glass holder (Narishige)
- Granulocyte-macrophage colony-stimulating factors (GM-CSF), murine, recombinant (Gibco, 13267–018)
- Four mice, six to eight weeks
- Macrophage maintenance media: RPMI medium 1640 with L-glutamine (Gibco, 11875–093), 10% calf serum, certified heat inactivated (Gibco, 10082–139), 5 mM Hepes–NaOH, 2% penicillin (100 U/ml)–streptomycin (100 μg/ml) (Gibco, 15070–030), 1% L-glutamine (Gibco, 25030–081), filter sterilized and pH adjusted to 7.3–7.5
- Phosphate-buffered saline (PBS) pH 7.4: 140 mM NaCl, 2.7 mM KCl, 10 mM Na_2HPO_4, 1.8 mM KH_2PO_4

Method

1. Kill mice and surgically remove both femurs. Remove the muscle and connective tissue, rinse with sterile PBS, and cut both ends from the bone. Cut only enough of the end of the bone for insertion of a 20 gauge needle, as the ends also have cell-containing marrow.
2. Using a 3 ml syringe and 20 gauge needle, carefully inject ~ 3 ml of macrophage maintenance media into the marrow space of each bone and collect the washed marrow from all eight bones in a 15 ml conical tube.
3. Centrifuge the bone marrow solution at 80 *g* for 5 min at 24 °C to pellet large contaminants. Collect the supernatant and centrifuge at 800 *g* for 10 min at 24 °C to pellet the cells.
4. Resuspend cell pellet from each mouse in 3 ml of macrophage

Protocol 2. *Continued*

maintenance media containing 3 μl of GM-CSF and count cells using a haemacytometer.

5. Plate cells in 6-well culture dishes containing a sterile 25 mm cover glass in the bottom of each well. The cells should be plated to a density of 2×10^6 cells/well by placing the appropriate number of drops of cell solution on each cover glass. Allow a minute for the cells to settle onto the glass then add 3 ml of macrophage maintenance media with GM-CSF to each well.
6. Place the 6-well culture dishes containing the bone marrow cells in the CO_2 incubator set at 37 °C. Incubate for seven days, changing the media every two to three days.
7. At day seven the macrophage cells should be fully differentiated and adherent to the cover glass and ready for induction of endosomal rocketing.

2.3 Culturing conditions for infective preparations of *Listeria*, *Shigella*, and vaccinia

Microscopic observations of actin-based motility require introduction and uptake of the competent pathogen by a host cell. The mechanisms of phagocytosis by the host cell, in the case of pathogenic bacteria, or host cell receptor binding and membrane fusion, in the case of vaccinia, are poorly understood. However, methods, determined empirically, have been developed for consistent and reproducible host cell infections by these pathogens.

Culturing of *Listeria* and *Shigella* is essentially the same and these procedures are described in *Protocols 3* and *4*. An exception is that because the genes required for invasiveness and actin-based motility are carried on extrachromosomal DNA in the form of a virulence plasmid in *Shigella*, *Shigella* cultures must be routinely screened for retention of this plasmid (7). Included in *Protocol 4* is the simple assay for determining the presence or absence of this plasmid by observing Congo red dye binding to those *Shigella* that are expressing genes from the virulence plasmid.

Although protocols are given below for preparation of vaccinia virus for infection of host cells, it should be cautioned that only laboratories that have been certified for culture of vaccinia should undertake these experiments. Staff from certified laboratories are already familiar with these common techniques for virus stock production. *Protocols 5* and *6* are for those researchers who may be contemplating motility experiments with vaccinia virus in collaboration with vaccinia-certified laboratories and who would like a brief background in the basic procedures. *Protocols 5* and *6* are based on material from refs 8 and 9 and the laboratory of Dr Richard C. Condit of the University of Florida.

Protocol 3. Infective preparation of the Gram-positive rod *Listeria monocytogenes* for infection of host cells

Equipment and reagents

- 100 mm Petri dishes
- 13 × 100 mm culture tubes (Sarstedt, 62.515.006)
- Frozen stock of *Listeria monocytogenes* strain 1043S serotype 1
- 30 °C and 37 °C incubator
- Brain heart infusion (BHI) media (Difco, 0037–17–8)
- BHI agar (Difco, 0418–17–7)
- Phosphate-buffered saline (PBS) pH 7.4

Method

1. From frozen stock of bacteria, streak a BHI agar plate and place in 30 °C incubator overnight. The next day pick a single colony, inoculate 10 ml of BHI media into a 13 × 100 mm culture tube, and place in 30 °C incubator overnight.
2. Determine the number of bacteria or colony-forming units (c.f.u.)/ml of culture by preparing serial dilutions in PBS, and spread 100 μl of each dilution on a BHI agar plate. Record the absorbance at A_{600} of the culture. Place the plates in the 30 °C incubator overnight and count the number of colonies the next day. Multiply the number of colonies on the plate by 10 (the 100 μl spread is 1/10th of a ml) and by the dilution factor to calculate the c.f.u./ml. An overnight culture should reach an A_{600} of ~ 0.8–1.2 and contain ~ 1.0×10^9 to 2.5×10^9 c.f.u./ml. The measured absorbance A_{600} of a culture can then be used as an approximate measure of the c.f.u./ml of the culture.
3. Pipette the calculated volume of bacterial culture to yield ~ 10^9 c.f.u. into a microcentrifuge tube and centrifuge at 5000 *g* for 5 min. Aspirate the media and resuspend the bacterial pellet in 1 ml of PBS. Repeat the centrifugation step twice to wash the bacteria and resuspend the bacterial pellet in 1 ml of PBS for a final bacteria concentration of ~ 10^9 c.f.u./ml.
4. The washed bacteria are now ready for inoculating host cells in culture and appropriate dilutions can be prepared based upon the required level of host cell infection.

Protocol 4. Infective preparation of the Gram-negative rod *Shigella flexneri* for infection of host cells

Equipment and reagents

- Frozen stock of *Shigella flexneri* wild-type serotype 5 strain M90T containing the native virulence plasmid
- 100 mm Petri dishes
- 37 °C incubator/shaker
- BHI agar (Difco, 0418–17–7)

Protocol 3. *Continued*

- Brain heart infusion (BHI) media (Difco, 0037–17–8)
- Phosphate-buffered saline (PBS) pH 7.4
- Tryptic soy broth (Difco)
- 13 × 100 mm culture tubes (Sarstedt, 62.515.006)
- Bacto agar (Difco)
- Congo red (Sigma, C6277)

Method

1. Prepare Congo red agar plates with tryptic soy broth, 1.5% Bacto agar (w/v), and 0.05% Congo red (w/v).
2. Screen for retention of the virulence plasmid in *Shigella* by streaking bacteria on Congo red agar plates and growing overnight at 37 °C. Bacterial colonies should be red, indicating retention of the 230 kilobase pair virulence plasmid. Red colonies can be picked and inoculated into BHI media and overnight cultures grown to prepare glycerol stocks. Glycerol stocks should be plated on Congo red agar plates for further verification of the *Shigella* stocks.
3. With verification of virulence plasmid-containing bacterial stock, streak BHI agar plates and grow colonies overnight in 37 °C incubator. Pick an individual colony and inoculate 10 ml of BHI for an overnight culture at 37 °C. Determine the c.f.u./ml of *Shigella* culture at a measured absorbance A_{600} by serial dilutions and plating as described in *Protocol 3* for *Listeria.*
4. On the day an infection is planned, dilute an overnight culture 1:50 and grow to mid-log phase with shaking in 37 °C incubator. At $A_{600} \approx \sim 0.6$–0.8 for the growing culture, pipette a calculated volume of bacterial culture to yield $\sim 10^9$ c.f.u. into a microcentrifuge tube and centrifuge at 5000 *g* for 5 min. Aspirate the media and resuspend the bacterial pellet in 1 ml of PBS. Repeat the centrifugation step twice to wash the bacteria, and resuspend the bacterial pellet in 1 ml of PBS for a final bacteria concentration of $\sim 10^9$ c.f.u./ml.
5. The washed bacteria are now ready for inoculating host cells in culture and appropriate dilutions can be prepared based upon the required level of host cell infection.

Protocol 5. Preparation of crude vaccinia virus stocks (infected cell lysates) for infection of host cells

Equipment and reagents

- Microtip sonicator (Ultrasonics, Model W-10)
- Temperature controlled, humidified CO_2 incubator set at 37 °C
- Frozen stock of vaccinia virus strain WR (ATCC, VR-1354)
- Confluent HeLa cell monolayers in 100 mm culture dishes (see *Protocol 1*)
- Phosphate-buffered saline (PBS) pH 7.4
- PBSAM pH 7.2: PBS, 0.01% BSA, 10 mM $MgCl_2$

Method

1. Thaw virus stock at room temperature and sonicate on ice for ~ 1 min with microtip sonicator or cup horn sonicator.
2. Prepare the viral inoculum by diluting the virus stock to 7×10^4 to 7×10^5 plaque-forming units (p.f.u.)/ml in PBSAM to give a multiplicity of infection (m.o.i.) of 0.01–0.1 in a 1.5 ml inoculum. The virus stock should be titred using a plaque assay before dilution (see *Protocol 6*).
3. Aspirate the media from the 100 mm dishes of HeLa cell monolayers and add 1.5 ml of viral inoculum to each dish. Place in 37 °C CO_2 incubator and adsorb the virus for 0.5–2 h with rocking every 15 min to spread the inoculum.
4. Remove the inoculum and replace with 10 ml of HeLa cell maintenance media (see *Protocol 1*). Return dishes to incubator for up to three days. Examine the cells daily for complete cytopathic effect (CCE) as indicated by rounding up of all the cells on the dish.
5. For isolation of the crude virus stock, scrape the cells from dishes that have attained a CCE without removing the media. Transfer the cell suspension from a single 100 mm dish to a 50 ml conical tube and centrifuge at 1800 *g* for 5 min at 4 °C. Remove the supernatant and resuspend the cell pellet in 1 ml of PBSAM for each 100 mm dish.
6. Lyse the cells by repeated freeze–thaw cycles (~ three cycles) using liquid nitrogen and thawing at room temperature. Check for complete cell lysis with a microscope using a × 40 objective. Compare pre- and post-lysis samples to accurately monitor cell lysis. If necessary, a microtip sonicator may be used to aid in lysing the cells.
7. Crude virus stock may be aliquoted, frozen, and stored at –70 °C. Titre the crude virus stocks by the plaque assay in *Protocol 6*.

Protocol 6. Plaque assay for determining the titre of vaccinia virus stocks

Equipment and reagents

- Microtip sonicator (Ultrasonics, Model W-10)
- Temperature controlled, humidified CO_2 incubator set at 37 °C
- Frozen stock of vaccinia virus strain WR (ATCC, VR-1354) or crude virus stock from *Protocol 5*
- Phosphate-buffered saline (PBS) pH 7.4
- Confluent HeLa cell monolayers in 60 mm culture dishes (see *Protocol 1*)
- PBSAM pH 7.2: PBS, 0.01% BSA, 10 mM $MgCl_2$
- Noble agar (Difco)
- Neutral red (0.1% in deionized water, filter sterilized)

Method

1. Prepare a nutrient agar overlay by autoclaving 2% noble agar in deionized water and maintaining at 56 °C. Prepare a 2 × maintenance

Protocol 6. *Continued*

media (2×MM) with 10% FBS, mix an equal volume of 2×MM with an equal volume of 2% noble agar, and place at 45 °C.

2. Thaw the virus stock and sonicate on ice for 1 min with microtip sonicator.
3. Prepare serial dilutions from 10^{-5} to 10^{-7} of virus stock in PBSAM. Crude virus stocks have a titre in the range of 10^8 to 10^9 p.f.u./ml, and plaques are best read from plates inoculated with 10^{-6} or 10^{-7} dilutions.
4. Remove the media from the confluent cell layers in the 60 mm dishes and inoculate with 0.5 ml of the serial dilution inoculum. Adsorb the virus for 0.5–2 h with rocking every 15 min to spread the inoculum.
5. Aspirate the inoculum and add 4 ml of nutrient agar overlay to each 60 mm dish and allow the agar to solidify. Incubate at 37 °C for two to three days and monitor plaque development with the microscope.
6. At day two or day three, depending on plaque development, add a second overlay of nutrient agar containing neutral red (0.005%). Incubate the dishes overnight and count the plaques the next day. Multiply the number of plaques counted times 2 (the inoculum is 0.5 ml), then multiply by the dilution factor for that dish to calculate the virus stock titre in p.f.u./ml.

2.4 Infection of PtK2 or HeLa cells by *Listeria* and *Shigella*

To fully exploit these intracellular pathogens as tools for investigating cytoskeletal processes and mechanisms, it is important to understand their life cycle. Many cell types are susceptible to infection by these pathogens including neutrophils, macrophages, epithelial cells, and endothelial cells. However, *in vivo* infection takes place primarily through epithelial cells lining the gastrointestinal tract (10). As shown in *Figure 1*, the first step in the infection process is attachment to the host cell membrane and phagocytosis by the host cell. *Listeria* uses proteins called internalins which enhance bacterial attachment and internalization (2, 11). Performing the same function in *Shigella* are the IpaB, IpaC, and IpaD proteins (12).

Once inside the host cell, the bacterium becomes enclosed in a subcellular compartment called the phagolysosome, a normally hostile and toxic environment for most bacteria. The low pH of this compartment, however, activates listeriolysin-O, a pore-forming toxin of *Listeria*, or a haemolysin of *Shigella* to lyse the phagolysosome, allowing escape of the bacterium into the cytoplasm. Within the cytoplasm, the bacteria double every 60 minutes. They become surrounded by actin, and approximately two hours later, actin filaments begin extending from one end of the bacterium, propelling the organism through the cytoplasm (10, 13). Many bacteria subsequently induce the formation of

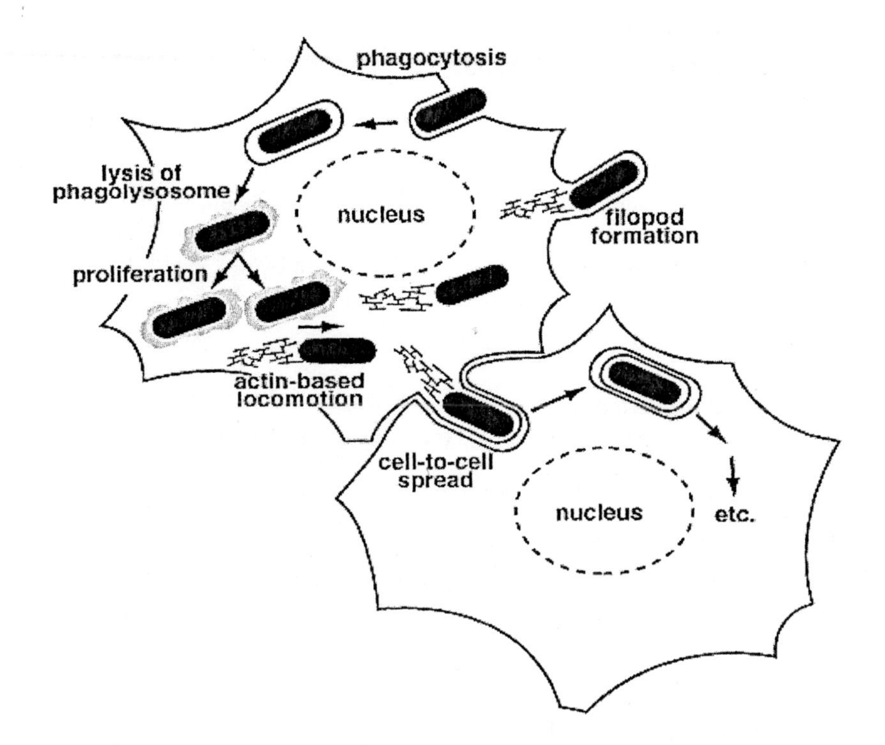

Figure 1. Intracellular life cycle of *Listeria* or *Shigella* in host cells. The bacterium is phagocytosed by the host cell and incorporated into a phagolysosome. The bacterium escapes from the phagolysosome and multiplies within the cytoplasm becoming surrounded by actin filaments. After sevèral hours the actin filaments polarize at one end of the bacterium forming the actin tail which propels the bacterium through the cytoplasm and ending at the periphery of the cell. Here the bacterium is propelled into a cell membrane protrusion, or filopod, and into the adjacent cell, where the cycle is repeated. (With permission from ref. 5.)

elongated membrane protrusions or filopods, which are stabilized actin filament bundles enveloped in membrane, each containing a bacterium at their tip. Adjacent cells ingest the filopods, and a characteristic double membrane phagolysosome can be observed in the new host. After avoiding the hostile extracellular environment of immunoglobulins, complement, and extracellular antibiotics, the life cycle of these pathogens begins over (14).

The method for infecting host cells with *Listeria* and *Shigella* is covered in *Protocol 7*. This method relies primarily on the ability of these pathogens to induce their own phagocytosis. The selection of host cell is important, as some cell types are more phagocytotic. We have found that the kangaroo rat PtK2 cell is readily infected with *Listeria* but less so with *Shigella*. On the other hand, HeLa cells phagocytose *Shigella* efficiently. These are general guidelines, and the researcher must determine the correct cell type for any given experiment.

For laboratory experiments, infections of host cells by these bacteria are facilitated by a short centrifugation step to bring the bacteria in contact with the cell membrane for efficient attachment and rapid internalization by phagocytosis. After an incubation period for completion of attachment and phagocytosis, the cells are washed to remove extracellular bacteria. The infected host cells, after an additional short incubation, are then ready for microscopic observation of intracellular bacterial motility.

Protocol 7. Infection of PtK2 or HeLa cells by *Listeria* or *Shigella*

Equipment and reagents

- Temperature controlled, humidified CO_2 incubator set at 37 °C
- Centrifuge with rotor for centrifuging microtitre plates
- *Listeria* or *Shigella* inoculum at ~ 10^9 c.f.u./ml in PBS from *Protocol 3* or *4*
- Phosphate-buffered saline (PBS) pH 7.4
- Semi-confluent PtK2 or HeLa cell monolayers in 35 mm glass-bottom microwell dishes (see *Protocol 1*)
- Gentamicin sulfate solution: 50 mg/ml (Sigma G1522)
- PtK2 or HeLa cell maintenance media without antibiotics (see *Protocol 1*)

Method

1. Remove the media from the cell monolayers in the 35 mm microwell dishes. Wash three times with PBS to remove all traces of the cell maintenance media containing antibiotic.
2. Replace the media with 2 ml of cell maintenance media without antibiotics and inoculate with bacterial inoculum prepared in *Protocols 3* or *4*. Note: the amount of inoculum will depend on the number of cells per dish, the desired level of infection, and the titre of the bacterial inoculum. The number of cells per semi-confluent dish can be counted by detaching the cells and counting the number of cells using a haemacytometer. A rule of thumb is that a confluent layer of cells on a 35 mm dish contains approx. 10^6 cells. The level of infection appropriate for a given experiment must be determined empirically. A good initial level of infection is 100 bacteria per host cell. The titre of the bacterial inoculum is determined in *Protocols 3* and *4* and is usually set at 10^9 c.f.u./ml. Therefore for a semi-confluent cell layer (2.5×10^5 cells) to be infected with 100 bacteria per cell requires an inoculum of 25 μl per 35 mm dish.
3. Immediately after inoculating the dish(es), centrifuge at 150 *g* at room temperature for 5 min using a swinging rotor designed for microtitre plates. Remove the dishes from the centrifuge and return to the CO_2 incubator for 1 h.
4. Aspirate the media containing the bacteria and wash the host cells at

least three times with PBS to remove extracellular bacteria. Add back fresh media containing 10 μg/ml, final concentration, of gentamicin sulfate to each dish to prevent further growth of extracellular bacteria.

5. Return the infected cells to the CO_2 incubator for 1–2 h for completion of phagocytosis and activation of bacterial motility. Check cells 1 h after the final wash step for level of infection and progression to motility with the microscope. Begin experiments at first indication of bacterial movement. Note: the time required before *Listeria* and *Shigella* begin to move differs. After the final wash and addition of gentamicin sulfate to the media, *Shigella* begins forming actin tails within 1–2 h. *Listeria,* however, may take up to 3–4 h after the final wash to begin actin-based motility. If performing this experiment for the first time, it is suggested that the bacteria be monitored for motility beginning 1 h after the final wash to determine the optimum time to motility for a set of experimental conditions. While monitoring the cells, if many bacteria are observed to be in filopods, the useful period for productive measurements of motility is over and it would be best to begin monitoring motility at an earlier time point.

2.5 Infection of host cells with vaccinia virus

A brief introduction into the life cycle of vaccinia will be given below, and the reader is encouraged to consult a thorough review (15).

The mode of infection by these large DNA viruses is very different from that taken by intracellular bacteria (compare *Figure 1* to *Figure 2*). Whereas the bacteria induce their own phagocytosis to gain entry, vaccinia enters by fusion of its external membrane with the host cell plasma membrane. It is believed that cell surface receptors and vaccinia membrane proteins mediate this fusion, but these specific components have not been clearly defined (15). Once inside the cytoplasm, the virion is uncoated and the viral cores are released. Components of these viral cores synthesize early mRNAs for translation of a number of proteins, including factors for DNA replication. DNA replication to produce progeny DNA follows, and transcription of intermediate genes encoding late transcription factors occurs. Translation of late gene mRNAs results in the production of viral structural proteins, enzymes, and transcription factors which will be included in the assembly of progeny virions. The viral particle goes through a complex maturation process in which all of the components are assembled into core progeny particles that are wrapped twice by membranes of the Golgi. The first wrapping takes place in the intermediate compartment of the *cis*-Golgi (16), forming the intracellular mature virion (IMV, see *Figure 2*). Formation of the intracellular enveloped virion (IEV) is the result of wrapping the IMV with an additional set of membranes derived from the *trans*-Golgi (17). Extracellular enveloped

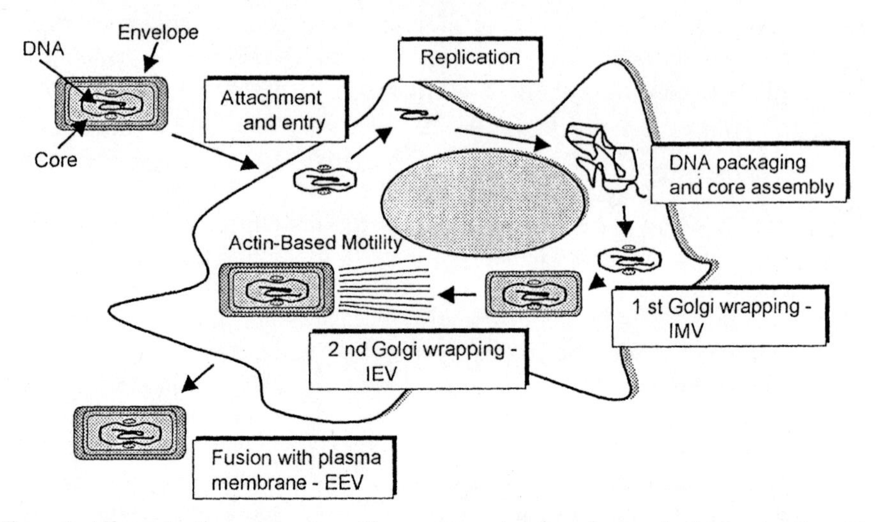

Figure 2. Life cycle of vaccinia virus. The enveloped virion attaches to the host cell and is internalized and uncoated, releasing the viral DNA and transcription apparatus into the cytoplasm. The viral DNA is replicated and transcribed, resulting in the production of viral progeny DNA, structural proteins, enzymes, and transcription factors that will be included in the progeny virion core particles. Maturation of virions takes place by wrapping of progeny cores by membranes derived from the *cis*-Golgi to form the intracellular mature virion (IMV) and then by a second wrapping by membranes derived from the *trans*-Golgi to form the intracellular enveloped virion (IEV). The IEV is capable of actin-based motility. The extracellular infectious form of the virus results from the fusion of the IEV with the plasma membrane to form the extracellular enveloped virion (EEV) and cell-associated enveloped virion (CEV). (Adapted from ref. 15.)

virions (EEV) and cell-associated enveloped virions (CEV) are formed when the IEV fuses through the host cell plasma membrane, shedding its last acquired membrane wrap for a new membrane composed of the host cell plasma membrane. Although all forms of the virus are infectious, only the IEV is competent to move by actin-based motility (18).

The procedure for infecting host cells grown in culture, see *Protocol 8*, relies on the very efficient attachment and membrane fusion of vaccinia. When using minimal inoculum volume, the centrifugation step required for efficient bacterial infection is not required for good viral infections. After introducing the viral inoculum, followed by a wash step, the time interval for formation of IEV is approximately three to five hours post-infection (h.p.i.). At this time, viral particles can be seen to move by actin assembly and continue for 10–12 h.p.i., whereupon numerous viral particles can be seen in filopods. For purposes of study of actin assembly, this represents the end of useful observations, although viral production continues for several days.

By electron microscopy the vaccinia virions appear as a rectangular box ~ 350 nm by 270 nm (19); consequently the virions are at the limit of resolution of the light microscope. Notwithstanding, a virus moving by actin assembly can be seen with a good × 63 or × 100 objective by phase-contrast imaging. This is because the formation of the actin tail behind the particle is of high contrast to the cellular background and can be readily differentiated from the surrounding cytoplasm. Therefore, vaccinia actin-based motility is tractable to study by the tools of video microscopy, such as microinjection, time-lapse imaging, and epifluorescence microscopy (20, 21). The use of these tools to study vaccinia actin-based motility is discussed in Section 3.

Protocol 8. Infection of host cells by vaccinia virus

Equipment and reagents

- Microtip sonicator (Ultrasonics, Model W-10)
- Temperature controlled, humidified CO_2 incubator set at 37 °C
- Titred frozen stock of vaccinia virus from *Protocol 5*
- Semi-confluent HeLa cell monolayers in 35 mm glass-bottom microwell dishes (see *Protocol 1*)
- Phosphate-buffered saline (PBS) pH 7.4
- PBSAM pH 7.2: PBS, 0.01% BSA, 10 mM $MgCl_2$

Method

1. Thaw the virus stock at room temperature and sonicate on ice for ~ 1 min with microtip sonicator or cup horn sonicator.
2. Prepare the viral inoculum by diluting the virus stock in PBSAM to give a m.o.i. of 10 in a 0.25 ml inoculum for each 35 mm microwell dish. (For example, if you have counted 1.5×10^6 cells/35 mm dish, and you require a m.o.i. of 10, then 1.5×10^7 p.f.u. is needed per 0.25 ml of inoculum. From a viral stock of 2×10^9 p.f.u./ml, 7.5 μl of stock would be added to 242.5 μl of PBSAM to make up the viral inoculum for each dish.)
3. Aspirate the media from the 35 mm dishes of HeLa cell monolayers and add 0.25 ml of viral inoculum to each dish. Place dishes in a 37 °C CO_2 incubator and adsorb the virus for 30 min with rocking every 10 min to spread the inoculum.
4. Remove the inoculum and wash the cells with PBSAM three times to remove the extracellular virus. Add 2 ml of HeLa cell maintenance media and return dishes to the incubator. Begin observations at the appropriate times. Note: generally, the host cells at 1 h.p.i. undergo rounding, and very little movement can be observed until 5 h.p.i. when many of the infected cells flatten out again. At 7 h.p.i. the greatest number of virions are moving by actin assembly, and this is usually the optimum time post-infection for studying actin-based motility by these particles.

2.6 Lanthanum/zinc treatment of mouse bone marrow macrophages to induce endosomal rocketing

As mentioned above, the rocketing endosome offers a simple model system for studying actin assembly in mammalian cells. However, how these structures are induced to form is far from understood. Modification of Ca^{2+}

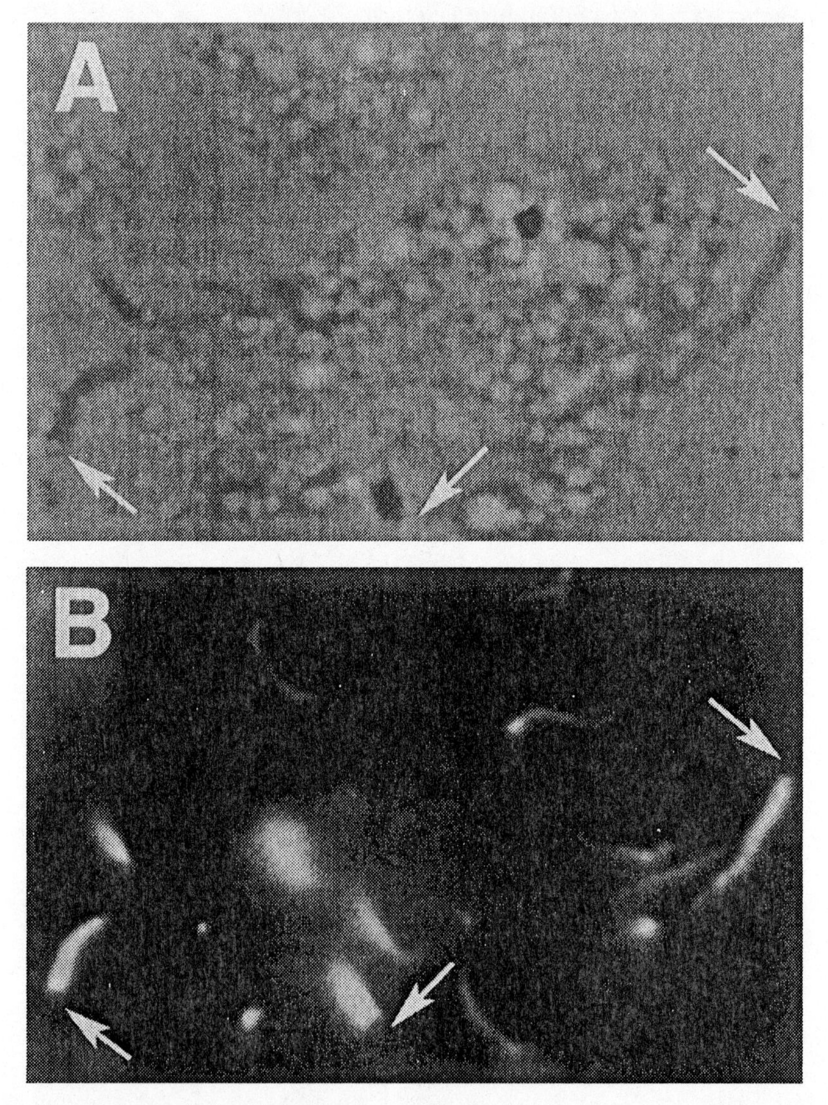

Figure 3. Lanthanum-induced endosomes. (A) Rocketing endosomes are readily seen by phase-contrast microscopy trailing an actin tail, as marked by the *arrows*. (B) Fluorescence image using bodipy-phalloidin to label polymerized actin. The *arrows* mark the interface of the membrane endosome with its actin tail.

flux in the cell by lanthanum ions has been proposed, but this hypothesis has not been tested. Endosome formation requires remodelling and re-forming of segments of internal host cell membranes, segments which must contain a membrane anchoring protein with a bound adapter protein, such as zyxin. Motor assembly can take place and actin polymerization can be initiated once a vesicle has formed with the adapter zyxin bound. Like *Listeria*, *Shigella*, and vaccinia, the membrane vesicles or endosomes which move by actin-based processes are also tractable to study by video microscopy (6). Rocketing endosomes are readily seen by light microscopy, ranging in size from 0.5 μm to 5 μm, and are often trailing relatively large phase-dense actin tails. Notice in *Figure 3A* the rounded endosome with the dark actin tail behind. In *Figure 3B* staining with fluorescently-labelled phalloidin, which is specific for filamentous actin confirmed that the endosome tails are composed of microfilaments.

Protocol 9 describes the method for inducing endosome formation and rocketing. The bone marrow macrophages are sequentially treated with lanthanum chloride and zinc chloride. After a short incubation of approximately 30 minutes, the endosomes form and begin to assemble dark clouds of actin at their periphery. When an asymmetric actin cloud is established, the actin tail structure forms and the endosome begins to rocket. Endosomes have been observed to move at the same rates as bacteria and viruses, in the range of 0.05 μm/sec to 0.45 μm/sec. Rocketing endosomes are stabilized through cell fixation, and localization studies can be done by antibody staining. In living cells the endosome is fragile and microinjection experiments have not been possible. Time-lapse imaging, however, can be usefully applied to endosomal rocketing studies in which experimental substances are applied exogenously in the cell media. See Section 3 for a discussion of time-lapse imaging, epifluorescence microscopy, and image analysis.

Protocol 9. Lanthanum/zinc treatment of mouse bone marrow macrophages to induce endosomal rocketing

Equipment and reagents

- Microincubator MS-200D (Narishige)
- Temperature controlled, humidified CO_2 incubator set at 37 °C
- 25 mm circular cover glass holder (Narishige)
- Differentiated bone marrow macrophage cells on cover glasses from *Protocol 2*
- Buffer A: 0.1 M NaCl in 15 mM Hepes–NaOH pH 7.2 (sterile)
- 0.1 M lanthanum chloride ($LaCl_3$) (Sigma)
- Macrophage maintenance media: RPMI medium 1640 with L-glutamine (Gibco, 11875–093), 10% calf serum, certified heat inactivated (Gibco, 10082–139), 5 mM Hepes–NaOH, 2% penicillin (100 U/ml)–streptomycin (100 μg/ml) (Gibco, 15070–030), 1% L-glutamine (Gibco, 25030–081), filter sterilized and pH adjusted to 7.3–7.5
- 0.1 M zinc chloride ($ZnCl_2$) (Sigma)

Method

1. Wash the differentiated bone marrow macrophage cells three times with PBS. Assemble the cover glass holder with cover glass and add

Protocol 9. *Continued*

990 μl of buffer A. Place the holder in the microincubator set at 37 °C on the microscope stage. To this add 10 μl 0.1 M $LaCl_3$ and incubate at room temp for 10 min.

2. To reduce cell damage or cell fragmentation during lanthanum treatment, remove the solution from the cells and add back 989 μl of buffer A. To this add 1 μl of 0.1 M $LaCl_3$ and 10 μl of 0.1 M $ZnCl_2$; incubate at room temp for 20 min.
3. Within 30 min the endosomes should begin to rocket, as seen in phase-contrast by dark, phase-dense actin tails behind the moving vacuole.

3. Observation, manipulation, and measurement of intracellular motility by video microscopy

3.1 Time-lapse phase-contrast imaging

The video microscope has contributed significantly to the study of intracellular movement of bacterial pathogens (22). With video microscopic computer imaging, the image of almost any specimen that can be resolved in a light microscope can be acquired, digitized, visualized live on a video monitor, and/or stored for later analysis and further refinement and enhancement. See ref. 23 for an excellent work on video microscopy, including the theory and applications. In the work described in this chapter, our emphasis is on phase-contrast as a method for contrast enhancement in light microscopy and on fluorescence microscopy. For purposes of time-lapse analysis and microinjection, images generated by phase-contrast have detail and resolution that is excellent for these applications. Other contrast generating methods such as dark-field, differential interference contrast, and polarizing microscopy are not covered here, yet the reader is encouraged to investigate the applicability of these for particular experiments.

Time-lapse imaging and analysis and microinjection are powerful experimental tools for assessing the mechanisms of cellular and intracellular motility. One tractable parameter of movement is velocity, and by time-lapse imaging velocity measurements can be readily recorded and analysed for monitoring a motile process. In the specific case of actin-based movement by intracellular pathogens, comparison of the speeds before and after an experimental treatment has been used to analyse the underlying mechanism. In these experiments specific peptide inhibitors, directed at cessation of bacterial motility, were microinjected into cells infected with *Listeria* or *Shigella*. Upon analysis, critical protein domains involved in the actin-based motility of *Listeria* and *Shigella* were identified (24, 25) (see *Table 1*).

In addition to speed measurements, dynamic structures that change in form over time can be analysed by time-lapse imaging. One such parameter in actin-based motility is actin filament tail length. Because the disassembly of the actin filament at one end is relatively constant, the changes in actin tail length over time can be used as an indirect measure of the rate of polymerization at the fast growing ends of the actin tail filaments. Direct comparison of filament length of the bacterial actin tail and bacterial speeds after an experimental treatment are also used to make inferences about the underlying mechanism.

Data for a time course study are collected either by direct frame grabbing using a time-lapse image acquisition function of the imaging software, or by VCR. Image acquisition by VCR tape can offer an inexpensive and unlimited format for image data storage. If the VCR is used to record a time series, frame grabbing from the videotape can be done later for acquiring images. It should be noted that, in computers of recent manufacture, memory for image acquisition and storage is not limiting and therefore some of the advantages of VCR tape no longer apply. In addition, recordable CD's now offer inexpensive data storage and high fidelity.

Metamorph is the image acquisition, hardware control, and image analysis software from Universal Imaging that we use in the computer controlling our video microscope. For time-lapse analysis a useful feature of *Metamorph* is the image stack. Here a stack is a collection of images or planes that is opened in the same window. A stack can be created with any set of two or more images, and the stack can be manipulated and stored as a single file. For time-lapse analysis a dedicated function is set in *Metamorph* for setting time-lapse acquisition parameters and activation of time-lapse acquisition. When the time-lapse function is activated, a stack is automatically begun and the entire series of images is acquired. Each stack representing an individual experiment can then be saved as a single file. Later objects or points in each image plane can be tracked in each successive plane and the distance per time interval logged into a spreadsheet file within *Metamorph* or linked to another spreadsheet program such as *Excel*. *Protocol 10* describes a method for time-lapse image acquisition using *Metamorph*.

Protocol 10. Image acquisition for a time-lapse study by digital imaging microscopy

Equipment and reagents

- Inverted microscope (Nikon Diaphot–TMD) with × 40, × 63, or × 100 objectives
- Cooled charge-coupled device (CCD) camera, 756 × 483 × 8-bit active picture elements (Hamamatsu, Model C5985)
- Computer/Windows 95 with video acquisition board (frame grabber) installed and configured
- Shutter driver (Uniblitz, Model D122)
- *Metamorph* computer image analyser (Universal Imaging)
- 25 mm circular cover glass holder (Narishige)
- Microincubator MS-200D (Narishige)
- Infected cell monolayers on 35 mm glass-bottom microwell dishes from *Protocol 7*

Protocol 10. *Continued*

Method

1. Infect cell monolayers growing on 35 mm glass-bottom microwell dishes, as described in *Protocol 7*.
2. Pre-heat the microincubator to 37 °C on the microscope stage and select microwell dish for observation.
3. Check that the shutter driver is functioning properly by selecting the appropriate shutter, acquiring an image, and observing for correct shutter opening and closing.
4. With the microscopic field selected and an image in-focus, begin image acquisition by selecting acquire time-lapse from pull down menu. Set beforehand the parameters for length of wait time between acquired images and total number of images to be acquired. Note: for bacteria moving at 0.05–0.2 μm/sec a time interval of 30 sec between acquired images is optimum for velocity measurements. The number of images acquired will depend on the experiment, but for most bacterial motility experiments using microinjection, a total of 12–14 images is sufficient.
5. Acquire an image stack for the time course. Acquisition may be paused at any time during the process for microinjection, addition of reagents, etc., and then resumed. Note: *Metamorph* saves an extensive data set automatically for each image, including time of acquisition, which can be retrieved, for later analysis. Please note: when analysing the data, if the time-lapse has been paused and restarted, be sure to recheck the time interval between image planes.
6. Save image stack to a file for later analysis—see Section 3.4.

Figure 4 is an example of a time-lapse image series from a microinjection experiment (24). Selected images from the entire image series (14 images taken at 30 second intervals) are displayed as a panel of four images from left to right, top to bottom. In each image a triangle has been superimposed to define a region within the cell shown that does not change over time. This reference triangle focuses the observer's attention on the area of interest and establishes a landmark from which to assess movement or lack of movement. Movement then is clearly demonstrated in (A) to (B) and directed cessation of movement by microinjection of a peptide inhibitor in (C) and (D). Observe this by selecting a bacterium in (A), at time 0, in reference to the sides of the triangle, and follow it through the time course from (A) to (D). Section 3.4 will discuss other equally effective methods for presenting data from time-lapse experiments.

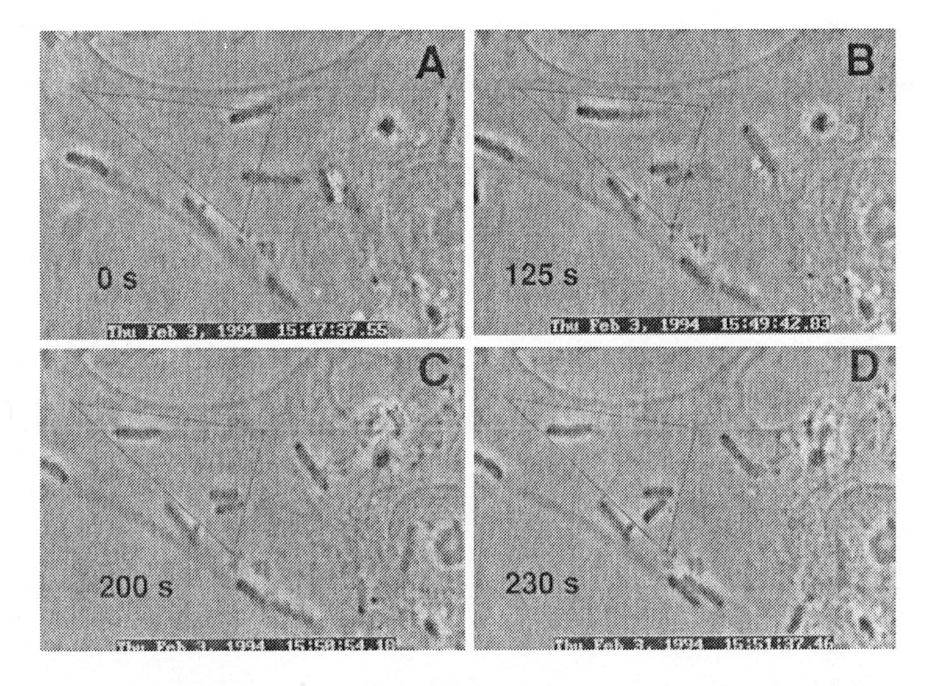

Figure 4. Time-lapse analysis of *Shigella* movement and actin rocket tail formation in PtK2 host cells before and after microinjection of the synthetic ActA peptide. Before injection the bacteria are seen to move at 0.12 μm/sec, and maximum tail length is 6.0 μm (A and B). After injection of an estimated intracellular concentration of 80 nM of ActA analogue (needle concentration 0.8 μM ActA peptide) at 160 sec, bacterial movement stops and the actin tails almost completely disappear (C and D). Solid bar, 10 μm. (With permission from ref. 24.)

3.2 Microinjection

Although the instrumentation for microinjection has been dramatically improved in recent years, microinjection as an experimental tool has not been widely accepted. The reasons for the under-utilization of this technique, in part, may be due to high initial investment costs in equipment, or the lack of training, and/or lack of understanding of the capabilities of the technique. For presenting within the cell a prescribed amount of membrane impermeable substance, it is unparalleled. Only by microinjection can one introduce a macromolecule to a specific cell and be confident of the final concentration within each cell treated. Introduction of a protein by transient overexpression from a transfected cDNA clone is limited by the inability of the experimenter to know accurately the amount of the target protein within the cell and the inability to limit the exposure to a specific cell. Microinjection gives the experimenter flexibility in the choice of cell or cells on a culture dish in which to administer the specific treatment and also the choice of leaving natural controls in the same microscopic field (uninjected cells) for direct comparison. This kind of experimental design is not always available in overexpression experiments or when using membrane permeable agents. Dose–response analysis can also be readily accomplished by microinjection for inhibition or activation studies.

The method of microinjection described uses constant flow at the needle tip and variable time of contact with the interior of the cell to deliver a volume of injected substance. Injecting by this method offers the experimenter more flexibility in judging the amount of substance delivered to the interior of the cell. Cell type, cell membrane characteristics, and the injected solution all can cause variability in cell-to-cell injections. Changing the time the needle is in contact with the cell when using constant flow at the needle tip can compensate for this. Pulse injections do not offer this option unless repeated injections to the same cell are performed, which is usually undesirable.

The technique of microinjection in theory is very simple, yet in practice can be difficult to master. *Protocol 11* describes a simple procedure for orienting the needle in the microscopic field. Positioning of the needle is for most people the most difficult part of the procedure, resulting in many broken needles and many frustrated experimenters. Also included in *Protocol 11* is the procedure for injecting a fluorescently-labelled standard protein to assess the effectiveness of the microinjection. *Protocol 12* describes a typical micro-injection experiment using a synthetic peptide at a known concentration for an inhibition study, co-injected with a fluorescently-labelled standard protein. In Section 3.4 methods are discussed for analysis and presentation of data recorded from this type of experiment.

Protocol 11. Preparation for microinjection and microinjection of a fluorescently-labelled protein

Equipment and reagents

- Inverted microscope (Nikon Diaphot–TMD) with × 40, × 63, or × 100 objectives
- Epifluorescence attachment (high intensity lamp, filter cassettes, fluor objectives)
- Micromanipulator (Eppendorf, Model 5171)
- Microinjector (Eppendorf, Model 5242)
- Cooled CCD camera, 756 × 483 × 8-bit active picture elements (Hamamatsu, Model C5985)
- Computer/Windows 95 with video acquisition board (frame grabber) installed and configured
- Shutter driver (Uniblitz, Model D122)
- *Metamorph* computer image analyser (Universal Imaging)
- 25 mm circular cover glass holder (Narishige)
- Microincubator MS-200D (Narishige)
- Glass capillary tubing: OD 1.5 mm, ID 1.17 mm, length 15 cm (Warner Mfg., GC150TF-15)
- Vertical pipette puller (Kopf, Model 720)
- Cell monolayers on 35 mm glass-bottom microwell dishes from *Protocol 7*
- Bovine serum albumin, Cascade Blue conjugate (25 mg/ml) (Molecular Probes, A-964)

Method

1. Pre-heat the microincubator to 37 °C on the microscope stage and select a microwell dish for observation. Bring the cell layer into focus. Objectives × 40, × 63, and × 100 can be used for a microinjection experiment, although the higher the magnification the more difficult it is to locate the needle initially.
2. Use the vertical pipette puller to prepare the microinjection needles from the glass capillary tubing. It is desirable to pull needles with tips that are not too long and/or thin. Adjustments of the puller can be made to allow pulling a needle tip of optimum size and shape. See the pipette puller manufacturer's suggestions for initial settings.
3. To assess the quality of the microinjection and the relative amount injected, prepare a series of dilutions of the fluorescently-labelled protein for needle concentrations in the range of 0.5 μg/ml to 4 mg/ml. Centrifuge the samples to be loaded in a microcentrifuge at full speed to pellet any insoluble material in the solution that can lead to clogging of the needle. Load the needle with protein solution (~ 2–5 μl) and attach to the pressurized needle holder. Be sure the carrier gas (N_2) of microinjector is flowing at the correct pressure. Set the outflow pressure in the needle between 20–100 kPa. This outflow pressure will be adjusted later after performing a series of test injections. Place the loaded needle in the micromanipulator rest and clamp in place.
4. Lower the needle, using the micromanipulator, into the media of the culture dish as soon after loading the needle as possible. Change the revolving turret on the ocular stand of the microscope to place the Bertrand lens in the image path. (The Bertrand lens gives a telescopic view of the back aperture of the objective.) Using this lens, guide the needle into view and slowly lower the needle until the image of the tip of the needle is seen projecting through the image of the annular phase ring of the objective. For correct needle placement at this stage refer to *Figure 5A*. Continue to lower and advance the needle until the diffraction pattern of the needle shows a splayed image as in *Figure 5B* that is offset from the centre as indicated. At this point, change back to the specimen image path with the ocular turret. Locate the image of the needle as a faint shadow above the cell layer. Slowly bring the needle down to just above the focal plane of the cell layer. Choose a cell or cells to test the needle before injecting the experimental cells. Adjust flow pressure based on the test cells to deliver enough fluid to cause a slight distension of the cell. It is recognized that on average, using these methods, the final concentration of the injected substance within the cell is 1/10 the needle concentration (26).

Protocol 11. *Continued*

5. For each protein concentration, inject 10–20 cells. By epifluorescence check the cells for success and overall consistency of the micro-injections. Fluorescence of the labelled protein should be observed in the treated cells. The intensity of the fluorescence should be uniform throughout the cell, and the relative intensity should correlate with the concentration of labelled protein injected.

Activation and/or inhibition studies using synthetic peptides can be used to investigate binding domains and binding interactions which may not be possible to study by any other means. Using a synthetic peptide made to the oligoproline repeat region of the ActA surface protein of *Listeria*, studies were undertaken to investigate the molecular mechanism of *Shigella* intra-cellular motility. The surface protein of *Shigella*, IcsA, was demonstrated to be necessary and sufficient for *Shigella* actin-based motility. Although IcsA has no primary amino acid sequence homology to ActA, we believed that the molecular mechanism used by both pathogens must be conserved. Micro-injection studies, summarized in *Table 1*, convincingly demonstrated that peptides synthesized to the ABM-1 sequence of *Listeria* and the ABM-2 sequence of VASP, were extremely effective inhibitors of *Shigella* motility at submicromolar concentrations. In addition, the inhibition effects were not

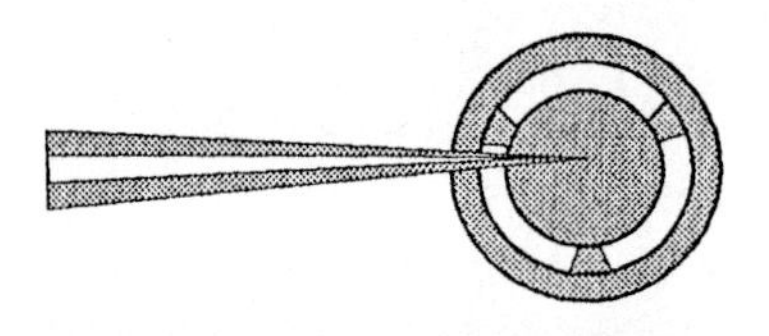

A. Bertrand lens view of needle entering from right, tip approximately 2 mm from cell surface

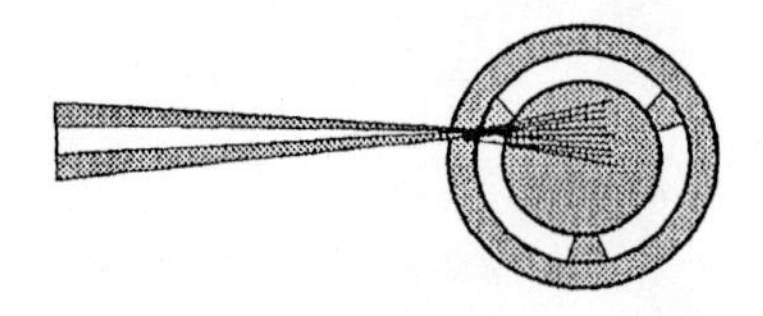

B. Bertrand lens view of needle tip approximately 0.1 mm from cell surface

Figure 5. Positioning of the needle for microinjection. A Bertrand lens on the microscope is used for the initial placement of the microinjection needle. The Bertrand lens gives a telescopic view of the back aperture of the objective. Using this view the needle can be guided to within 0.1 mm of the cell surface, after which the Bertrand lens is removed and the needle is focused using the objective.

Table 1. Effects of microinjected peptides on *Shigella* intracellular motility[a]

Additions	Intracellular concentration	Pre-injection velocity	Post-injection velocity	Post-injection/ pre-injection	P value
ActA peptide CFEFPPPPTDE	80 nM	0.06 ± 0.03 (n = 47)	0.004 ± 0.01 (n = 85)	0.07	< 0.001
Profilin	80 nM	0.14 ± 0.04 (n = 16)	0.12 ± 0.06 (n = 21)	0.85	N.S.[b]
ActA peptide and profilin	80 nM/80 nM	0.09 ± 0.07 (n = 16)	0.30 ± 0.11 (n = 33)	3.33	< 0.001
	20 nM/20 nM	0.13 ± 0.05 (n = 15)	0.17 ± 0.08 (n = 12)	1.31	N.S.
Poly-L-proline and profilin	2.5 μM/10 μM	0.14 ± 0.08 (n = 29)	0.06 ± 0.11 (n = 45)	0.43	< 0.001
MAP-2 peptide VKSKIGSTDNIKYZPKGG	10 μM	0.15 ± 0.05 (n = 17)	0.15 ± 0.07 (n = 44)	1.00	N.S.

[a] *Table 1* modified from ref. 24.
[b] N.S. = not significant.

observed with any of the other control peptides used. The results in *Table 1* pointed to a shared mechanism of actin-based motility and the importance of the ABM-1 and ABM-2 domains (23). *Table 1* has also been presented as an example of how to summarize microinjection data. Note in the fifth column the post-injection values are divided by the pre-injection values and that ratio is listed for each treatment. A value of the ratio greater than 1.00 indicates an increase and a value of less than 1.00 indicates a decrease in bacterial speed after microinjection. This is used to direct the reader to a single value to evaluate the experiments and allows the data to be more easily understood.

Protocol 12. Microinjection of synthetic peptides for inhibition of bacterial motility

Equipment and reagents

- Inverted microscope (Nikon Diaphot–TMD) with × 40, × 63, or × 100 objectives
- Epifluorescence attachment (high intensity lamp, filter cassettes, fluor objectives)
- Micromanipulator (Eppendorf, Model 5171)
- Microinjector (Eppendorf, Model 5242)
- Cooled CCD camera (Hamamatsu, Model C5985)
- Computer/Windows 95 with video acquisition board (frame grabber) installed and configured
- *Metamorph* computer image analyser (Universal Imaging)
- Shutter driver (Uniblitz, Model D122)
- 25 mm circular cover glass holder (Narishige)
- Microincubator MS-200D (Narishige)
- Cell monolayers on 35 mm glass-bottom microwell dishes from *Protocol 7* infected with *Listeria* or *Shigella*
- Bovine serum albumin, Cascade Blue conjugate (25 mg/ml) (Molecular Probes, A-964)
- PBS pH 7.4
- Synthetic peptides in PBS (100 μM stock) pH 7.4

Protocol 12. *Continued*

Method

1. Set up microscope and microinjection system as in *Protocol 11*. Load needle and position above cell layer as described. Microinject test cells and adjust the pressure.
2. Set time-lapse acquisition parameters to acquire 12 images for each injection experiment with a 30 sec wait time between image acquisition.
3. Locate a cell in the microscopic field that contains moving bacteria. In the case of *Listeria*, the actin tail trailing behind can identify moving bacteria. For *Shigella*, the moving bacteria do not always have a readily observable tail. In these cases, observe movement by looking for a cleared region of cytoplasm trailing the bacteria, or acquire two or more timed images and look for displacement of the bacterium over the time period.
4. Begin the time-lapse acquisition. At acquired image 4 or 5, pause the program and inject the cell with the selected agent. Resume the time-lapse and continue to completion. During the entire series readjust the focus to maintain a clear image of the bacteria and cell.
5. After completion of the time-lapse, change to the epifluorescent mode and observe the extent of the microinjection.
6. Continue with microinjections until at least three good microinjection time-lapse series have been acquired and recorded for each treatment and control.
7. Analyse data by statistical and graphing methods—see Section 3.4.

3.3 Immunolocalization of key components of cellular movement

Defining a molecular mechanism need not be limited to use of time-dependent parameters. Localization of cellular constituents by epifluorescence microscopy and immunofluorescence is a powerful technique that has been used extensively and with great success by many researchers. Immunofluorescence exploits the high affinity and selectivity of an antibody for its antigen and conjugated fluorescent chromophores to generate localized light sources within the specimen. The video microscope with an image intensifier and CCD camera or cooled CCD camera now permits the capture of images produced from these extremely low-light sources. With the availability of fluorescently-labelled antibodies, virtually any antigenic substance can be localized with a specific antibody. The broad panel of commercially available antibodies also has greatly expanded the possibilities for the researcher who

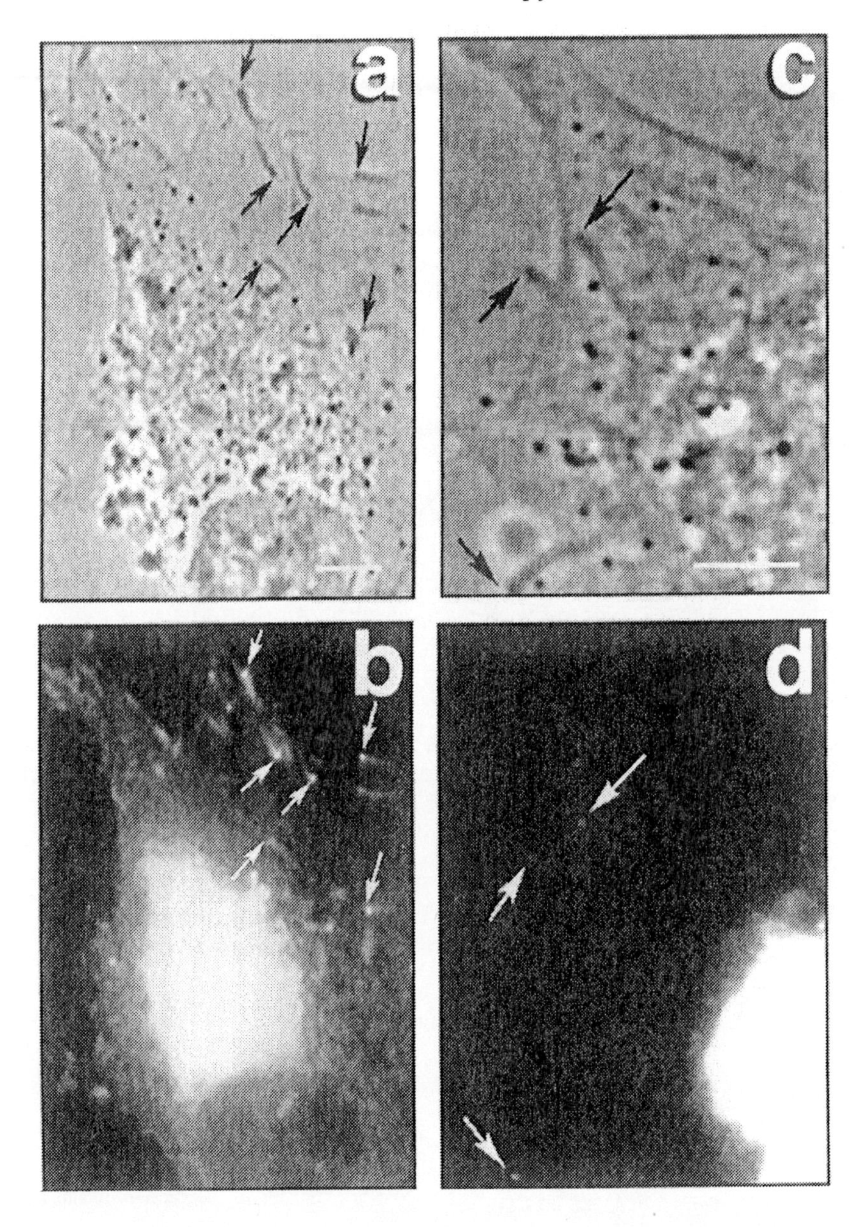

Figure 6. Phase-contrast and epifluorescent micrograph of vaccinia-infected HeLa cells. (a and c) Phase-contrast micrograph of HeLa cells infected with vaccinia virus for 7 h. (b) Immunofluorescence micrograph, using a anti-VASP antibody. *Arrows* point to the virus–actin tail interface. Viral factories are also intensely stained. (d) Hoechst dye staining for double-stranded DNA locates motile viral particles (*arrows*) and viral factories in the cytoplasm. (With permission from ref. 21.)

may have generated his or her own antibody to a specific antigen but who needs to localize known proteins for defining cellular location.

The methods in *Protocol 13* are just one procedure of many that are available that describe the preparation of living tissue for fixation, permeabilization, and antibody staining. In this protocol the cell monolayers are fixed, permeabilized, and stained in place in the microwell dishes. This method affords great flexibility. For example, cells that have been used in a microinjection experiment in microwell dishes can be directly fixed, avoiding the need to grow cells for immunofluorescence experiments on separate cover glasses. Obviously, methods of extraction that require acetone or other solvents that can attack polystyrene cannot be used in this procedure, but for most of the cytoskeletal proteins that have been studied in our laboratory such treatments are not necessary for good fixation and permeabilization. Although *Protocol 13* has been used in our laboratory with excellent results on a number of different cell types and microbes, the reader is encouraged to refer to those published methods that have been used with success on their particular cell type or organism.

Figure 6 demonstrates the use of this protocol in a co-localization experiment on HeLa cells infected with vaccinia virus. In the left panel *Figures 6a* and *6c* are the phase-contrast images in which the vaccinia can be see as dark spots trailing a thin dark filament, the actin tail. When these same cells are stained with a polyclonal antibody to the cytoskeletal protein VASP the region just behind the virion and some of the actin tail is fluorescent (*Figure 6b*). A Hoechst dye, which stains double-stranded DNA, labels the virion DNA (*Figure 6d*). Each virion identified by Hoechst staining is coincident with the same area of fluorescence identified by the VASP antibody.

Protocol 13. Immunolocalization: fixation, permeabilization, and staining of cell monolayers in microwell dishes

Equipment and reagents

- 12 mm circular cover glass
- Cell monolayers, infected or uninfected, growing on 35 mm glass-bottom microwell dishes from *Protocol 1*
- 3.7% formalin in PBS pH 7.4
- 0.2% Triton X-100 in PBS pH 7.4
- 50 mM NH_4Cl in PBS pH 7.4
- Blocking buffer: 1% BSA, PBS pH 7.4
- PBS pH 7.4
- Primary antibodies
- Secondary antibodies: anti-mouse IgG or anti-rabbit IgG conjugated to fluorescein isothiocyanate (FITC) (Sigma; F2883, F6005), anti-mouse IgG or anti-rabbit IgG conjugated to tetramethylrhodamine isothiocyanate (TRITC) (Sigma; T7782, T5268)
- Bodipy 581/591 phalloidin (Molecular Probes, B3416) or fluorescein phalloidin (F432)
- Fluormount-G (Fisher, 100–01)

Method

1. Remove the media from the cell monolayer and wash three times with PBS. Fix the cells directly in the microwell dish by adding 1 ml of 3.7% formalin and incubate for 20 min at room temperature.

2. Aspirate the fixation solution and wash three times with PBS. Permeabilize the cell membranes with 1 ml of 0.2% Triton X-100 and incubate for 5 min at room temperature.
3. Remove the permeabilization solution and wash three times with PBS. Inactivate unreacted aldehydes by treating with 1 ml of 50 mM NH_4Cl for 5 min at room temperature.
4. Wash the cell layer three times with PBS and add 1 ml of blocking buffer. Incubate 1 h at room temperature or overnight at 4 °C.
5. Prepare the primary antibody solution by diluting the primary antibody into blocking buffer. For immunofluorescence, a good starting ratio of primary antibody to blocking buffer is 1:200. If the fluorescence signal is too low, the antibody concentration may need to be increased.
6. Remove the blocking buffer and add 250 μl of primary antibody solution per microwell dish. Place a circle of filter paper cut to fit the top of the microwell dish and wet with water to create a moisture chamber that will prevent drying. Incubate the fixed cells with the primary antibody for 1 h at room temperature. This incubation can be done at 4 °C overnight if necessary.
7. Prepare the secondary antibody solution as above by diluting the stock in blocking buffer. A good starting ratio of secondary antibody to blocking buffer is 1:500. Wash with 2 ml of blocking buffer for a total of three times.
8. Incubate the cells with secondary antibody for 1 h at room temperature. After incubation, wash the cells with 2 ml of PBS for a total of three times.
9. If the cells are to be stained for F-actin using fluorescently-labelled phalloidin, prepare the stain by evaporating 10–25 μl of labelled phalloidin stock to dryness and resuspending the pellet in 500 μl of 0.2% Triton X-100 in PBS. Remove the last wash and incubate the cell layer with 250 μl of the phalloidin solution for 20 min at room temperature. After the incubation step wash the cells with 2 ml of PBS for a total of three washes.
10. Remove the last wash and add 50 μl of Fluormount-G, anti-quenching preservative directly to the cell layer. Place a 12 mm cover glass on top and press out air bubbles. The cells are now ready to be visualized by epifluorescence microscopy. At this time it is also possible to store the prepared cells for later microscopy by placing the cover on the microwell dish, re-wetting the filter paper in the top of the dish, and sealing the cover to the dish by wrapping with Parafilm. Wrap the dishes in aluminium foil and store at 4 °C.

3.4 Image analysis and data presentation

A high quality micrograph taken on scientific 35 mm photographic emulsion through a microscope port can have detail and clarity that is unsurpassed in digitized images from even the best 8-, 12-, or even 16-bit CCD cameras. A typical digitized image acquired from a 756 × 483, 8-bit photodiode array contains 0.35 megabytes of information, while the 35 mm negative contains approximately 40 megabytes of information. Despite this large difference in information storage, the digitized image from a high quality video camera is far superior as a tool for image data analysis, manipulation, storage, and, in most cases, presentation.

Clearly, image digitization simplifies and speeds up the storage and retrieval of image information, but in addition, the digitized image is in a format that allows for mathematical manipulation. The numeric values of the X-coordinate, Y-coordinate, and greyscale define the position and intensity of the signal output from each picture element or pixel in an array of pixels which specifies the image as seen by the camera. The digitized signal from each pixel in an image can be manipulated mathematically. This gives rise to many of the image processing and enhancements that are now available. One such manipulation known as thresholding allows setting a minimum and/or maximum greyscale range to the image to aid in differentiation between regions of significance. Other manipulations use arithmetic operations on individual pixel output to convert images to 'skeletonized' forms or for processing stacks for image overlays. A third form of image processing is convolution. In image convolution the pixel and its adjacent pixels are arithmetically manipulated to generate a filter or mask which, when applied to the image, may be used to enhance differences in regions of low contrast or smooth regions of high contrast. Some of these convolution masks are the sharpen filter, low pass and median pass filters, and the unsharpen filter. Finally, mathematical operations can be applied to detect edges or boundaries of an object by measuring the slope or gradient of greyscale intensities in its vicinity. An abrupt change in the slope of intensities can define an edge or boundary. Singly or together these manipulations can be useful in motion analysis where it is desirable to clearly differentiate objects from frame-to-frame within a stack. This aids in accurate distance measurements.

The intent here has not been to present a detailed discussion of image processing or to give specific examples, but to introduce the reader to the most simple image manipulations that may be applied for optimum data retrieval from acquired images. It is suggested that the manual of the imaging software be consulted for detailed descriptions for applying filters and other mathematical manipulations.

Figures 4, *7*, and *8* are given as examples of data presentation. These figures are the result of microinjection experiments of PtK2 cells infected with *Shigella* (24). *Figure 8* is an example of a composite image created for em-

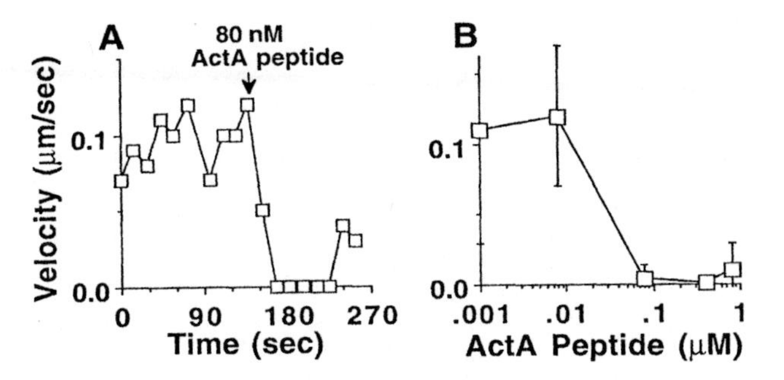

Figure 7. (A) Time plot of a microinjection experiment: velocity of a single *Shigella* bacterium in a PtK2 cell before and after microinjection of the ActA analogue (estimated intracellular concentration 80 nM). The *arrow* marks the time point at which the peptide was introduced. (B) Dose–response analysis of a microinjection experiment; effect of varying intracellular concentrations of the ActA analogue on *Shigella* intracellular velocity. Horizontal axis is in a log scale. Intracellular concentrations of 8, 80, 400, 800 nM were studied. Bars represent the standard deviation of the mean for 30–80 velocity determinations per concentration. (With permission from ref. 24.)

phasis of movement. *Figures 4* and *7* support the conclusion that a peptide inhibitor synthesized to the ABM-1 sequence of *Listeria* ActA is a potent inhibitor of *Shigella* actin-based motility. In this presentation the data from acquired images from a series of microinjection experiments in which this peptide inhibitor was used is presented in three different ways. First, direct presentation of images from a time-lapse experiment was shown in *Figure 4* as a dramatic way to display this experimental result. Secondly, a time plot of bacterial movement of the microinjection experiment in *Figure 4* is given in *Figure 7A*. Time plots can be very effective in graphically representing movement. As seen in *Figure 7A* the bacterial speed is plotted versus time with the time of microinjection of the inhibitor marked by an *arrow*. Measurements of speed are taken by using the *Track Points* function in *Metamorph*. Using this function distances are recorded in an image stack by tracking user specified points from image plane to image plane. The distances are logged and speed calculated based on the acquisition time recorded for each image. Quantification of the information from the experiment in which the images in *Figure 4* were obtained allows the readers to make their own decisions as to whether the speeds observed were within a normal range and over what time interval, specifically, the inhibition was observed. Although not shown here, time plots of actin tail length can also be analysed and presented as a measure of motility and actin dynamics.

Thirdly, dose–response analysis is another appropriate method to quantify microinjection experiments. For these experiments, the treatment is administered over a range of concentrations and the effects measured for each concentration. The data are statistically analysed and graphed as in *Figure 7B*. Here bacterial speeds are plotted versus the concentration of the peptide

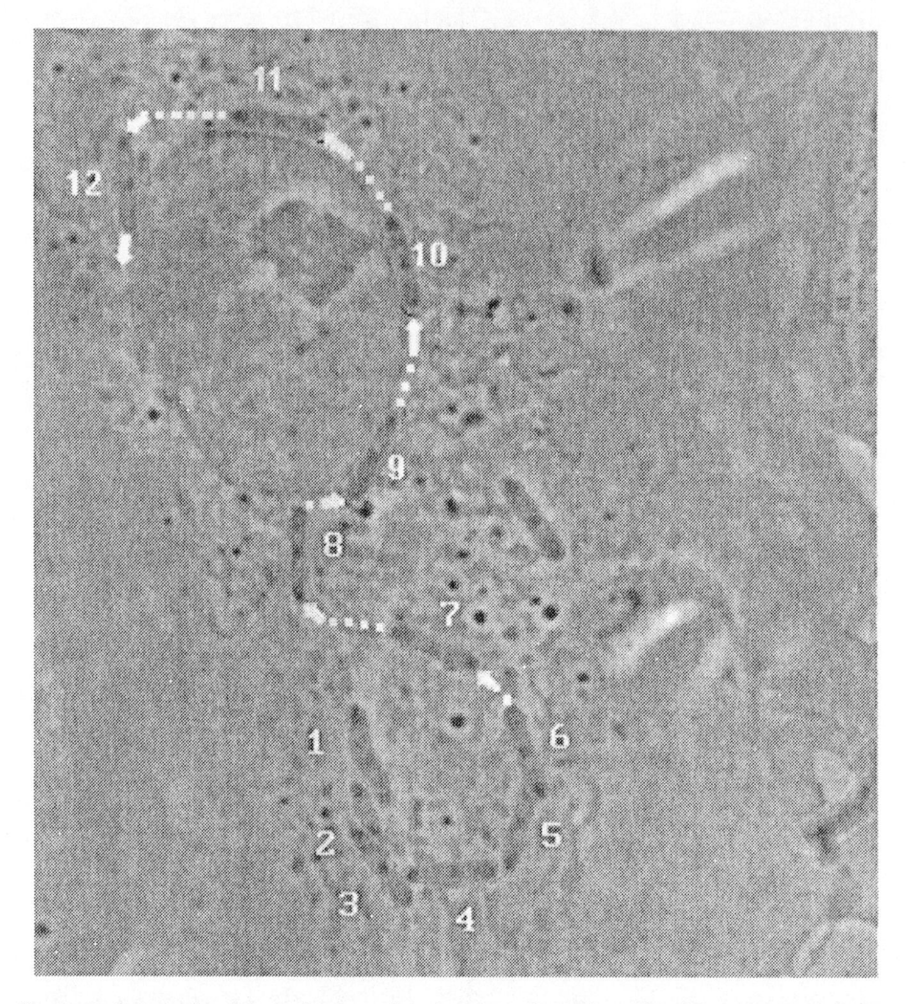

Figure 8. Composite image from a time-lapse series of a microinjection experiment measuring *Shigella* motility in a PtK2 cell before and after injection of a mixture of profilin and the ActA analogue. Depicted are the path and distances covered by a single bacterium before and after microinjection. Images show the position of the bacterium at 30 second intervals and are numbered sequentially. The cell was microinjected with the solution mixture between images 3 and 4 of the composite. After microinjection, note the progressive increase in the distance travelled by the bacterium after each time interval. The moving bacterium is 5 μm in length. (With permission from ref. 24.)

inhibitor injected. It can be seen that the inhibitory effects of the ActA peptide are concentration dependent. A lower intracellular concentration (8 nM) of the peptide fails to inhibit, while higher intracellular concentrations (400–800 nM) consistently block intracellular movement of *Shigella*.

A composite image is given in *Figure 8* to demonstrate an unconventional method of image data presentation. A PtK2 cell infected with *Shigella* was microinjected with a mixture of the actin-binding protein profilin and the synthetic ActA peptide. To our surprise this mixture was not inhibitory but caused the bacteria to move at speeds significantly greater than normal. As one way to convincingly portray these results, we chose to create this image. The figure is composed from a time-lapse series of 12 images taken at 30 second intervals recorded during the microinjection experiment. The background for the composite was chosen from the first image of the series. The path of the bacteria over time was depicted by superimposing the image of the moving bacteria from each successive image on the background. Numbers next to the individual bacterium correspond to the image of the time-lapse series. The injection takes place between image 3 and 4. Notice how dramatically the distance moved by the bacterium increases over each time interval. Data displayed in this way are visually powerful and, as a supplement to statistical analysis, time plots, and dose–response graphs, can convincingly present experimental results.

4. Concluding remarks

The procedures in this chapter have been presented as a guide for designing a strategy for identifying and/or characterizing key components in cellular processes and, specifically processes of cell motility. The intent was not to be exhaustive in covering individual techniques, but to provide a more general guide to enable the reader to examine experimental studies in cell biology with an eye for applying one or more of these methods. It should be clear now that video microscopy, microinjection, and immunolocalization are powerful tools, at a useful and appropriate resolution, for the study of many aspects of living systems. Although we have presented the methods here based on our work in actin-based motility, we hope this chapter may find use as an outline of the logic and experimental procedures for discovering novel aspects of cell signalling, regulation, and cell motility.

References

1. Tapon, N. and Hall, A. (1997). *Curr. Opin. Cell Biol.*, **9**, 86.
2. Cossart, P. and Lecuit, M. (1998). *EMBO J.*, **17**, 3797.
3. Kang, F., Laine, R. O., Bubb, M. R., Southwick, F. S., and Purich, D. L. (1997). *Biochemistry*, **36**, 8384.
4. Finlay, B. B. and Cossart, P. (1997). *Science*, **276**, 718.
5. Southwick, F. S. and Purich, D. L. (1994). *BioEssays*, **16**, 885.
6. Southwick, F. S., Heuser, J. E., Beckerle, M. C., Shen, P., and Purich, D. L. (1997). *Mol. Biol. Cell*, **8**, 169a.
7. Maurelli, A. T., Blackmon, B., and Cirtiss, R. (1984). *Infect. Immun.*, **43**, 397.
8. Condit, R. C. and Motyczka, A. (1981). *Virology*, **113**, 224.

9. Joklik, W. K. (1962). *Virology*, **18**, 9.
10. Dabiri, G. A., Sanger, J. M., Portnoy, D. A., and Southwick, F. S. (1990). *Proc. Natl. Acad. Sci. USA*, **87**, 6068.
11. Gaillard, J. L., Berche, P., Frehel, C., Gouin, E., and Cossart, P. (1991). *Cell*, **65**, 1127.
12. Menard, R., Sansonetti, P., and Cossart, C. (1994). *EMBO J.*, **13**, 5293.
13. Tilney, L. G. and Portnoy, D. A. (1989). *J. Cell Biol.*, **109**, 1597.
14. Southwick, F. S. and Purich, D. L. (1996). *N. Engl. J. Med.*, **334**, 770.
15. Moss, B. (1996). In *Fields virology*, 3rd edn (ed. B. N. Fields, D. M. Knipe, P. M. Howley, *et al.*), Chapter 38, p. 2637. Lippencott-Raven, Philadelphia.
16. Sodeik, B., Doms, R. W., Ericsson, M., Hiller, G., Machamer, C. E., van't Hof, W., *et al.* (1993). *J. Cell Biol.*, **121**, 521.
17. Schmelz, M., Sodeik, B., Ericsson, M., Wolfee, E. J., Shida, H., Hiller, G., *et al.* (1994). *J. Virol.*, **68**, 130.
18. Mathew, E., Sanderson, C. M., Hollinshead, M., and Smith, G. L. (1998). *J. Virol.*, **72**, 2429.
19. Dubochet, J., Adrian, M., Richter, K., Garces, J., and Wittek, R. (1994). *J. Virol.*, **68**, 1935.
20. Cudmore, S., Cossart, P., Griffiths, G., and Way, M. (1995). *Nature*, **378**, 636.
21. Zeile, W. L., Condit, R. C., Lewis, J. I., Purich, D. L., and Southwick, F. S. (1998). *Proc. Natl. Acad. Sci. USA*, **95**, 13917.
22. Fung, D. C. and Theriot, J. A. (1998). In *Motion analysis of living cells* (ed. D. R. Soll and D. Wessels), Chapter 8, p. 157. Wiley-Liss, New York.
23. Inoue, S. and Spring, K. R. (1997). In *Video microscopy: the fundamentals*, 2nd edn. Plenum Press, New York and London.
24. Zeile, W., Purich, D. L., and Southwick, F. S. (1996). *J. Cell Biol.*, **133**, 49.
25. Southwick, F. S. and Purich, D. L. (1994). *Proc. Natl. Acad. Sci. USA*, **91**, 5168.
26. Graessmann, G. and Graessmann, A. (1986). In *Microinjection and organelle transplantation techniques* (ed. J. E. Celis, A. Graessmann, and A. Loyton), Chapter 1, p.10. Academic Press, New York.

8

Structural studies on tubulin and microtubules

KENNETH H. DOWNING

1. Introduction

Microtubules (MTs) are involved in a wide variety of essential functions in eukaryotic cells, ranging from intracellular transport to cell motility and maintenance of cell shape. The identification of microtubules and their activities started a quest to understand the mechanisms behind their functions and the interactions among various proteins that make them possible, and this quest continues today. In early work, mainly done by light and electron microscopy (for a review, see ref. 1), the distribution of microtubules within the cell showed that they were an important component of the cytoskeleton, and their association with chromosomes in mitotic cells hinted at the critical role they play in cell division. Much of this work was done with sections of fixed and embedded material, where not only the distribution of MTs was revealed, but also the basic structure. It was seen that MTs were hollow cylinders made up of about 13 parallel protofilaments. Visualization of MT structure at higher resolution, and even determination of the structure of the protein subunit, has relied on electron microscopy. This chapter focuses on the methods used to derive structural information on microtubules and their complexes with drugs and other proteins by electron microscopy.

Microtubules can be highly dynamic structures within the cell. Much of the knowledge about the dynamic properties comes from light microscopy, which has confirmed that different arrays of MTs exist in the cell at different times in the cell cycle. For example the construction of the spindle apparatus during mitosis, and its disassembly as chromosomes separate, involve rapid polymerization and depolymerization of MTs. In addition, video-enhanced microscopy shows dynamic behaviour of individual MTs as they grow and shrink to explore the cell interior, for example, searching for a chromosome to which to attach. This dynamic behaviour is also seen *in vitro*, both with individual MTs and in bulk solution. Understanding the mechanisms of these activities has been an active field of research for scientists from a wide range of disciplines.

The timing of MT assembly and disassembly appears to be controlled by various proteins within the cell that are themselves regulated in a cell cycle-specific manner. Interfering with these processes by disrupting the binding of these other proteins or binding of ligands that stabilize or destabilize MTs can disrupt the cell cycle. This property has made MTs an effective target for a number of anti-proliferative drugs, most notably Taxol®, which is effective against several types of cancers. A number of laboratories and pharmaceutical companies are engaged in trying to understand the mechanisms of action of drugs known to bind to tubulin and in searching for new anticancer drugs. Electron microscopy can play a role in this effort using the methodology that led to determination of the structure of tubulin.

The transport of organelles along MTs involves proteins known as *motor molecules*, which use energy derived from ATP hydrolysis to move toward one end or the other of MTs. The study of complexes of these motors on MTs is currently a particularly active area utilizing electron microscopy.

The dynamic properties of MTs as well as their interactions with many other proteins and drugs have led to much interest in the structure of tubulin, the main component of MTs. Understanding the structure of the protein, and eventually the structure of the various complexes, will not only answer basic questions about biophysics and cell biology, but may also lead to ways to manage and direct these interactions. The strong tendency of tubulin to polymerize has so far made it difficult to obtain crystals suitable for X-ray diffraction. In the absence of direct structural data, many studies have been aimed at understanding tubulin and MT properties using methods that range from proteolysis of the protein to visualization by electron microscopy of its various polymer forms.

With the recent determination of tubulin's structure by electron crystallography (2), there are new possibilities for understanding the molecular basis of MT properties and interactions in more detail. It is also now possible to obtain medium resolution, 3D reconstructions of MTs complexed with motors and other ligands. In this chapter we focus on the methodologies using electron microscopy which are in use to obtain structural data at the molecular level both about tubulin itself and about MTs and their complexes with natural and therapeutic ligands. These include high resolution crystallographic studies based on two-dimensional crystals of the protein and on medium to high resolution studies of intact microtubules and motor molecules by cryo-electron microscopy.

1.1 Developments in electron microscopy

The earliest applications of electron microscopy (EM) in microtubule studies used thin section techniques to investigate the distribution of MTs within the cell, showing interactions of MTs with an organizing centre, chromosomes, and other organelles. Following the development of procedures to isolate

MTs and, later, tubulin, negative staining was used to obtain better images of individual MTs, including three-dimensional (3D) reconstructions as techniques for working with helical specimens developed (3). To go beyond the resolution of stained specimens, techniques were developed to embed specimens in media which prevent dehydration in the vacuum of the microscope. Preservation was first achieved by glucose embedding, followed shortly by embedding in vitreous ice. In parallel with new techniques for specimen preparations, instrumentation advances included development of stable cold stages that allowed obtaining high resolution information from frozen specimens. Continuing advances in image recording and processing techniques have also helped to make it possible to retrieve data to high resolution.

Radiation damage caused by the electron beam presents a fundamental limit to the exposure that can be given to a specimen. The limited exposure in turn limits the signal-to-noise ratio (SNR) in the image, and thus the resolution that can be achieved in images of single molecules. This limit can, however, be overcome by averaging images of many equivalent views of the molecule. The essence of electron crystallography is that a crystal provides many images of the same molecule in precisely defined positions and orientations, allowing straightforward combining of the images. The periodic, repeating structure of MTs also presents a conceptually simple way of combining images of different subunits, although the number of subunits is not nearly as great in a microtubule image as in a crystal. Thus the resolution so far achieved in MT reconstructions has not yet reached the same level as with crystals.

2. Microtubules

The basic structure of a MT is shown schematically in *Figure 1*. Heterodimers of α and β tubulin associate head-to-tail to form protofilaments which then associate laterally to form the wall of the MT. Each tubulin monomer has a molecular weight of around 50 000 Da and is represented by a ball of about 50 Å diameter. Most MTs *in vivo* contain 13 protofilaments, and in these MTs the protofilaments run parallel to the MT axis. As indicated in the figure, the α and β monomers are arranged to follow a shallow helical path around the MT wall. Since this path rises by three monomers around the MT, this is known as a *three-start* helix. Note that arranging dimers on such a three-start helix (or any helix with an odd start number) requires that there be a mismatch, or seam, somewhere around the helix where the continuous path of one type of monomer is broken. Thus a 13 protofilament MT is not strictly a helical structure.

Some MTs *in vivo*, and many formed *in vitro*, contain different protofilament numbers, from as low as 9 to as high as 18. The lateral contacts between protofilaments in all of these MTs, though, appear to be essentially

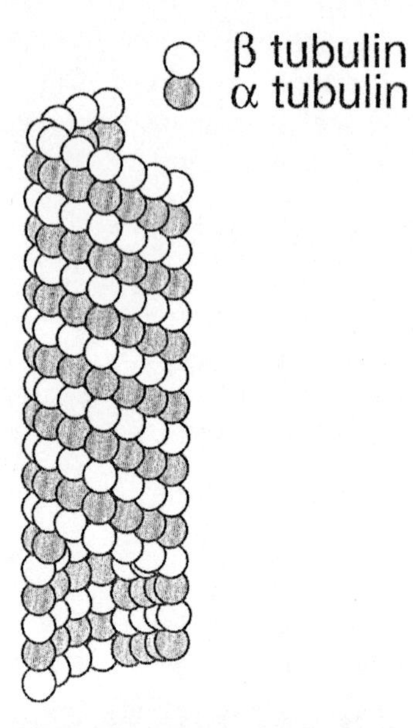

Figure 1. Schematic diagram of a microtubule. The basic structural unit is the αβ tubulin heterodimer, which forms protofilaments that associate in parallel to form the closed wall of the microtubule. The α and β subunits are indicated by different shading. This model is of a 13 protofilament, three-start MT, the most common type formed under most conditions. At the bottom end, the front part is cut away to show the seam that runs along the back side.

the same. One can think of a surface lattice of the MT as the planar structure that would be rolled up to produce the cylinder. Differences in protofilament number correspond to different widths of this surface, with the same internal structure. Only small distortions are required in this surface lattice to accommodate different numbers of protofilaments in order to make the monomers line up along the shallow helical path. With different distortions, different helix start numbers can be produced. For any given number of protofilaments, there is one start number that will require the smallest distortion and should correspond to the most frequently observed helical structure for the MT (4). The basic premise of this prediction has been confirmed by EM (4), and more recent investigations reveal the strong conservation of the interprotofilament interaction (5). The distortions of the surface lattice produce a distinct pattern that can be used to identify both the protofilament number and helix start number in cryo-EM images of MTs (6).

Within a MT the protofilaments all run in the same direction, giving a distinct polarity to the MT. In the cell most MTs originate at a MT organizing centre, where the minus ends are anchored, while the plus ends usually orient toward the cell periphery. In neurons, for example, the plus ends point away from the cell body. In the mitotic spindle the minus ends are anchored at the spindle pole, while the plus ends connect to the chromosomes at the kinetochore. The polarity is important in determining which direction various motors travel in the cell. Most members of the kinesin family of motors move towards the plus end, while dynein and a few kinesin-like motors move toward the minus end.

The attachment of ligands to the ends of MTs is at present poorly understood, but should become an area of intense study as EM techniques develop to visualize interactions of proteins at higher resolution. For example, γ tubulin is located at the minus end of the MT, but it is not yet clear whether γ tubulin forms a cap at the end and how other proteins may be involved in anchoring the end. The connection of the plus end to the kinetochore may be even more complicated, as there appear to be mechanisms by which energy is used to pull the chromosome along the MT as it disassembles from the plus end.

Microtubule polymerization can be induced under a variety of conditions, most (but not all) of which require GTP. Most commonly polymerization is induced in the presence of GTP at around 37 °C. Glycerol, sucrose, or DMSO is occasionally added to promote polymerization. Taxol and related drugs can induce polymerization even in the absence of GTP. Non-hydrolysable GTP analogues can also be used, since GTP hydrolysis is not required for the polymerization process. Indeed, the MT appears to be put into a metastable state following hydrolysis, and if all of the hydrolysable GTP is converted to GDP the MT falls apart. This disassembly process has been captured and visualized in the EM by freezing MTs under conditions in which they have been induced to disassemble rapidly, revealing a curved conformation of the protofilament that is favoured in the GDP state (7).

Two protocols for MT formation are given here as examples of the variety of conditions that have been used in previous work.

Protocol 1. Polymerization of microtubules induced by glycerol[a]

Reagents

- Buffer: 0.1 mM Mes, 1 mM EGTA, 0.1 mM EDTA, 1 mM GTP, 0.5 mM $MgCl_2$ pH 6.4
- Tubulin stored in 8 M glycerol

Method

1. Thaw an aliquot of tubulin on ice.
2. Suspend pellet in buffer; add $MgCl_2$ to 6 mM.
3. Incubate at 37 °C for 30 min to polymerize.
4. Centrifuge at 100 000 *g*, 37 °C for 10 min to pellet the MTs.
5. Suspend pellet in buffer plus 4 M glycerol. Incubate on ice for 30 min.
6. Centrifuge at 100 000 *g*, 4 °C for 10 min to remove denatured protein.

Protocol 1. *Continued*

7. To the supernatant add $MgCl_2$ to 6 mM. Incubate for 30 min at 37 °C. After 10–30 min the reaction will have reached a steady state.

[a] Adapted from ref. 8. The initial polymerization and depolymerization steps select tubulin which is assembly competent and not aggregated by denaturation, respectively. These steps can be omitted under many conditions when non-quantitative work is being performed.

Protocol 2. Polymerization of microtubules induced by Taxol

Reagents

- G-PEM buffer: 80 mM Pipes, 1 mM EGTA, 1 mM GTP, 1 mM $MgCl_2$ pH 6.8
- Taxol: 1 mg/ml in DMSO
- Tubulin: 10 mg/ml in G-PEM buffer (Cytoskeleton)

Method

1. Thaw a 25 μl aliquot of tubulin on ice.
2. Dilute tubulin with 75 μl of G-PEM buffer.
3. Add Taxol to 15 μM. Incubate for 30 min at 37 °C.

2.1 Sample preparations for electron microscopy

Microtubules can be stabilized by the addition of Taxol or related drugs after they have formed. In the absence of stabilizing ligands, the continued polymerization and depolymerization will eventually deplete the GTP, leading to loss of intact MTs. In addition, cooling the solution, such as may occur due to evaporation from the small droplets used for EM preparations, may induce depolymerization. Thus Taxol is often added to maintain the MT sample, if it was not already added to produce the MTs from the beginning.

Negative staining of samples for observation in the electron microscope provides a very rapid method for visualizing structures, although the resolution with stains is generally limited to around 20–50 Å. This resolution is more than sufficient for visualizing the subunits along protofilaments. A simple protocol for negative staining is given here; again, there is some latitude in the exact procedure, depending on what level of detail is required. This simple procedure produces adequate results for a wide range of applications, for example in visualizing MTs and their protofilaments. Refinements of the basic procedure are discussed in many EM techniques books.

The development of cryotechniques has provided a tremendous advance in available resolution with the EM. The basic principle is to preserve the specimen under conditions which retain the native, hydrated state. Substituting the water by glucose or other sugars provides sufficient hydrogen bonds that proteins retain their native conformation. Glucose embedding can be done with essentially the same procedure as for negative staining, with a 0.5–2% glucose solution substituted for the uranyl acetate. A further refinement of the embedding procedure is given in *Protocol 7*.

Protocol 3. Visualization of MTs by negative staining

Equipment and reagents

- Carbon-coated EM grids[a]
- Uranyl acetate solution: 1–4% in H_2O
- MT solution at a protein concentration of 1–5 mg/ml

Method

1. Place 5 μl of MT suspension on carbon-coated grid. Allow to adsorb for 30 sec.
2. Rinse in distilled water for 10 sec.
3. Add 10 μl uranyl acetate. Blot on filter paper.

[a]The grids are often glow-discharged before use to improve the uniformity of the stain distribution.

More frequently, the specimen is frozen in a layer of water that is thin enough to see through in the EM. Conventional freezing, though, involves a volume change as the water crystallizes, which can cause serious distortions to the protein. To overcome this problem, techniques have been developed to freeze the sample fast enough that the water vitrifies rather than crystallizes. Plunging the specimen into liquid ethane produces such rapid cooling that crystallization is avoided. This method can vitrify samples as much as one micrometre deep into solid material. For the types of EM samples such as MTs that are at most a few hundred Ångstroms thick, plunge freezing is a reliable method for preserving the specimen structure.

Plunge freezing also provides such a rapid cooling rate that dynamic processes can be stopped with a time resolution on the order of milliseconds. While this possibility has been used to examine such rapid phenomena as transitions in the conductance of membrane channels (9), it is also very useful for slower events, such as the depolymerization of MTs, that are still too fast to stop by most other techniques (7).

Frozen hydrated specimens are generally prepared on holey carbon films. The ideal hole size is around 1–2 μm, a size that will roughly fill the micrograph at a magnification around × 20 000–50 000. The films are made hydrophilic so that as the solution is blotted, a meniscus of liquid remains in the hole. Some trials are required to blot just the right amount to leave the correct liquid thickness.

While the use of a cold stage in the EM is desirable for glucose-embedded samples in order to reduce effects of radiation damage, it is essential with frozen hydrated specimens. The vitrified water must be held below about –150 °C in order to prevent crystallization.

Protocol 4. Plunge freezing in liquid ethane

Equipment and reagents

- Plunge freezing apparatus such as shown in *Figure 2*
- Holey carbon films on EM grids
- MT solution at 0.1–1 mg/ml
- Liquid nitrogen
- Ethane gas

Method

1. Treat the grids to make them hydrophilic, usually by glow-discharge.
2. Fill the liquid nitrogen and cool the ethane cup.
3. Condense ethane in the inner container until it is filled with liquid.
4. Mount the grid in the tweezers above the ethane cup.
5. Put 2–5 μl of sample on grid.
6. Blot the bulk of the liquid from the grid, leaving a very thin layer.
7. Release the trigger to drop the grid into the ethane.

2.2 Images of microtubules

Figure 3 illustrates the typical appearance of a frozen-hydrated MT preparation. Although the contrast is much lower than with stained specimens, the level of preservation is so much better that features such as the protofilaments are much more clearly resolved.

The Fourier transform of an MT image, such as shown in *Figure 3b*, contains a small number of spots compared to transforms of many other helical samples such as flagella or actin filaments. However, the information content is sufficient to reconstruct a three-dimensional structure, as discussed below. In addition, the positions of some of the low-order spots can be used to determine the hand of the protofilament twist.

The three-dimensional structure of the MT is also well preserved within the volume of the ice, as opposed to the flattening and distortion that frequently occur with negative stain, and densities of the top and bottom halves of the MT are superimposed in the image. In MTs that have a twist in the protofilaments (i.e. those with other than 13 protofilaments), this superposition produces a moiré pattern of fringes along the MT. There is one beat of this moiré pattern each time the top and bottom protofilaments superimpose along the MT, thus providing an easy measure of how much the protofilaments twist. This moiré pattern provides information that can be used to interpret the number of protofilaments and the helix start number. As shown

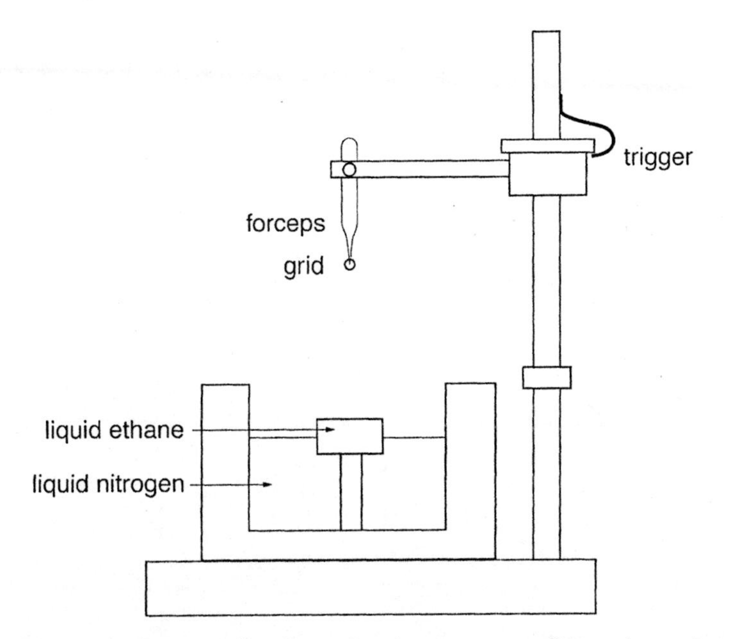

Figure 2. Schematic diagram of a plunge freezing apparatus. Ethane is condensed in a cup that is cooled by liquid nitrogen. In some designs, a heater is put in the cup to keep the ethane from freezing, since its melting point is somewhat above the temperature of liquid nitrogen. Otherwise, if the ethane does begin to freeze it can be melted by poking a metal object into the frozen part just before the grid is plunged. The trigger that releases the tweezers holding the grid is usually controlled by a foot pedal, freeing both hands for blotting the grid. The figure shows a plunger driven by gravity. Other implementations use a spring or compressed air to drive the grid into the ethane even faster.

in *Figure 4*, the top and bottom of the helix produce separate sets of diffraction spots in the Fourier transform of the image, as long as there is some twist to the protofilaments. The knowledge that the shallow helix of monomers is left-handed allows the determination of which spots are associated with each half of the MT and thus the hand of the twist.

3. Motor molecules

The interaction of motor molecules with MTs is an essential function in the cell, and understanding the mechanisms of motor movement is a fascinating biophysical problem. ATP binding and hydrolysis are involved in motor binding and the subsequent conformational changes of the motor that move the motor and its cargo along the MT. With the availability through molecular cloning of substantial quantities of purified motor domains from several types of motor molecules, several groups have studied the structure of the motor–MT complex by EM (10–13). The goal, of course, is to understand the mech-

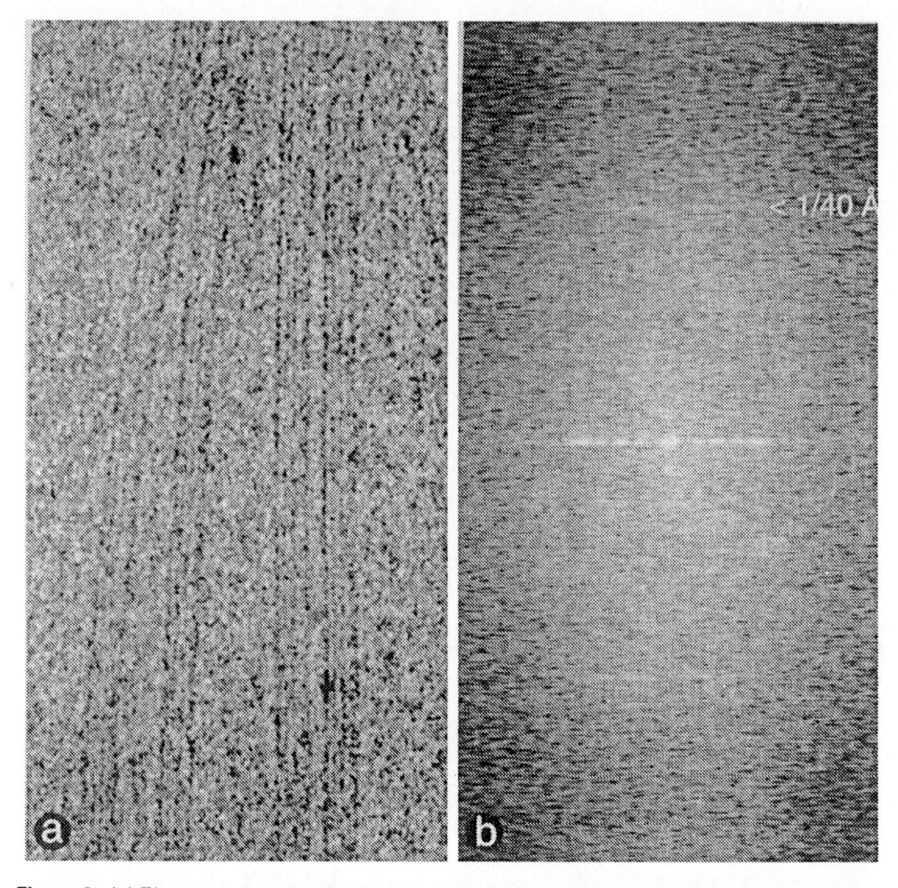

Figure 3. (a) Electron micrograph of a frozen-hydrated microtubule preparation. All of the MTs in this field contain 13 protofilaments. Protofilaments are seen running along the length of the MTs, although the subunits along the protofilaments are difficult to discern in this image. (b) Fourier transform of one of the MTs in (a). The prominent spots along the equator (the horizontal axis at the centre of the pattern) arise from diffraction by the protofilament spacing, while the spots at a distance 1/40 Å up from the equator arise from the separation of monomers along the protofilaments.

anism of binding and movement at the atomic level. This goal can be approached by obtaining sufficient resolution in reconstructions of the complex to allow fitting of atomic structures of tubulin and the motor proteins together. In addition, visualizing the complex in different nucleotide states of the motor can provide information about the movements involved. We give here an example of a protocol for forming the complex on an EM grid, which is subsequently plunge frozen for study in the microscope.

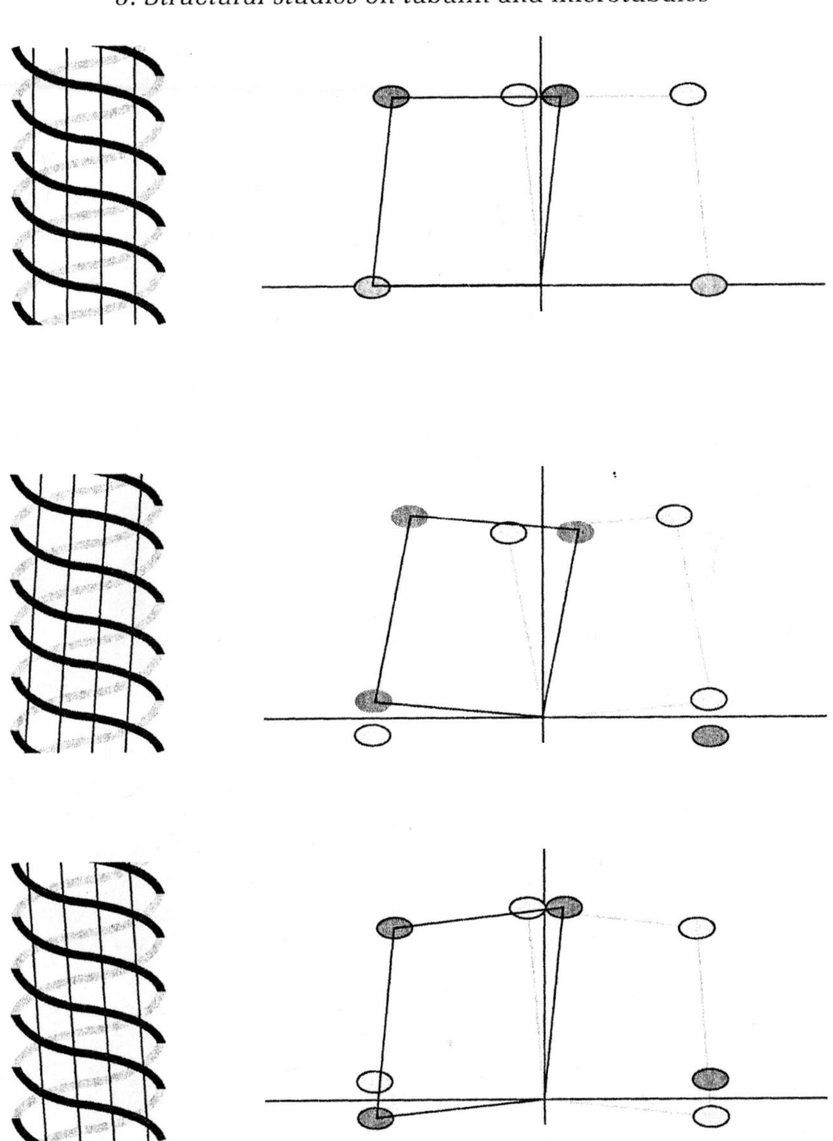

Figure 4. Indexing MT diffraction patterns. Schematic diagrams of three helical patterns are shown, corresponding to three-start helices with no protofilament twist, right-handed twist, and left-handed twist. The front and back surfaces of the helices are indicated by dark and light shading, respectively. Only protofilaments along the front surface are indicated. Adjacent to each helix diagram is a schematic of its Fourier transform, with shading of diffraction spots corresponding to whether they arise from the front or back surface of the helix.

Protocol 5. Preparation of kinesin or ncd bound to microtubules[a]

Equipment and reagents

- Holey carbon films on EM grids, glow-discharged
- MT solution at 2–3 mg/ml
- ncd or kinesin at 5–10 mg/ml
- Buffer with desired nucleotide: ATP, AMPPNP, etc.

Method

1. Apply 3 μl of MT solution to the grid. Allow to adsorb for 3–5 min.
2. Rinse the grid with buffer.
3. Add 3 μl of the motor solution, and allow to react for 3–5 min.
4. Blot and either stain as in *Protocol 3* or plunge into liquid ethane as in *Protocol 4*.

[a] Adapted from ref. 14.

4. 3D reconstruction of helical structures

Each image in the EM is a two-dimensional projection of a three-dimensional object. In order to reconstruct the 3D structure one must somehow collect projections at various angles through the specimen. A helical structure naturally provides a full-range of projection angles in a single image, with the object (for example the tubulin dimer) arranged in regular orientations around the helix. Thus helical objects are particularly well suited to 3D reconstructions. The mathematics of the reconstruction resembles that of conventional crystallography, although somewhat different functions are used to reconstruct the object. In the same way that the diffraction pattern of a crystal is related to the crystal itself by a Fourier transform, with the diffraction spot amplitudes representing coefficients of sine and cosine terms, the transform of a helical pattern gives coefficients of Bessel functions that describe the 3D structure. Several software packages are available and currently in use by electron microscopists for helical reconstruction. Some of these are described in a journal issue devoted to discussion of software for EM reconstructions (15, 16).

Another technique that has been developed for reconstructing three-dimensional volumes is known as 'back projection'. The basic principle of this method is that densities in a series of projections through the specimen at different angles are projected back into the 3D volume. This method requires a set of projections at different angles, which can be obtained in the electron microscope by tilting the specimen. Back projection is particularly well suited to visualizing structures of non-periodic structures, and is thus the method of choice in tomography, in applications ranging from EM in cell biology to whole-body scanners. The method is also appropriate for helical or near-helical structures such as microtubules. In a helical structure the angle of projection varies continuously along the helix axis. Thus the equivalent of a tilt series is obtained by examining a series of adjacent areas along the length of the structure. This has a particular advantage with many MTs, where the presence of a seam interrupts the helical structure, but the regular twist along the axis still provides the series of projections.

A true helical structure can be reconstructed with either helical or back projection methods, but the presence of a seam requires the use of back projection. At the resolution so far achieved in MT reconstructions, there is little distinction between the α and β tubulin monomers. On the other hand, motors bind predominantly to the β monomer, greatly enhancing the visibility of the surface lattice. Back projection reconstructions of MTs decorated with motor molecules have shown the seam in MTs, confirming the basic premise of the accommodation model (17, 18).

5. Tubulin structure

The importance of tubulin has inspired intense efforts to obtain crystals suitable for study by X-ray diffraction, but the tendency of tubulin to polymerize into linear forms, as well as its inherent tendency to degrade rapidly, has so far prevented success. Under various conditions tubulin forms rings, spirals, hoops, and sheets, in addition to the various sizes of MTs (19). Each of these forms appears to be based on the same protofilament structure, indicating the dominance of the longitudinal interactions between dimers.

The tendency of microtubules to align into parallel bundles, on the other hand, has provided the ability to collect some structural data from fibre X-ray diffraction patterns (20). Information from such patterns confirms the axial repeat of monomers along the MT axis and gives an indication of secondary structure, but does not provide sufficient data for constructing an atomic model of the protein, as has been done with several other helical systems.

One of the tubulin polymer forms is an arrangement in which the protofilaments are antiparallel, rather than the parallel arrangement of MTs. This antiparallel alignment removes the tendency to form cylinders, and instead leads to formation of two-dimensional, crystalline sheets of tubulin (21, 22). This form was discovered shortly after the first demonstration that electron microscopy could be used to visualize molecular structures at sufficient detail to resolve secondary structures (23), and several groups began to apply electron crystallography to the tubulin sheets (24, 25). These early efforts provided some further insights into the behaviour of MTs, although the resolution did not extend much beyond 20 Å.

The instrumentation and methodology of electron crystallography developed quite significantly over the next two decades following the first studies of the tubulin sheets. Taking advantage of these improvements, the sheets were used in the first determination of the structure of tubulin (2). The sheets also provide a basis for studying tubulin–drug interactions. The first structure was solved with Taxol bound to the sample in order to stabilize the sheets, and visualization of the Taxol binding site sparked nearly as much interest as the tubulin structure itself. Similar work should be useful in visualizing other drugs that stabilize MTs and sheets. Binding of motor molecules and other proteins to the sheets may also provide a way to obtain resolution higher than can currently be obtained on complexes with intact MTs.

The sheets are formed under polymerizing conditions with the addition of zinc ions. As with the formation of MTs, there is some latitude in the conditions, although growth of the largest sheets requires adjustment of conditions to match each batch of tubulin. A sample protocol is as follows.

Protocol 6. Formation of zinc-induced crystalline sheets of tubulin[a]

Reagents

- Incubation buffer: 80 mM Mes, 200 mM NaCl, 3 mM GTP, 1.25 mM $MgSO_4$, 1.25 mM $ZnSO_4$, 0.025 mg/ml pepstatin (Sigma) pH 5.9
- Tubulin at 10 mg/ml in G-PEM buffer (Cytoskeleton)

Method

1. Thaw a 25 μl aliquot of tubulin on ice.
2. Add an equal volume of incubation buffer.
3. Incubate at 30 °C for 10–24 h.

[a] Adapted from ref. 26. NaCl is not essential in forming the sheets, but allows longer incubation times without tubulin degradation, which in turn produces the largest sheets.

As with MTs discussed above, embedding the sample in vitreous ice or glucose retains the native, hydrated state, and thus gives the possibility of obtaining structural data to atomic resolution. Glucose is known to promote formation of MTs, and appears to slightly destabilize the sheets. Tannic acid has been used in studies of microtubules since the early days, as well as in embedding EM samples for high resolution work (27). The best preparations of the tubulin sheets so far have used a mixture of tannin and glucose (26). A problem that has often been encountered in high resolution studies of proteins by EM arises from variability in the surface properties of the carbon support films. The films age in often unpredictable ways which can affect how well the crystals are preserved. Carbon films are formed by evaporation onto mica and then floated off onto a water surface for mounting on the grids. In an effort to make the preparations more reproducible, the following protocol was developed for mounting the sample on the side of the carbon that had been in contact with the mica before any exposure to atmosphere.

Protocol 7. Preparation of glucose/tannin-embedded tubulin sheets

Equipment and reagents

- Carbon film on mica
- EM grids, 300 mesh
- Tubulin sheets at around 5 mg/ml
- Tannic acid: 1% (w/v), filtered and adjusted to pH 5.9
- Glucose: 1% (w/v)

Method

1. Place about 1 ml of the tannin solution in a small crystallizing dish.
2. Cut a square of carbon-coated mica about 3 mm on edge.
3. Immerse the mica under the tannin so that the carbon floats off.
4. Bring a grid up from below the carbon film to pick it up from the surface. A small lens of around 2 μl of the tannin solution will adhere to the grid under the carbon.
5. Turn the grid over, inject 2 μl of the tubulin solution into the tannin droplet, and mix gently.
6. Inject 3 μl of the glucose solution, mix, and withdraw about 5 μl with the pipette.
7. Blot the grid on filter paper.

6. 3D structure determination from 2D crystals

The data obtained from crystalline specimens in the electron microscope includes both electron diffraction patterns and high resolution images. As in X-ray crystallography, diffraction patterns provide amplitudes of the structure factors that are used to compute a density map. The images provide a direct measurement of structure factor phases, as discussed below. Several of the software packages currently used for processing data from 2D crystals are described in a journal volume devoted to descriptions of EM-related software (see, for example, ref. 28.). First we discuss analysis of electron diffraction patterns.

6.1 Diffraction

Figure 5a shows an electron diffraction pattern of a crystalline tubulin sheet. This pattern was recorded on a CCD camera in an electron microscope operating at 400 kV. The spacing of the diffraction spots indicates the unit cell

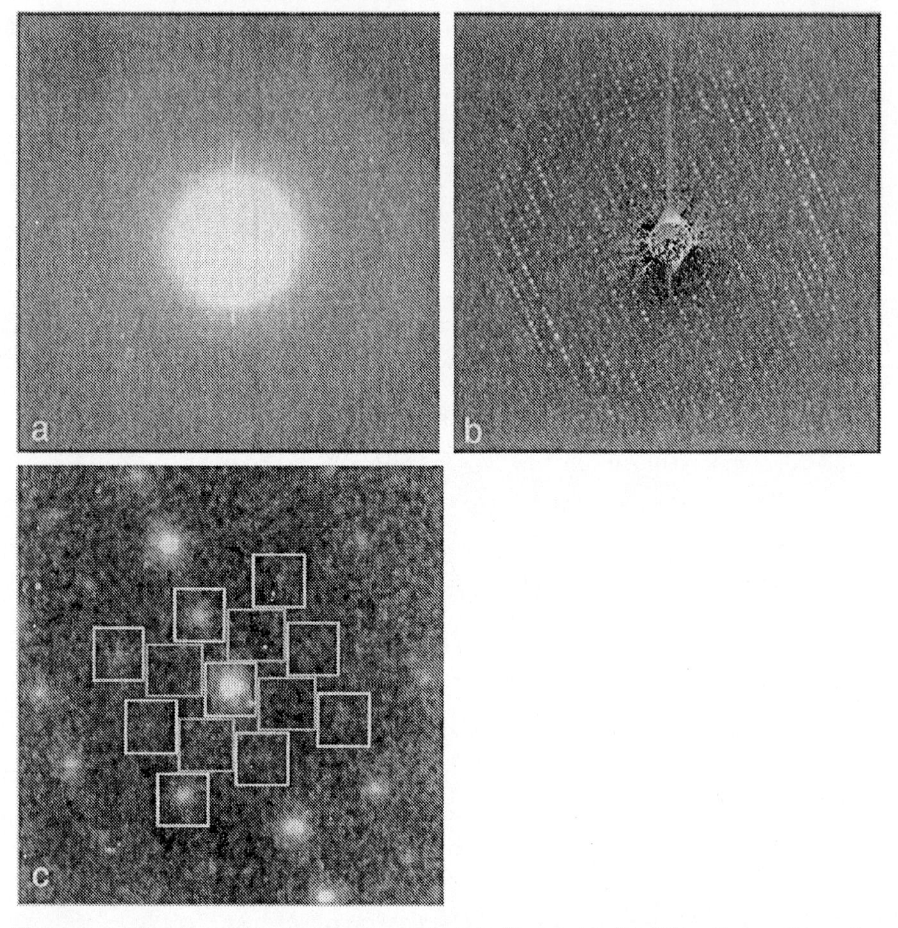

Figure 5. (a) Electron diffraction pattern of a tubulin crystal. (b) Diffraction pattern after subtraction of the radially symmetric component. (c) Spot (white) and background (grey) boxes used for determining intensities.

size of 80 × 92 Å, and spots can be seen out to somewhat beyond 3 Å near the edge of the pattern.

Because of the high linearity of the CCD, the signal is proportional to the electron intensity. When diffraction patterns are recorded on photographic film, the large dynamic range of the patterns generally exceeds the linear range of the emulsion and a correction needs to be made to convert the optical density on the film to electron intensity. Inelastic scattering and scattering from the amorphous support film contribute to the large halo around the centre of the pattern. This halo and the large dynamic range make it difficult to display the raw pattern so that the spots can be seen in both the low and high resolution ranges. The bright streak through the centre arises from saturation and blooming of the signal on the CCD due to the large over-exposure of the central region that is required to obtain good measurements of diffraction spots in the outer regions of the pattern.

Removal of the radially symmetric background in the pattern reveals a much better impression of the spots themselves. The radial background is determined by summing all intensities as a function of radius in the pattern, excluding the diffraction spots. Subtracting the radial pattern gives the pattern shown in *Figure 5b*, where the spots can now be seen much more clearly. There is still a locally varying background around the spots, which must be compensated in computing the spot intensities.

The diffraction spot intensities are measured by summing all intensities within a box whose size is chosen to be large enough to include the entire spot. The local background is measured in the area between spots, usually in four or six boxes as shown in *Figure 5c* (29).

6.2 Images

In order to avoid excessive radiation damage, images of unstained protein crystals must be recorded with an exposure that is so low that details are not visible in the image. The Fourier transform of the image provides a representation of the information in the image that corresponds to a diffraction pattern. The required data are, in fact, extracted from these diffraction spots in the computed Fourier transform. Because the transform is computed as an array of complex numbers, both amplitudes and phases are available. The amplitudes correspond to those of the electron diffraction pattern, while the phases are (aside from some offset discussed below) the values needed to construct a map of the specimen. Amplitudes from images are subject to degradation by effects that are not present in electron diffraction patterns, and are generally not as accurate as those that can be obtained by electron diffraction. Factors such as the defocus used in recording the image and the coherence of the beam affect the amplitudes in ways that can be described analytically and thus corrected. However, it is found that when the crystals are large enough to allow recording electron diffraction patterns, better measurements are obtained by diffraction than from the images. Phases determined from the images, on the other hand, can be remarkably precise. The ability to extract accurate phases from the images provides a distinct advantage of electron crystallography and produces reconstructions that are directly interpretable even at intermediate resolution.

A number of image processing steps are involved in extracting data from the images (30). Aside from computation of the Fourier transform, one generally needs to compensate for distortions in the image that have the effect of blurring the diffraction spots and thus decreasing the signal and apparent resolution. Since the procedures used in this processing illustrate some interesting general principles, it is useful to follow some of the steps. These operations can be applied to images of stained or unstained specimens, and because the results are visually clearer with the high contrast of a stained specimen, we use a negatively stained tubulin image to illustrate this section.

Figure 6a shows an image of a stained tubulin crystal. Although by most standards the contrast in this image is low, the crystal lattice is visible, but little detail can be seen beyond the presence of lattice lines. A simple operation reduces the non-periodic noise in the image and gives a much clearer view of the crystal. All of the periodic information content of the image is represented in the diffraction spots in the Fourier transform (*Figure 6b*), so masking the transform to zero out all values except those in the spots removes

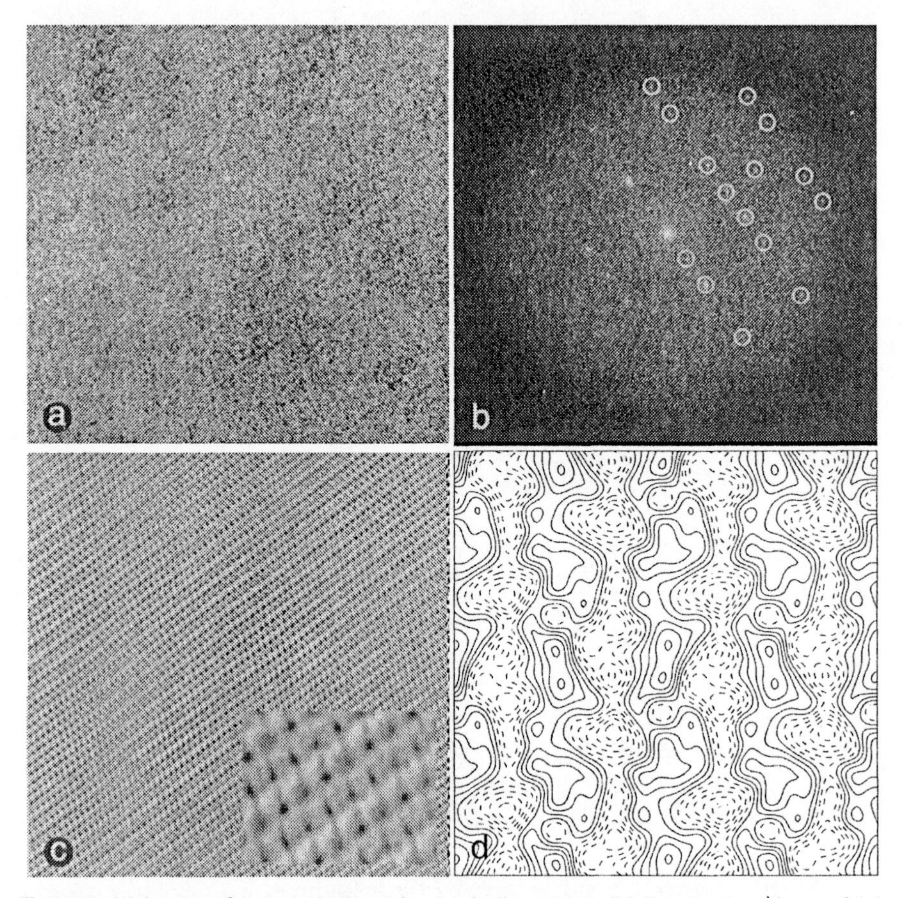

Figure 6. (a) Image of a negatively stained tubulin crystal. (b) Fourier transform of (a). Diffraction spots are circled in one-half of the pattern, indicating the mask that is used in computing a filtered image. (c) Filtered image obtained by inverse Fourier transformation of the transform in (b) after masking. (d) Contour plot of a density map obtained from negatively stained tubulin sheets. The protofilaments run vertically with adjacent protofilaments antiparallel. Solid lines enclose protein. The α and β monomers have distinctly different shapes as seen in these stained preparations.

the non-periodic component. Distortions and other imperfections in the crystal spread out the diffraction spots, so it is necessary to choose the mask parameters appropriate to the image.

An inverse Fourier transform of the masked transform gives the filtered image in *Figure 6c*. Now the unit cell is clearer, as shown in the magnified inset, although the distortions in the image lead to some variation in the appearance within the image.

A further step can be taken to remove the image distortions. By cross-correlating the filtered image with a small area containing a few unit cells, cut out from the centre of the image, the position of each unit cell in the image can be found. The correlation pattern contains a peak at the position of each unit cell, and these peaks can be identified by a computer program. Another program identifies the displacements of the peak locations from where they would be in the ideal, undistorted lattice, and computes a function that is used to shift the image so that each unit cell is put back onto its 'ideal' position. The Fourier transform of this 'unbent' image then has diffraction spots that are significantly sharper than in the original image, with a resultant improvement in signal-to-noise ratio. The results of the data extraction are ultimately limited by the SNR, so improving the SNR generally leads to an improvement in resolution. This step has proven to be essential in determining structures to high resolution. In this type of work the areas of crystals that are processed are so large that even relatively small distortions produce large spreading of the diffraction spots, making it difficult to extract high resolution data before correcting the distortions.

The resulting data from a single image can be used to produce a density map of the specimen, shown in *Figure 6d*. This map represents a projection of the densities within the 3D specimen onto a plane. In order to have the image properly centred, it is necessary to shift the *phase origin* of the data in the transform (a theorem of Fourier transforms states that a position shift in image space corresponds to a phase shift in Fourier space). When merging data from different images, it is also essential that each image be shifted to the proper phase origin.

6.3 Three-dimensional reconstruction

The three-dimensional structure is determined by combining data from many different views of the sample. In work with 2D crystals, these views are obtained by tilting the specimen in the microscope. With stained specimens it is possible to record a tilt series from a single specimen area, but with unstained specimens only a single image is recorded from each crystal. Data from different views are merged in Fourier space, building up a three-dimensional representation of the transform of the 3D object. Each individual view of the object represents a projection of the specimen, and the Fourier transform of each projection corresponds to the data on one plane through

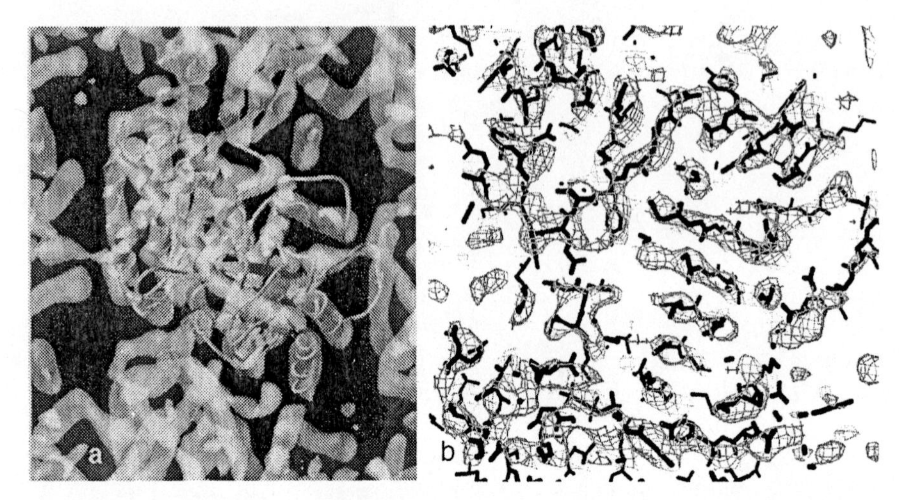

Figure 7. (a) Solid surface representation of a density map of a 3D reconstruction of tubulin at a resolution of 6.5 Å. The surface has been made transparent and a ribbon diagram of the tubulin structure is included. All of the main secondary structure elements of the final model can be identified in the 6.5 Å map. (b) Wire frame display of a section of the 3.7 Å resolution map of tubulin, showing part of the atomic model built into the density.

the origin of the 3D Fourier transform of the object. One needs to be able to determine the tilt angle and the direction of the tilt axis for the image, which together define the corresponding plane in the Fourier transform, so that data from different views are properly merged. Once a 3D density map has been calculated, standard X-ray crystallographic methods are used to construct and refine the atomic structure. One of the significant advantages of obtaining accurate phases from the images is that the structure can be interpreted with confidence even at moderate resolution and before refinement has begun.

In displaying 3D density maps one has the choice of using a solid surface or wire frame representation. The solid surface (which can also be made semi-transparent) is particularly useful with medium resolution structures, while the wire frame is generally more convenient when the resolution is high enough to produce a highly convoluted surface and when actually constructing an atomic model within the density map. Examples of both types of displays are shown in *Figure 7*.

7. Drug binding studies

The atomic model of tubulin was determined using samples which had been stabilized with Taxol. Taxol is an effective anticancer drug, and its mode of action has been the subject of numerous studies. Thus visualizing the Taxol binding site has provided important insights into its binding, as well as understanding of how certain tubulin mutations are related to Taxol resistance in cell culture and possibly *in vivo* (31). It will be informative to study other drug interactions by the same methods. It has been found so far that all the drugs that destabilize microtubules also destabilize the zinc-induced sheets, so these drugs are not candidates for crystallographic study. However, it can be expected that most of the MT-stabilizing drugs will also stabilize the sheets, so that electron crystallography may be a viable way to obtain structural information on interactions with these drugs.

It would, of course, be possible to record a full set of diffraction and image data on each drug of interest and compute the three-dimensional map in order to identify the drug in the 3D density map. However, another technique from X-ray crystallography may provide a simpler way to study these systems. Difference Fourier analysis is often used to examine small changes in structure, such as the addition of a small ligand to a proteins. Making a small change in the structure will make correspondingly small changes to the structure factors. To a first approximation, one could assume that either the amplitudes or the phases were unperturbed, and simply measure the other. A difference Fourier map can then be computed from the changes in the one data set. It is substantially easier to collect and process a set of electron diffraction data, and difference Fourier maps based on electron diffraction have been used, for example, to study conformational changes in bacteriorhodopsin through its photocycle (32, 33). One potential drawback is that the accuracy of measuring diffraction amplitudes must be very high, a problem that has yet to be fully overcome with the tubulin sheets.

Numerous software packages have been developed by structural biologists for modelling of protein–ligand interactions. It appears likely that these docking programs will be helpful in understanding the interactions of tubulin and drugs that cannot be studied by crystallographic methods. Hypotheses derived from this type of work can be tested against biochemical and biophysical data, as well as through the use of site-directed mutagenesis.

8. Protein–protein interactions

Mutagenesis studies have already been used to gain insights on the interaction of tubulin with other proteins. The general approach in this work is to change charged amino acids on the protein surface to neutral residues, usually alanine, either singly or in small clusters, under the assumption that charge interactions will have a large effect in determining the strength of the interactions. Site-directed mutation of kinesin, in the region of the protein thought to interact with microtubules, was followed by analysis of kinesin binding properties (34). The magnitude of the effect of each mutation maps out the expected strength of interaction between the two proteins, defining a hypothetical footprint of one protein on the other. This information strongly constrains models for the orientation of kinesin when bound to a MT.

Similar studies have been done with tubulin site-directed mutants. Alanine scans of both α and β tubulin have been done in yeast (35, 36), with characterization of the resultant phenotypes. Using a two-hybrid system that detects interactions between two proteins *in vivo*, the mutant strains were also tested for interaction with Alf1, a member of the tubulin-specific chaperonin complex (37), and with Bim1, a microtubule-associated protein in yeast (36). In both cases a binding footprint was clearly mapped out on the tubulin surface.

References

1. Porter, K. R. (1980). In *Microtubules and microtubule inhibitors* (ed. M. De Brabander and J. De May), p. 555. Elsevier/North-Holland Biomedical Press.
2. Nogales, E., Wolf, S. G., and Downing, K. H. (1998). *Nature*, **391**, 199.
3. Amos, L. A. and Klug, A. (1974). *J. Cell Sci.*, **14**, 523.
4. Chrétien, D., Kenney, J. M., Fuller, S. D., and Wade, R. H. (1996). *Structure*, **4**, 1031.
5. Chrétien, D., Flyvbjerg, H., and Fuller, S. F. (1998). *Eur. Biophys. J.*, **27**, 490.
6. Chrétien, D. and Wade, R. (1991). *Biol. Cell*, **71**, 161.
7. Mandelkow, E.-M., Mandelkow, E., and Milligan, R. A. (1991). *J. Cell Biol.*, **114**, 977.
8. Lu, Q. and Luduena, R. F. (1994). *J. Biol. Chem.*, **269**, 2041.
9. Unwin, N. (1995). *Nature*, **373**, 37.
10. Hirose, K., Lockhart, A., Cross, R. A., and Amos, L. A. (1995). *Nature*, **376**, 277.
11. Hoenger, A., Sablin, E. P., Vale, R. D., Fletterick, R. J., and Milligan, R. A. (1995). *Nature*, **376**, 271.
12. Kikkawa, M., Ishikawa, T., Wakabayashi, T., and Hirokawa, N. (1995). *Nature*, **376**, 274.
13. Arnal, I., Metoz, F., DeBonis, S., and Wade, R. H. (1996). *Curr. Biol.*, **6**, 1265.
14. Sosa, H., Dias, D. P., Hoenger, A., Whittaker, M., Wilson-Kubalek, E., Sablin, E., *et al.* (1997). *Cell*, **90**, 217.
15. Carragher, B., Whittaker, M., and Milligan, R. A. (1996). *J. Struct. Biol.*, **116**, 107.
16. Owen, C. H., Morgan, D. G., and DeRosier, D. J. (1996). *J. Struct. Biol.*, **116**, 167.
17. Sosa, H., Hoenger, A., and Milligan, R. A. (1997). *J. Struct. Biol.*, **118**, 149.
18. Wade, R. H., Meurer-Grob, P., Metoz, E., and Arnal, I. (1998). *Eur. Biophys. J.*, **27**, 446.
19. Mandelkow, E. and Mandelkow, E.-M. (1989). In *Cell movement: kinesin, dynein and microtubule dynamics*, p. 23. Alan R. Liss, Inc.
20. Mandelkow, E., Thomas, J., and Cohen, C. (1977). *Proc. Natl. Acad. Sci. USA*, **74**, 3370.
21. Larsson, H., Wallin, M., and Edstrom, A. (1976). *Exp. Cell Res.*, **100**, 104.
22. Gaskin, F. and Kress, Y. (1977). *J. Biol. Chem.*, **252**, 6918.
23. Henderson, R. and Unwin, P. N. (1975). *Nature*, **257**, 28.
24. Amos, L. A. and Baker, T. S. (1979). *Nature*, **279**, 607.
25. Crepeau, R. H., McEwen, B., Dykes, G., and Edelstein, S. J. (1977). *J. Mol. Biol.*, **116**, 301.
26. Nogales, E., Wolf, S. G., Zhang, S. X., and Downing, K. H. (1995). *J. Struct. Biol.*, **115**, 199.

27. Kühlbrandt, W. and Downing, K. H. (1989). *J. Mol. Biol.*, **207**, 823.
28. Crowther, R. A., Henderson, R., and Smith, J. M. (1996). *J. Struct. Biol.*, **116**, 9.
29. Baldwin, J. and Henderson, R. (1984). *Ultramicroscopy*, **14**, 319.
30. Henderson, R., Baldwin, J. M., Downing, K. H., Lepault, J., and Zemlin, F. (1986). *Ultramicroscopy*, **19**, 147.
31. Giannakakou, P., Sackett, D. L., Kang, Y.-K., Zhan, Z., Buters, J. T. M., Fojo, T., *et al.* (1997). *J. Biol. Chem.*, **272**, 17118.
32. Subramaniam, S., Gerstein, M., Oesterhelt, D., and Henderson, R. (1993). *EMBO J.*, **12**, 1.
33. Vonck, J. (1996). *Biochemistry*, **35**, 5870.
34. Woehlke, G., Ruby, A. K., Hart, C. L., Ly, B., Hom-Booher, N., and Vale, R. D. (1997). *Cell*, **90**, 207.
35. Reijo, R. A., Cooper, E. M., Beagle, G. J., and Huffaker, T. C. (1994). *Mol. Biol. Cell*, **5**, 29.
36. Richards, K., Nogales, E., Anders, K., Scaife, C., Downing, K. H., Schwartz, K., *et al.* (in preparation).
37. Feierbach, B., Nogales, E., Downing, K. H., and Stearns, T. (1999). *J. Cell Biol.*, **144**, 113.

9

Regulation of intermediate filament dynamics: a novel approach using site- and phosphorylation state–specific antibodies

HIROYASU INADA, KOH-ICHI NAGATA, HIDEMASA GOTO, and MASAKI INAGAKI

1. Introduction

Intermediate filaments (IFs) are major components of the cytoskeleton and nuclear envelope in most types of eukaryotic cells (1, 2). On the basis of primary sequence homology, cell- and tissue-specific expression, or the exon-intron structure of the genes encoding each subunit, one can classify all IF proteins as one of the following (3, 4): Type I, acidic keratins in epithelia; Type II, neutral/basic keratins in epithelia; Type III, desmin in muscle cells, vimentin in mesenchymal cells and most types of cultured cells, glial fibrillary acidic protein (GFAP) in glial cells, and peripherin in peripheral neurons; Type IV, neurofilament (NF) subunits such as NF-L, NF-M, and NF-H in neurons, and α-internexin in the central nervous system; Type V, lamins which assemble into a nuclear lamina in all eukaryotic cells; Type VI, nestin in neuroectodermal stem cells.

Although IFs were thought to be relatively stable compared to other cytoskeletons such as microtubules and actin filaments, it is now widely accepted that IFs undergo dynamic reorganization of filament structures through site-specific phosphorylation. We, for the first time, proposed the notion that site-specific phosphorylation induces disassembly of IF using vimentin filament *in vitro* (5). Thereafter, it has become increasingly evident that site-specific phosphorylation of IF proteins alters filament structure. Several kinases are apparently responsible for IF phosphorylation *in vitro* (6–9).

We next wanted to determine the physiological significance of IF protein phosphorylation *in vivo* and to monitor *in vivo* phosphorylation states of various IF proteins. Therefore, we developed specific antibodies that recognize the phosphorylated, but not non-phosphorylated, amino acid residues of IF proteins (10, 11). Using a series of these antibodies, the spatiotemporal phosphorylation states of various IF proteins could be visualized. This new method demonstrates another important advantage, that antibodies can

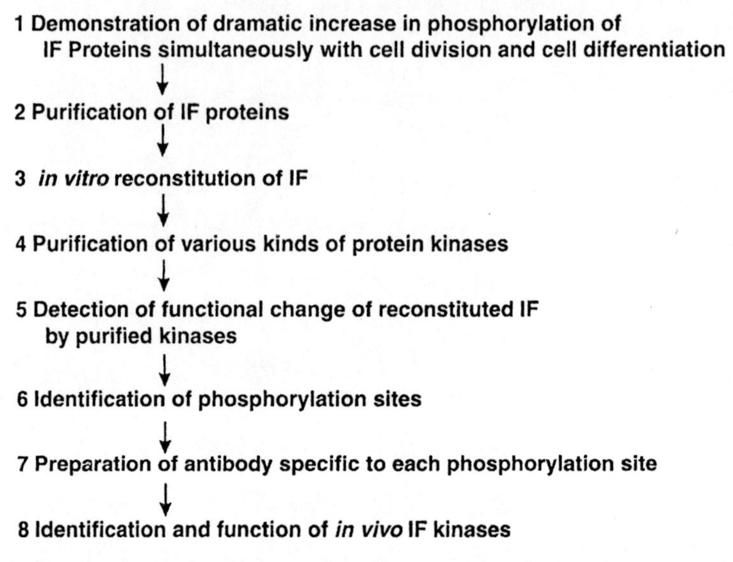

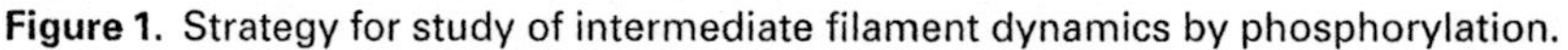

Figure 1. Strategy for study of intermediate filament dynamics by phosphorylation.

monitor the localized kinase activity responsible for IF phosphorylation on a specific serine and/or threonine residue. Information on the subcellular distribution of protein kinases has been obtained using immunocytochemistry with antibodies against enzymes, or biochemical methods with subcellular fractionation. However, neither method is completely satisfactory for determining the intracellular organization of kinase activity. Use of an immunocytochemical approach facilitated depiction of the distribution of enzymes. However, co-localization by immunochemical methods does not always mean that the catalytic activity is co-localized, since kinases often require cofactors or modification for activation. On the other hand, biochemical fractionation paved the way in examining enzyme activity *in vitro*. However, the activity *in vivo* may differ, since it is frequently difficult to isolate intact organelles, mainly due to technical limitations (see Chapter 4 for a further discussion). Development of the site- and phosphorylation state-specific antibody has overcome these difficulties and with use of this new tool the multi-IF kinases were seen to be activated at different time schedules during the cell cycle. Moreover, specific phosphorylation of IF proteins by various kinases showed spatially different distribution patterns in mitotic cells (10, 12, 13). Analysis using the site- and phosphorylation state-specific antibodies we developed is now an established approach, and production of antibodies against such phosphoproteins is now feasible (7).

In this chapter we describe in detail procedures for *in vitro* and *in vivo* studies done to analyse IF function through phosphorylation. Emphasis will be placed on protocols which we routinely use in our laboratory. These procedures, although established for IF proteins, should be equally applicable to other proteins. *Figure 1* shows the strategy we used to identify phosphorylation-related intermediate filament dynamics.

2. Regulation of IFs by site-specific phosphorylation *in vitro*

The first direct evidence supporting IF regulation by phosphorylation was obtained for vimentin in *in vitro* studies (5). Vimentin filaments reconstituted *in vitro* undergo complete disassembly when phosphorylated by purified protein kinase A (A kinase) or protein kinase C (C kinase). Subsequently, a similar *in vitro* disassembly induced by phosphorylation was noted for almost all major IF proteins, such as vimentin (5, 14–24), GFAP (20, 21), desmin (14, 22, 23, 25), keratin (26, 27), α-internexin (28), NF-L (29–33), and lamin (34–39). Interestingly, analysis of tryptic phosphopeptide maps showed that sites of phosphorylation by protein kinases differ, which in turn suggests that phosphorylation by protein kinases is site-specific.

IF proteins are composed of an amino-terminal head, a central rod, and carboxyl-terminal tail regions (2). Phosphorylation sites responsible for the disassembly of vimentin (18, 24, 40–43), GFAP (20, 21, 44), desmin (22, 23, 25, 45), keratin 8 (46), and NF-L (31, 33, 47) are located in the head domain, not in the rod or tail domain. Together with data obtained by mutagenesis analysis that the head domain plays an important role in IF assembly (48–51), these results indicate that phosphorylation of the head domain is responsible for disassembly of these IFs. In this section, we describe one approach to detect IF phosphorylation sites *in vitro*.

2.1 Chromatographic purification of IF proteins

As a first step for *in vitro* studies, we obtain purified IF proteins. Since IFs are among the most insoluble structures of the cytoskeleton, purification and maintenance of IF proteins in a soluble state under non-denaturing conditions can be difficult (52). To purify an IF protein from insoluble cytoskeletal fraction or a bacterial inclusion body preparation, we use urea as a denaturant followed by two steps of chromatography. A procedure for isolating vimentin from inclusion bodies of *E. coli* is given in *Protocol 1*. We use pET vector (Stratagene) as an expression system in *E. coli* that is selectable by ampicillin resistance and inducible by isopropyl β-D-thiogalactoside (IPTG). This method can be used for purification of other IF proteins.

Protocol 1. General purification protocol of vimentin expressed in *E. coli*

Reagents

- LB medium: 20 g tryptone, 10 g yeast extract, 20 g NaCl per 2000 ml
- Lysis buffer: 0.6 M KCl, 1% (v/v) 2-mercaptoethanol (2-ME), 1% (v/v) Triton X-100, 10 mM $MgCl_2$, 2 mM EGTA, 2 mM phenylmethylsulfonyl fluoride (PMSF), 1 μg/ml leupeptin in PBS
- Buffer A: 2 mM EGTA, 1% 2-ME, 2 mM PMSF, 1 μg/ml leupeptin in PBS
- Buffer B: 25 mM Tris–HCl pH 7.5, 8 M urea, 2 mM EGTA, 100 mM 2-ME, 1 mM PMSF
- Buffer C: 25 mM Tris–HCl pH 7.5, 7 M urea, 2 mM EGTA, 100 mM 2-ME, 1 mM PMSF
- Buffer D: 20 mM formic acid pH 4.0, 7 M urea, 2 mM EGTA, 100 mM 2-ME, 1 mM PMSF
- Buffer E: 10 mM Tris–HCl pH 8.8, 7 M urea, 2 mM EGTA, 50 mM 2-ME, 1 mM PMSF
- Buffer F: 10 mM Tris–HCl pH 8.8, 2 mM EGTA, 50 mM 2-ME, 1 mM PMSF

Method

1. Inoculate 50 ml LB medium containing 100 μg/ml ampicillin with a single colony from an agar culture plate. Incubate overnight at 37 °C with vigorous shaking.
2. Inoculate 2000 ml LB medium containing 100 μg/ml ampicillin with 20 ml of the 50 ml overnight culture described above. Incubate at 37 °C with vigorous shaking until the culture reaches 0.6–0.7 OD 600 nm.
3. Induce protein expression by adding 2 ml of 1 M IPTG (final concentration is 1 mM) and incubate at 37 °C for 4 h with vigorous shaking.
4. Harvest cells by centrifugation at 8000 *g* for 10 min at 4 °C, and decant the supernatant. Freeze in dry ice for at least 15 min. Cells can be stored at –80 °C for weeks.
5. Thaw rapidly and completely resuspend the cells in 500 ml of lysis buffer. Sonicate the cells on ice until the solution is no longer viscous (e.g. 1 min burst for five times, on ice).
6. Centrifuge the homogenate at 8000 *g* for 30 min at 4 °C, and discard the supernatant.
7. Wash the resulting pellet in 100 ml of buffer A and centrifuge at 8000 *g* for 30 min at 4 °C. Repeat this step three times. The pellets can be resuspended easily after removal from the tubes by dispersing them with several strokes of a glass homogenizer. The supernatant should be clear after the last wash.
8. After solubilization of the inclusion body pellet with 50 ml of buffer B, stirring overnight at room temperature, dialyse the sample in buffer C at 4 °C. All the purification procedures described below are performed at 0–4 °C.

9. Centrifuge the sample at 100 000 *g* for 60 min at 4°C, and collect the supernatant.
10. Apply the urea-solubilized sample to a DEAE–cellulose column (2.0 × 20 cm) equilibrated with buffer C. Wash the column with three column volumes of the same buffer, at a flow rate of 25 ml/h.
11. Elute with a linear gradient between 0–250 mM NaCl (total volume, 600 ml) and collect 6 ml fractions. Analyse vimentin content of the fraction using SDS–PAGE. Vimentin was eluted at around 70 mM NaCl.
12. Pool the vimentin-rich fractions and dialyse against buffer D.
13. Apply the sample to a CM–cellulose column (2.0 × 20 cm) equilibrated with buffer D. Wash the column with three column volumes of the same buffer, at a flow rate of 20 ml/h.
14. Elute with a linear gradient between 0–300 mM NaCl (total volume, 600 ml) and collect 6 ml fractions. Analyse the vimentin content of the fraction using SDS–PAGE and determine the vimentin concentration of each fraction by the method of Bradford using bovine serum albumin (BSA) as a standard. Vimentin was eluted at around 150 mM NaCl.
15. Pool the most enriched vimentin fractions (at least 2–3 mg/ml) and dialyse overnight against 2000 ml of buffer E at 4°C with one buffer change.
16. Determine the protein concentration of vimentin again and adjust it to 2.5–3 mg/ml. The yield of recombinant vimentin as determined by Bradford's method is in the order of 50–100 mg/2000 ml of bacterial culture.
17. The final protein preparation is stored in aliquots, snap frozen in liquid N_2, and kept at –80 °C until use.
18. Before use, dialyse the preparation extensively to remove urea. Thaw an aliquot rapidly and dialyse overnight against buffer F at 4°C. Collect soluble vimentin by centrifugation at 15 000 r.p.m. in a microcentrifuge for 60 min at 4 °C. Determine the soluble vimentin concentration again by the method of Bradford using BSA as a standard.

2.2 Phosphorylation assay

In vitro phosphorylation assay can be useful to verify the putative *in vivo* IF kinases. Before the stoichiometric phosphorylation experiments, we recommend determining the optimal conditions for phosphorylation of purified IF protein by each protein kinase. IF molecules can form an unusual ribbon-like structure (wider than 10 nm) at a lower pH or at a high concentration of divalent cations (such as Mg^{2+} or Ca^{2+}) *in vitro*. The rate of IF phosphorylation was found to be reduced under these conditions (5, 53). Thus, we recommend that IF protein phosphorylation reactions are performed at pH 7.0–7.5 in a lower concentration of Mg^{2+} (< 3 mM).

Assembly and morphology of IF under *in vitro* conditions were found to be affected by IF protein concentration, ionic strength, and temperature (53). Thus, phosphorylation reactions for soluble IF protein are recommended to be performed under the following conditions:

(a) IF protein concentration: 100–200 μg/ml.

(b) NaCl concentration: 0–30 mM.

(c) Temperature: around 25 °C.

For example, vimentin (150 μg/ml) was phosphorylated by incubation with 0.5 μg/ml of cdc2 kinase, 0.1 mM [γ-^{32}P]ATP, 1.5 mM $MgCl_2$, 25 mM Tris–HCl pH 7.5, at 25 °C, or by incubation with 5 μg/ml of the catalytic subunit of protein kinase A, 0.1 mM [γ-^{32}P]ATP, 0.3 mM $MgCl_2$, 25 mM Tris–HCl pH 7.5, at 25 °C, or by incubation with 5 μg/ml of protein kinase C, 0.1 mM [γ-^{32}P]ATP, 0.3 mM $MgCl_2$, 50 μg/ml phosphatidylserine, 2.5 μg/ml diacylglycerol, 25 mM Tris–HCl pH 7.0, at 25 °C, or by incubation with 5 μg/ml of CaM kinase II, 0.1 mM [γ-^{32}P]ATP, 0.1 mM $CaCl_2$, 40 μg/ml CaM, 3 mM $MgCl_2$, 25 mM Tris–HCl pH 7.5, at 25 °C, or by incubation with 9 μg/ml of GST-Rho-kinase, 0.1 mM [γ-^{32}P]ATP, 0.4 mM $MgCl_2$, 25 mM Tris–HCl pH 7.5, at 25 °C.

When phosphorylation reactions using filaments prepared *in vitro* are performed, the following procedure is recommended:

(a) Soluble IF protein (protein concentration < 1 mg/ml) is polymerized by incubation in 25 mM Tris–HCl pH 7.0–7.5, containing 50–150 mM NaCl at 37 °C for 30–60 min, or at 4 °C for 24–48 h.

(b) The prepared polymerized sample is ready for phosphorylation assay, under optimal conditions. The concentration of IF protein, divalent cations, and the pH are important (see above).

Then, check to see if the purified IF protein is phosphorylated by each protein kinase, at a reasonable rate and stoichiometry, under the determined conditions.

2.3 Effect of phosphorylation on IF filaments structure *in vitro*

Under physiological conditions, the majority of IF proteins, e.g. GFAP and vimentin, form 10 nm filaments *in vitro* (54), which are considered to be assembled by the following steps (55–63):

(a) Two IF molecules interact to form a parallel α-helical coiled-coil dimer structure.

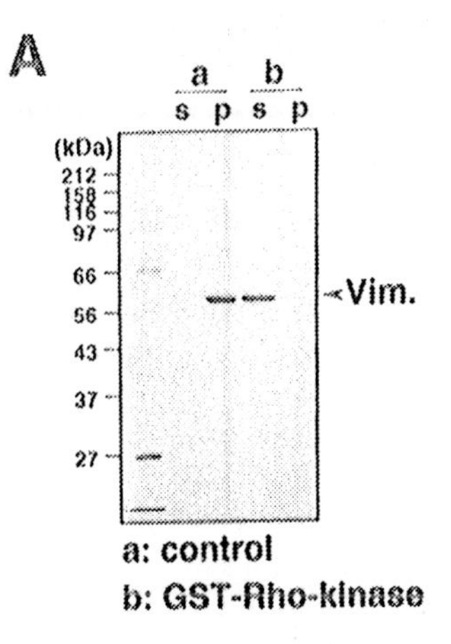

a. Control

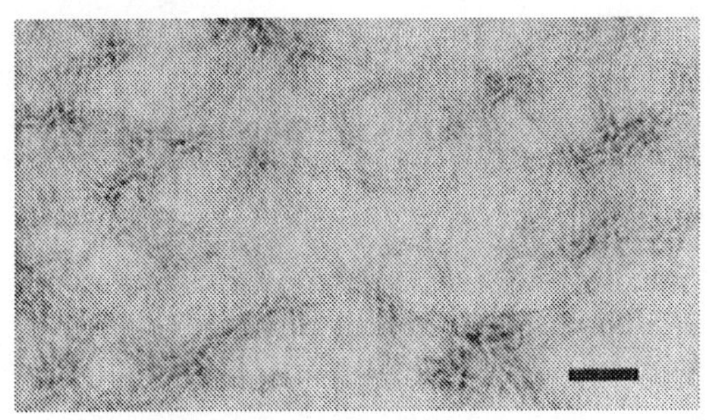

b. GST-Rho-kinase

Figure 2. Effect of phosphorylation by GST-Rho-kinase on the filament-forming potential of vimentin. (A) Fractionation of soluble vimentin and vimentin filaments. After vimentin was phosphorylated with (b) or without (a) GST-Rho-kinase for 60 min at 25°C, the samples were incubated with 100 mM NaCl for further 60 min at 37°C and subjected to high speed centrifugation (12 000 *g*). The supernatant (s) and the precipitate (p) were then subjected to SDS–PAGE. (B) Electron microscopy of negatively stained vimentin samples was performed. (Modified with permission, from ref. 24.)

(b) Two dimers form antiparallel protofilaments as a tetramer which characterizes the non-polarized structure.

(c) The protofilaments assemble into a 10 nm filament structure.

In the case of GFAP, vimentin, and desmin, the head domain is positively charged due to numerous arginine residues. This positive charge may be important for assembly of such IF proteins. Regulation of IF assembly by phosphorylation at the head domain may relate to regulation of the electric charge in the domain (53, 64, 65).

An *in vitro* method of IF reconstitution by changing ionic strength (pH 7.0–7.5) (66, 67) and analytical methods of examining IFs by centrifugation and by negative stain electron microscopy are illustrated in *Protocol 2*. The objective is to analyse the effect of vimentin phosphorylation on filament formation. *Figure 2* shows effects of vimentin phosphorylation on filament-forming ability.

Protocol 2. *In vitro* reconstitution of IF filaments by changing ionic strength

Equipment and reagents

- 3MM paper (Whatman)
- 0.4% (w/v) sucrose
- 2% (w/v) uranyl acetate

A. *In vitro reconstitution*

1. Incubate soluble vimentin, with or without vimentin kinase, under optimal conditions for phosphorylation.
2. Add 1 M NaCl (final concentration is 100–150 mM). Incubate at 37 °C for 60 min.

B. *Analysis of IF filament formation with centrifugation*

1. Centrifuge the sample (see *Protocol* 2A) at 12 000 *g* for 60 min at 4 °C. Separate and collect the supernatant and the precipitate.
2. Subject to SDS–PAGE. Polymerized IF protein (IF filaments) is detected in the precipitate.

C. *Analysis of IF filament formation with electron microscopy (negative staining)*

All procedures are performed at room temperature.

1. Place one drop of the samples (see *Protocol* 2A) onto carbon-coated mesh.
2. Wait for about 10 sec and carefully remove the solution using 3MM paper.
3. Place one drop of 0.4% sucrose.
4. Wait for about 10 sec and carefully remove the solution using 3MM paper.

5. Place one drop of 2% uranyl acetate.
6. Wait for about 10 sec and carefully remove the solution using 3MM paper.
7. Dry at room temperature and analyse by electron microscopy. These samples can be stored for one week at room temperature.

2.4 Detection of phosphorylation sites

Site-specific phosphorylation of IF protein(s) on serine and threonine leads to alteration of the filament structure, both *in vitro* and *in vivo*. Therefore, identification of phosphorylated amino acid residue(s) within the IF molecule during cellular events is of great importance to understand mechanisms by which cellular IF reorganization is regulated. As a first step, we attempted to identify *in vitro* phosphorylation sites of IF proteins by various protein kinases.

Most studies done to examine phosphorylation sites of proteins utilize [^{32}P]phosphate as a readily detectable reporter group. Although a series of non-radioactive methods has been developed (68), we describe here the method using [^{32}P]phosphate.

The phosphorylation assay is performed in the presence of [γ-^{32}P]ATP. Phosphorylated IF protein is first treated with an appropriate protease. The fragmented peptides are separated by chromatography. We often use reverse-phase and/or ion exchange high-performance liquid chromatography (HPLC) to separate the peptides (see *Protocol 3*).

(a) Determine the amino acid sequence of each radioactive peptide and compare it with the reported amino acid sequence of the IF protein.

(b) Determine the phosphorylated residue(s) of each peptide by phospho-amino acid analysis (69). Phosphorylation site(s) can be determined if each peptide has a single serine or threonine residue.

(c) If a peptide has multiple serine or threonine residues, amino acid sequence analysis after ethanethiol treatment is useful, since this treatment specifically converts phosphoserine and phosphothreonine into *S*-ethylcysteine and β-methyl-*S*-ethylcysteine, respectively (68, 70). Mutational analysis is also useful for identifying of phosphorylation sites.

Protocol 3. Fragmentation of phosphorylated vimentin

Reagents

- 5% and 30% (w/v) trichloroacetic acid (TCA)
- Buffer A: 100 mM Tris–HCl pH 8.0, 8 M urea
- Lysyl-endopeptidase (Wako Pure Chemical)
- L-1-tosylamide-2-phenyl-ethyl chloromethyl ketone (TPCK)-treated trypsin (Sigma)

Protocol 3. *Continued*

Method

1. Phosphorylate vimentin (150 μg in 1 ml reaction mixture) with [γ-^{32}P]ATP at the proper stoichiometry, under optimal conditions for the kinase used.
2. Remove non-reactive [γ-^{32}P]ATP by precipitating vimentin with TCA. Add 500 μl of 30% ice-cold TCA into 1 ml of the reaction mixture (final concentration is 10%) and mix immediately. Place the tube on ice for at least 1 h.
3. Collect the TCA precipitate by centrifugation at 8000 *g* in a microcentrifuge for 10 min at 4°C. Carefully remove the supernatant (the pellet may be visible at this stage).
4. Wash the precipitate with 1 ml of 5% ice-cold TCA and centrifuge again at 8000 *g* in a microcentrifuge for 5 min at 4°C. Carefully remove the supernatant.
5. Wash the precipitate with 1 ml of diethyl ether and centrifuge again at 8000 *g* in a microcentrifuge for 5 min at 4°C. Carefully remove the supernatant. Repeat this step twice.
6. Air dry (do not lyophilize) the pellet, which can be stored at –20°C.
7. Dissolve the precipitate with 50 μl of buffer A by gentle shaking for 1–2 h at room temperature.
8. Add 50 μl of dH_2O to dilute (final 50 mM Tris–HCl pH 8.0, 4 M urea).
9. Incubate with 5 μg lysyl-endopeptidase for 2 h at 30°C.
10. After incubation, centrifuge at 8000 *g* in a microcentrifuge for 30 min at 4°C. Then, transfer the supernatant (no pellets) to a fresh tube.
11. Add 1 ml of 0.1% trifluoroacetic acid (TFA) and filter the sample with a 0.2 μm filter. Subject to reverse-phase HPLC. The digested sample is usually fractionated on a Zorbax C8 column equilibrated with 5% (v/v) 2-propanol:acetonitrile (7:3) containing 0.1% TFA. In our laboratory, peptides are eluted with a 60 min linear gradient of 5–50% (v/v) 2-propanol:acetonitrile at a flow rate of 0.8 ml/min, followed by a further 10 min linear gradient of 50–80% (v/v) 2-propanol:acetonitrile at a flow rate of 0.8 ml/min. Since the head domain of vimentin contains no lysine residues, digestion of vimentin with lysyl-endopeptidase generates about a 12 kDa fragment (consisting mainly of the head domain) and smaller ones. Thus, a single radioactive peak is obtained.
12. Lyophilize the radioactive fraction(s), which can be stored at –20°C. Resuspend the radioactive head domain of vimentin in 400 μl of 50 mM Tris–HCl pH 7.5. Treat with 1:50 (w/w) TPCK-treated trypsin for 8 h at 37°C. Incubate with TPCK-treated trypsin identically for an additional 8 h.

13. Centrifuge at 8000 *g* in a microcentrifuge for 30 min at 4°C. Transfer the preparation (no pellets) to a fresh tube.
14. Add 1 ml of 0.1% TFA and pass the sample through a 0.2 μm filter. Then perform reverse-phase HPLC.
15. Lyophilize each radioactive peptide, which can be stored at –20°C. Each peptide is used for amino acid sequence analysis and phosphoamino acid analysis.

3. Cell biology of IFs

Monitoring *in vivo* IF phosphorylation is important for a better understanding of IF reorganization in cells. Labelling of cells with radioactive phosphate has been a widely used strategy to monitor *in vivo* IF phosphorylation (8). However, using this method it is difficult to determine the spatiotemporal distribution of phosphorylated IF proteins in cells.

It is also of great importance to identify protein kinase(s) responsible for *in vivo* IF phosphorylation. *In vitro* phosphorylation studies revealed putative IF kinases. However, not all of these kinases can phosphorylate IFs *in vivo*. It has been reported that treatment with protein kinase activators, such as dibutyryl cAMP for A kinase, phorbol ester for C kinase, and Ca^{2+} ionophore for CaM kinase II, leads to cellular IF phosphorylation, but the possibility that other protein kinases are activated directly or indirectly by these stimuli must be determined. Therefore, it is essential to determine whether the phosphorylation sites of an IF protein in stimulated cells are the same as the sites phosphorylated *in vitro* by purified kinases using two-dimensional analysis of tryptic phosphopeptides (17, 71). However, this method is laborious, and it is not always easy to obtain sufficient consistent data. Indeed, the phosphopeptide analysis provides clear evidence only when additional kinases are not involved in IF phosphorylation *in vivo*.

To overcome these difficulties, we developed site- and phosphorylation state-specific antibodies for IF proteins, using phosphorylated peptides or synthetic phosphopeptides as antigens for immunization (10). Immunocytochemical analysis using these antibodies has the advantage that one can monitor the spatiotemporal IF phosphorylation in a single cell. These antibodies can also be utilized as an alternative method for identification of *in vivo* IF kinase(s) (7).

3.1 Preparation of site- and phosphorylation state-specific antibodies for phosphopeptides

3.1.1 Design of site-specific antibodies for phosphopeptides

We were the first to use a phosphopeptide as an antigen for the production of an antibody that recognizes phosphorylation of a protein at a specific site (10,

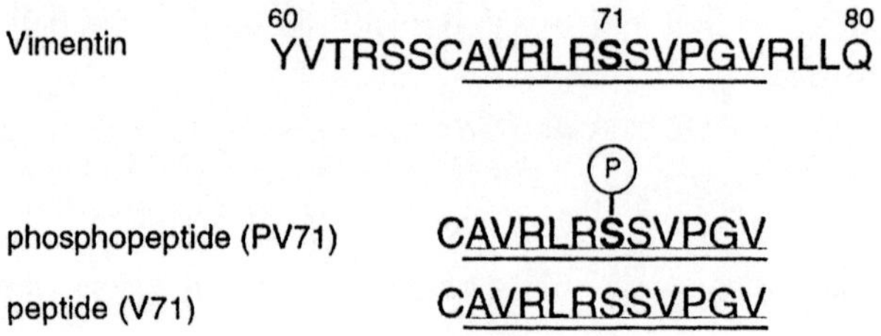

Vimentin 60 71 80 YVTRSSCAVRLRSSVPGVRLLQ

phosphopeptide (PV71) CAVRLRSSVPGV (P)

peptide (V71) CAVRLRSSVPGV

Figure 3. Design of site- and phosphorylation state-specific antibody (GK71) for serine 71 of vimentin.

11). The phosphopeptides containing phosphoserine or phosphothreonine residues had been identified as phosphorylation sites by *in vitro* studies. The production of a site- and phosphorylation state-specific antibody by a phosphopeptide has the advantage that the phosphorylation site has been previously recognized as a possible epitope (7, 10, 11, 72–94).

The following is a brief description of methods we use to prepare site- and phosphorylation state-specific antibodies. Since five or six amino acid residues constitute an antigen epitope recognized by an antibody molecule, a peptide consisting of a phosphoserine or phosphothreonine and its flanking sequences of five amino acids (total 11 amino acids) is designed. We introduced a cysteine residue at the N- or C-terminal of the synthetic peptide and coupled it to the carrier protein, keyhole limpet haemocyanin (KLH), using maleimidobenzoic acid *N*-hydroxysuccinimide ester (MBS). KLH, one of the most commonly used carrier proteins, is effective for the antigen presentation required for antibody production (95). A synthetic peptide designed to elicit a site- and phosphorylation state-specific antibody against serine 71 of vimentin is shown in *Figure 3*.

3.1.2 Production and purification of site- and phosphorylation state-specific antibodies

We used two adjuvant systems; Freund complete/incomplete adjuvant system (Difco Laboratories) and RIBI adjuvant system (RIBI ImmunoChem Res. Inc.) for production of antibodies. We show here a protocol using the RIBI adjuvant system. See manufacturers' protocols or references for other adjuvant systems. Additionally, for immunization (*Protocol 4*) and purification (*Protocol 5*) of site- and phosphorylation state-specific antibodies, the purity of synthetic peptides is critical.

Protocol 4. Polyclonal antibody production

Equipment and reagents

- RIBI adjuvant (RIBI Immunochem. Res. Inc.)
- 0.22 μm Millex-GP filter (Millipore)
- Synthetic phosphopeptide: KLH conjugated, as an antigen (Peptide Institute Inc.)

Method

1. Obtain pre-immune serum (about 1 ml) from each rabbit prior to the initial injection.
2. Immunize rabbits with 1 ml total (0.05 ml intradermal in six sites, 0.3 ml intramuscular into each hind leg and 0.1 ml subcutaneous in neck region) of peptide-conjugate (100–200 μg peptide) in RIBI adjuvant.
3. For booster injections of the conjugate (100–200 μg peptide), give RIBI adjuvant four weeks after first injections.
4. After 10–14 days from the first boost, obtain serum samples (about 50 ml) from each rabbit. Incubate samples at 37 °C for 1 h and overnight at 4 °C. Centrifuge the samples at 1000 g for 30 min at 4 °C and collect the supernatants. Incubate at 56 °C for 30 min and pass through a 0.22 μm filter. Store in aliquots at –80 °C.
5. Purify (see *Protocol 6*) and screen the samples, using ELISA (see *Protocol 8*). Positive sera that appear to be specific for the phosphopeptide can be identified.
6. If the serum sample is positive, a second booster injection of the conjugate (100–200 μg peptide) is given and serum can thus be obtained after another two weeks.

Various methods can be used to purity antibodies (72, 95, 96). Since site- and phosphorylation state-specific antibodies require strong specificity, we use an immunoaffinity method for antibody purification. The major advantage of this method is its unique potential to isolate specific antibodies from a mixed pool. Advantages and disadvantages of immunoaffinity purification are as follows:

Advantages:

- yields pure antibody
- yields specific antibody

Disadvantages:

- expensive
- multiple steps
- requires pure antigen
- may inactivate antibody

A summary of the immunoaffinity purification methods we use is as follows. In the first step, phospho- and non-phosphopeptides are covalently attached to a solid phase matrix. There are many different protocols for covalently binding antigens to a solid phase. We use FMP (2-fluoro-1-methylpyridinium toluene-4-sulfonate)-activated cellurofine. After the preparation of phospho- and non-phosphopeptides-matrix, the mixed antibody pool is bound to the phosphopeptide-matrix and contaminating antibodies are removed by washing. In the next step, the antibody is released into the eluate, using stringent elution conditions. Next, the antibody is bound to the non-phosphopeptide-matrix. Finally, the site- and phosphorylation state-specific antibody is collected in flow-through fractions.

Protocol 5. Preparation of affinity matrix

Equipment and reagents

- FMP-activated cellurofine (Seikagaku Co.)
- Synthetic phosphopeptide (Peptide Institute Inc.)
- Synthetic non-phosphopeptide (Peptide Institute Inc.)
- Coupling buffer: 50 mM $NaHCO_3$–Na_2CO_3 pH 8.0–10.0
- Blocking buffer: 50 mM Tris–HCl pH 8.0, 0.1 M ethanolamine
- Elution buffer: 0.1 M glycine–HCl pH 2.5, 10% (v/v) ethylene glycol
- Wash buffer: 20 mM Tris–HCl pH 7.5, 1 M NaCl, 1% (v/v) Triton X-100
- TBS: 20 mM Tris–HCl pH 7.5, 0.15 M NaCl
- Stock buffer: 20 mM Tris–HCl pH 7.5, 0.15 M NaCl, 20 mM β-glycerophosphate, 0.2% (w/v) NaN_3

Method

1. Prepare a solution of antigen (phosphorylated and non-phosphorylated peptides, about 2 mg) in 10 ml coupling buffer in a 15 ml tube. Check the pH of the solution, which must be 8.0–10.0.
2. Add 2 ml of the FMP-activated cellurofine.
3. Mix gently overnight at 4°C on a rocker or shaker.
4. Wash with 10 ml coupling buffer by invert mixing. Centrifuge at 150 g for 2 min at 4°C and remove the supernatant carefully. Repeat this step twice.
5. Wash with 10 ml blocking buffer by invert mixing. Centrifuge at 150 g for 2 min at 4°C and remove the supernatant carefully. Repeat this step twice.
6. Add 10 ml blocking buffer and mix gently overnight at 4°C on a rocker or shaker.
7. Wash with 10 ml dH_2O by invert mixing. Centrifuge at 150 g for 2 min at 4°C and remove the supernatant carefully. Repeat this step twice.
8. Wash with 10 ml elution buffer by invert mixing. Centrifuge at 150 g for 2 min at 4°C and remove the supernatant carefully. Repeat this step twice.
9. Wash with 10 ml dH_2O by invert mixing. Centrifuge at 150 g for 2 min at 4°C and remove the supernatant carefully. Repeat this step twice.
10. Wash with 10 ml wash buffer by invert mixing. Centrifuge at 150 g for

2 min at 4°C and remove the supernatant carefully. Repeat this step twice.

11. Wash with 10 ml TBS by invert mixing. Centrifuge at 150 g for 2 min at 4°C and remove the supernatant carefully. Repeat this step twice. Separate the matrix into 1 ml fractions. The antigen–matrix can be stored at 4°C for several months. If the matrix is stored for further use, equilibrate it with stock buffer.

Protocol 6. Purification of antibody

Reagents

- TBS (see *Protocol 5*)
- Wash buffer (see *Protocol 5*)
- Elution buffer (see *Protocol 5*)
- Neutralizing buffer: 1 M Tris–HCl pH 8.5
- Buffer A: 20 mg/ml BSA in TBS
- Stock buffer (see *Protocol 5*)

Method

1. Mix 5 ml of the serum (see *Protocol 4*) and 5 ml of TBS with the phosphorylated peptide–matrix (see *Protocol 5*). Each step described below is performed at 4°C.
2. Agitate gently overnight at 4°C, which should be sufficient to keep the matrix in suspension.
3. Transfer the matrix to a suitable column (for example, Muromac® mini column, Muromachi Kagaku Kogyo Kaisha, Ltd.).
4. Wash the column with 10 ml of ice-cold TBS.
5. Wash the column with 10 ml of ice-cold wash buffer.
6. Wash the column with 10 ml of ice-cold TBS.
7. Elute with elution buffer using a stepwise elution (500 μl each, about five times).
8. Collect each fraction in a separate tube, containing 37.5 μl of neutralizing buffer. Mix the eluates gently and constantly.
9. Check each tube for the presence of the antibody using SDS–PAGE.
10. Collect the antibody-rich fraction (about two fractions, 1 ml) and add 50 μl of buffer A.
11. Dialyse for 4 h or more against two changes of 200 ml TBS at 4°C.
12. Mix the antibody with 1 ml of the non-phosphorylated peptide–matrix (see *Protocol 5*) in a 15 ml tube.
13. Agitate gently overnight at 4°C, which should be sufficient to keep the matrix in suspension.
14. Transfer the matrix to a suitable column (e.g. Muromac® mini column).

Protocol 6. *Continued*

15. Collect the flow-through. Wash the matrix with 500 μl of TBS and collect the flow-through fraction which contains the site- and phosphorylation state-specific antibody.
16. To store for further use, wash columns (phosphorylated and non-phosphorylated peptides–matrix) with ice-cold TBS, wash, and elution buffers. Wash with ice-cold TBS and equilibrate with ice-cold stock buffer. Store at 4 °C.

After this step of purification, we first perform an enzyme-linked immunosorbent assay (ELISA) to screen for site- and phosphorylation state-specific antibody production. The ELISA is a simple and sensitive assay (96). Microtitre plates are coated with the phosphorylated peptide and non-phosphorylated peptide. Antibodies are selected for further analysis. *Figures 4A* and *4B* show the specificity of site- and phosphorylation state-specific antibody (GK71) against phospho- and non-phosphopeptide by ELISA. In *Figure 4C*, the specificity of GK71 was also examined by Western blot analysis. GK71 reacted with vimentin phosphorylated by Rho-associated kinase but not with non-phosphorylated vimentin or vimentin phosphorylated by protein kinase A, protein kinase C, cdc2 kinase, or CaM kinase II (*Figure 4C*).

Protocol 7. Preparation of plates for ELISA

Reagents

- Buffer A: 0.1 M Na_2HPO_4–NaH_2PO_4 pH 7.4
- Blocking buffer: 10 mM Na_2HPO_4–NaH_2PO_4 pH 8.0, 5% (w/v) BSA, 5% (w/v) sucrose, 0.1% (w/v) NaN_3

Method

1. Dilute peptides to 1 μg/ml with buffer A. Check pH of the mixture, which must be 7.0–7.5.
2. Dispense 60–70 μl into each well of a Corning 96-well microtitre plate and incubate overnight at 4 °C.
3. Remove contents of the wells and wash each well four times with 200–250 μl of PBS.
4. Dispense 300–350 μl of blocking buffer into each well and incubate the preparation overnight at 4 °C or for 2–4 h at 37 °C.
5. Dry the wells for 1–2 h at room temperature and store the plates at 4 °C.

Protocol 8. Examination of specificity of purified antibody using ELISA

Reagents

- Buffer A: 1% (w/v) BSA, 1% (w/v) sucrose, 0.1% (w/v) NaN_3 in PBS
- Primary antibody: site- and phosphorylation state-specific antibody (obtained by *Protocol 6*)
- Reaction solution: 4 mg *o*-phenylenediamine, 500 μl methanol, 9.5 ml dH_2O, 3.4 μl H_2O_2, freshly prepared
- Buffer B: 10 mM Na_2HPO_4–NaH_2PO_4 pH 8.0, 0.1 M NaCl, 1% (w/v) BSA, 0.1% (w/v) *p*-hydroxy-phenylacetic acid, 0.025% (w/v) thimerosal
- Secondary antibody: HRP (horseradish peroxidase)-conjugated anti-rabbit Ig antibody (Sigma)

Method

1. Dilute primary antibody 10- to 10 000-fold with buffer A.
2. Dispense 50 μl of each antibody sample into wells and incubate for at least 45 min at 37 °C.
3. Remove contents of the wells and wash five times with 100–150 μl of PBS.
4. Dispense 100 μl of a 1000-fold dilution of the secondary antibody in buffer B. Incubate for at least 45 min at 37 °C.
5. Remove contents of the wells and wash five times with 100–150 μl of PBS.
6. Add 100 μl of reaction solution and incubate 10–30 min at room temperature.
7. Stop the reaction with 100 μl of 2 M H_2SO_4 and measure the colour development at 492 nm.

3.1.3 Preparation of monoclonal antibodies for phosphopeptide

Site- and phosphorylation state-specific monoclonal antibodies are prepared using standard techniques (95, 96). We use BDF1[(C57BL/6 × DBA2)F1] mice for immunization because we find that F1 hybrid mice produce a larger amount of antibodies against IF phosphopeptides than do other strains.

3.2 Visualization of phosphorylation state of IF proteins and identification of IF kinases *in vivo*

The question arises as to why the phosphopeptides of IF readily elicit antibodies specific not only for the phosphopeptide but also for the native phosphoprotein. Our reasoning is as follows. The site of phosphorylation is mainly located in the head domain, which is essential for filament formation. This domain has the same turn structure found in the synthetic peptide/phosphopeptide. Such structural homology ensures that an antibody against

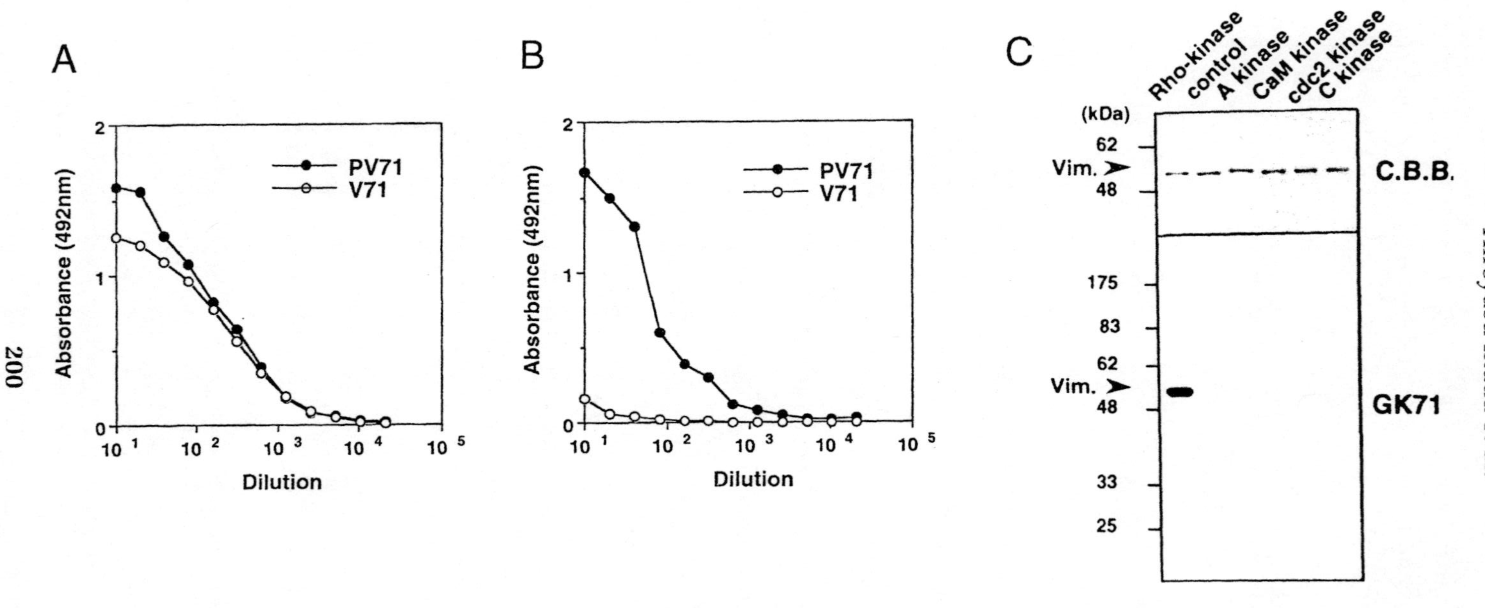

Figure 4. Specificity of site- and phosphorylation state-specific antibody (GK71). Specificity of unpurified (antisera, A) and purified (B) GK71 for phospho- (PV71) and non-phosphopeptide (V71) by ELISA. (C) Specificity of GK71 analysed by Western blotting. Vimentin (Vim) was non-phosphorylated (control) or phosphorylated by GST-Rho-kinase (Rho-kinase), catalytic subunit of protein kinase A (A kinase), CaM kinase II, cdc2 kinase, or protein kinase C (C kinase), respectively. (Modified with permission, from ref. 24.)

phosphopeptides of IF can recognize not only antigen phosphopeptide but also phospho-IF proteins. In addition, immunocytochemical detection of IF phosphorylation and of IF kinase activity is facilitated using this technique, since there is an extensive network of IFs (10 nm filaments) in the cytoplasm.

Among *in vitro* phosphorylation sites are those specifically phosphorylated by a single kinase (18, 20–25, 32, 33, 36, 37, 39–45, 47, 76, 97–106). For example, Ser33, Ser55, Ser71, and Ser82 on vimentin are site-specific for C kinase (40, 41), Cdc2 kinase (18, 42), Rho-kinase (24), and CaMK II (43), respectively. Specific sites such as these serve as pertinent substrates for detection of *in vivo* phosphorylation of IF by these kinases. Using antibodies recognizing the phosphorylation of distinct specific sites on vimentin, together with other methods , we concluded that C kinase, Cdc2 kinase, Rho-kinase, and CaMK II, as *in vivo* vimentin kinases, phosphorylate vimentin during cell signalling and the cell cycle. For example, we recently developed a site- and phosphorylation state-specific antibody (GK71) that recognizes vimentin phosphorylated at Ser71 by Rho-associated kinase (*Figure 4*). Immunocytochemical studies using GK71 revealed that Rho-associated kinase phosphorylates vimentin specifically in the cleavage furrow (*Figure 5*). Vimentin phosphorylated by Rho-associated kinase lost its ability to form filaments *in*

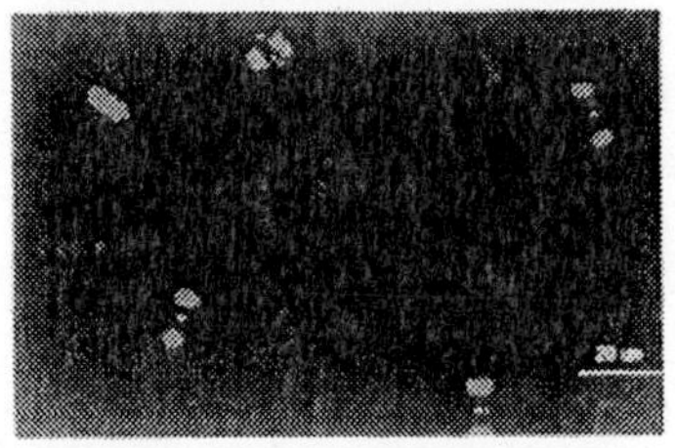

Figure 5. Immunofluorescence micrographs of mitotic U251 human glioma cells stained with the antibody 1B8 or GK71. The antibodies 1B8 and GK71 recognize vimentin and vimentin Ser71 phosphorylated by Rho-associated kinase, respectively. Confocal microscopic images of U251 cells stained with 1B8 (A) or GK71 (B and C) (green). Bars, 20 μm (A and B); 10 μm (C). (Modified with permission, from ref. 24.)

vitro (*Figure 2*). Therefore, we propose that Rho-associated kinase has a role in governing regulatory processes in assembly–disassembly and turnover of vimentin filaments at the cleavage furrow during cytokinesis (24).

In cell localization studies, one of the problems in immunocytochemistry is fixation. Various methods are used to fix cells (95); we use a formaldehyde and methanol fixation method (see *Protocol 9*). Another problem is quality of antibodies. For purification several steps are required, and antibodies may be inactivated. If purified antibodies are inactive for immunocytochemistry, the protocol must be changed and other immunization or purification methods tried.

Protocol 9. Immunocytochemistry using a site- and phosphorylation state-specific antibody

Reagents

- 3.7% (v/v) formaldehyde in PBS
- Buffer A: 1% (w/v) BSA, 1% (w/v) sucrose, 0.1% (w/v) NaN_3 in PBS
- Primary antibody: site- and phosphorylation state-specific antibody (obtained by *Protocol 6*)
- 0.5 μg/ml of propidium iodide (PI) (Sigma)
- Secondary antibody: FITC (fluorescein isothiocyanate)-labelled anti-rabbit Ig antibody (Biosource)
- 0.5 μg/ml of 4′,6-diamidine-2′-phenylindole-dihydrochloride (DAPI) (Boehringer Mannheim)

Method

1. Prior to cell staining, cells are grown on glass coverslips (or slides).
2. Fix cells with 3.7% ice-cold formaldehyde (4 °C) for 10 min.
3. Wash the coverslip gently twice in ice-cold PBS (4 °C) for 10 min.
4. Treat the fixed cells with methanol (–20 °C) for 10 min.
5. Wash the coverslip gently twice in ice-cold PBS (4 °C) for 10 min.
6. Place each coverslip, which is on a layer of Parafilm, in a Petri dish containing water-saturated paper.
7. Dilute the primary antibody with buffer A (at first, we test 1:2 dilution of the starting materials). Place the antibody gently on the coverslip, and incubate for 2 h at 37 °C or overnight at 4 °C.
8. Wash twice in PBS for 10 min at room temperature.
9. Dilute the FITC-labelled secondary antibody with PBS (at first, we test 1:100 dilution of the starting materials), and place it gently on the coverslip.
10. Incubate for 1 h at 37 °C in the dark.
11. Wash twice in PBS for 10 min at room temperature.
12. If necessary, incubate with 0.5 μg/ml of DAPI or 0.5 μg/ml of PI for 10 min at room temperature and wash twice in PBS for 10 min.
13. Mount the coverslip and analyse by fluorescence microscopy.

This technique is useful to monitor the spatiotemporal organization of kinase activities. Using this strategy, we consider that one can observe regulation of enzymatic activity in a restricted area of cells. This technique should be applicable to obtain specific antibodies against other proteins, which can detect activities of various kinases in the membrane and nucleus. In addition, monitoring unknown kinase activity may be feasible, using this or more developed methods.

4. Conclusion

It is known that IF reorganization plays an important role when cells undergo morphological changes. The IF structure is regulated by site-specific phosphorylation and dephosphorylation of constituent proteins. We have established a method to produce site- and phosphorylation state-specific antibodies for phospho-IF proteins. These antibodies are useful tools to analyse site-specific IF phosphorylation *in vivo*, as well as to identify IF kinases. The site- and phosphorylation state-specific antibodies we have described here are expected to have a wide application for analysing various cellular events.

Acknowledgements

This work was supported in part by Grants-in-Aid for Scientific Research and Cancer Research from the Ministry of Education, Science, Sports, and Culture of Japan, Japan Society of the Promotion of Science Research for the Future, special coordination funds from the Science and Technology Agency of the Government of Japan, and a grant from Bristol-Myers-Squibb. We thank K. Kuromiya for secretarial assistance, and M. Ohara for critique of the manuscript.

References

1. Lazarides, E. (1980). *Nature*, **283**, 249.
2. Steinert, P. M. and Roop, D. R. (1988). *Annu. Rev. Biochem.*, **57**, 593.
3. Klymkowsky, M. W. (1995). *Curr. Opin. Cell Biol.*, **7**, 46.
4. Fuchs, E. and Weber, K. (1994). *Annu. Rev. Biochem.*, **63**, 345.
5. Inagaki, M., Nishi,Y., Nishizawa, K., Matsuyama, M., and Sato, C. (1987). *Nature*, **328**, 649.
6. Inagaki, M., Matsuoka, Y., Tsujimura, K., Ando, S., Tokui, T., Takahashi, T., *et al.* (1996). *BioEssays*, **18**, 481.
7. Inagaki, M., Inagaki, N., Takahashi, T., and Takai, Y. (1997). *J. Biochem.*, **121**, 407.
8. Evans, R. M. and Fink, L. M. (1982). *Cell*, **29**, 43.
9. Celis, J. E., Lasen, P. M., Fey, S. J., and Celis, A. (1983). *J. Cell Biol.*, **97**, 1429.
10. Nishizawa, K., Yano, T., Shibata, M., Ando, S., Saga, S., Takahashi, T., *et al.* (1991). *J. Biol. Chem.*, **266**, 3074.

11. Yano, T., Taura, C., Shibata, M., Hirono, Y., Ando, S., Kusubata, M., *et al.* (1991). *Biochem. Biophys. Res. Commun.*, **175**, 1144.
12. Matsuoka, Y., Nishizawa, K., Yano, T., Shibata, M., Ando, S., Takahashi, T., *et al.* (1992). *EMBO J.*, **11**, 2895.
13. Inagaki, N., Ito, M., Nakano, T., and Inagaki, M. (1994). *Trends Biochem. Sci.*, **19**, 448.
14. Inagaki, M., Gonda, Y., Matsuyama, M., Nishizawa, K., Nishi, Y., and Sato, C. (1988). *J. Biol. Chem.*, **263**, 5970.
15. Evans, R. M. (1988). *FEBS Lett.*, **234**, 73.
16. Tokui, T., Yamauchi, T., Yano, T., Nishi, Y., Kusagawa, M., Yatani, R., *et al.* (1990). *Biochem. Biophys. Res. Commun.*, **169**, 896.
17. Chou, Y. H., Bischoff, J. R., Beach, D., and Goldman, R. D. (1990). *Cell*, **62**, 1063.
18. Kusubata, M., Tokui, T., Matsuoka, Y., Okumura, E., Tachibana, T., Hisanaga, S., *et al.* (1992). *J. Biol. Chem.*, **267**, 20937.
19. Wyatt, T. A., Lincoln, T. M., and Pryzwansky, K. B. (1994). *J. Biol. Chem.*, **266**, 21274.
20. Inagaki, M., Gonda, Y., Nishizawa, K., Kitamura, S., Sato, C., Ando, S., *et al.* (1990). *J. Biol. Chem.*, **265**, 4722.
21. Tsujimura, K., Tanaka, J., Ando, S., Matsuoka, Y., Kusubata, M., Sugiura, H., *et al.* (1994). *J. Biochem.*, **116**, 426.
22. Geisler, N. and Weber, K. (1988). *EMBO J.*, **7**, 15.
23. Kusubata, M., Matsuoka, Y., Tsujimura, K., Ito, H., Ando, S., Kamijo, M., *et al.* (1993). *Biochem. Biophys. Res. Commun.*, **190**, 927.
24. Goto, H., Kosako, H., Tanabe, K., Yanagida, M., Sakurai, M., Amano, M., *et al.* (1998). *J. Biol. Chem.*, **273**, 11728.
25. Inada, H., Goto, H., Tanabe, K., Nishi, Y., Kaibuchi, K., and Inagaki, M. (1998). *Biochem. Biophys. Res. Commun.*, **253**, 21.
26. Yano, T., Tokui, T., Nishi, Y., Nishizawa, K., Shibata, M., Kikuchi, K., *et al.* (1991). *Eur. J. Biochem.*, **197**, 281.
27. Ku, N.-O. and Omary, M. B. (1997). *J. Biol. Chem.*, **272**, 7556.
28. Tanaka, J., Ogawara, M., Ando, S., Shibata, M., Yatani, R., Kusagawa, M., *et al.* (1993). *Biochem. Biophys. Res. Commun.*, **196**, 115.
29. Hisanaga, S., Gonda, Y., Inagaki, M., Ikai, A., and Hirokawa, N. (1990). *Cell Regul.*, **1**, 237.
30. Nakamura, Y., Takeda, M., Angelides, K. J., Tada, K., and Nishimura, T. (1990). *Biochem. Biophys. Res. Commun.*, **169**, 744.
31. Gonda, Y., Nishizawa, K., Ando, S., Kitamura, S., Minoura, Y., Nishi, Y., *et al.* (1990). *Biochem. Biophys. Res. Commun.*, **167**, 1316.
32. Guan, R. J., Hall, F. L., and Cohlberg, J. A. (1992). *J. Neurochem.*, **58**, 1365.
33. Hashimoto, R., Nakamura, Y., Goto, H., Wada, Y., Sakoda, S., Kaibuchi, K., *et al.* (1998). *Biochem. Biophys. Res. Commun.*, **245**, 407.
34. Peter, M., Nakagawa, J., Doree, M., Labbe, J.-C., and Nigg, E. A. (1990). *Cell*, **61**, 591.
35. Dessev, G., Lovcheva-Dessev, C., Bischoff, J. R., Beach, D., and Goldman, R. (1991). *J. Cell Biol.*, **112**, 523.
36. Peter, M., Heitlinger, E., Hander, M., Aebi, U., and Nigg, E. A. (1991). *EMBO J.*, **10**, 1535.
37. Hocevar, B. A., Burns, D. J., and Fields, A. P. (1993). *J. Biol. Chem.*, **268**, 7545.

38. Fields, A. P., Pettit, G. R., and May, W. S. (1988). *J. Biol. Chem.*, **263**, 8253.
39. Peter, M., Sanghera, J. S., Pelech, S. L., and Nigg, E. A. (1992). *Eur. J. Biochem.*, **205**, 287.
40. Ando, S., Tanabe, K., Gonda, Y., Sato, C., and Inagaki, M. (1989). *Biochemistry*, **28**, 2974.
41. Geisler, N., Hatzfeld, M., and Weber, K. (1989). *Eur. J. Biochem.*, **183**, 441.
42. Chou, Y.-H., Ngai, K. L., and Goldman, R. (1991). *J. Biol. Chem.*, **266**, 7325.
43. Ando, S., Tokui, T., Yamauchi, T., Sugiura, H., Tanabe, K., and Inagaki, M. (1991). *Biochem. Biophys. Res. Commun.*, **175**, 955.
44. Nakamura, Y., Takeda, M., Aimoto, S., Hojo, H., Takao, T., Shimonishi, Y., *et al.* (1992). *J. Biol. Chem.*, **267**, 23269.
45. Kitamura, S., Ando, S., Shibata, M., Tanabe, K., Sato, C., and Inagaki, M. (1989). *J. Biol. Chem.*, **264**, 5674.
46. Ando, S., Tokui, T., Yano, T., and Inagaki, M. (1996). *Biochem. Biophys. Res. Commun.*, **221**, 67.
47. Sihag, R. K. and Nixon, R. A. (1991). *J. Biol. Chem.*, **266**, 18861.
48. Traub, P., Scherbarth, A., Wiegers, W., and Shoeman, R. L. (1992). *J. Cell Sci.*, **101**, 363.
49. Raats, J. M. H., Pieper, F. R., Egberts, W. T. M. V., Verrijp, K. N., Ramaekers, F. C. S., and Bloemendal, H. (1990). *J. Cell Biol.*, **111**, 1971.
50. Hatzfeld, M., Dodemont, H., Plessmann, U., and Weber, K. (1992). *FEBS Lett.*, **302**, 239.
51. Herrmann, H., Hofmann, I., and Franke, W. W. (1992). *J. Mol. Biol.*, **223**, 637.
52. Coulombe, P. A. and Fuchs, E. (1990). *J. Cell Biol.*, **111**, 153.
53. Inagaki, M., Nakamura, Y., Takeda, M., Nishimura, T., and Inagaki, N. (1994). *Brain Pathol.*, **4**, 239.
54. Rueger, D. C., Huston, J. S., Dahl, D., and Bignami, A. (1979). *J. Mol. Biol.*, **135**, 53.
55. Soellner, P., Quinlan, R. A., and Franke, W. W. (1985). *Proc. Natl. Acad. Sci. USA*, **82**, 7929.
56. Shoeman, R. L. and Traub, P. (1993). *BioEssay*, **15**, 605.
57. Robson, R. M. (1989). *Curr. Opin. Cell Biol.*, **1**, 36.
58. Stewart, M. (1993). *Curr. Opin. Cell Biol.*, **5**, 3.
59. Geisler, N., Schünemann, J., and Weber, K. (1992). *Eur. J. Biochem.*, **206**, 841.
60. Stewart, M., Quinlan, R. A., and Moir, R. D. (1989). *J. Cell Biol.*, **109**, 225.
61. Frank, W. W., Winter, S., Schmid, E., Soellner, P., Hammerling, G., and Achstatter, T. (1987). *Exp. Cell Res.*, **173**, 17.
62. Quinlan, R. A., Cohlberg, J. A., Schiller, D. L., Hatzfeld, M., and Franke, W. W. (1984). *J. Mol. Biol.*, **178**, 365.
63. Klymkowsky, M. W., Maynell, L. A., and Nislow, C. (1991). *J. Cell Biol.*, **114**, 787.
64. Inagaki, M., Takahara, H., Nishi, Y., Sugawara, K., and Sato, C. (1989). *J. Biol. Chem.*, **264**, 18119.
65. Quinlan, R. A., Moir, R. D., and Stewart, M. (1989). *J. Cell Sci.*, **93**, 71.
66. Huiatt, T. W., Robson, R. M., Arakawa, N., and Stromer, M. H. (1980). *J. Biol. Chem.*, **255**, 6981.
67. Renner, W., Franke, W. W., Schmid, E., Giesler, N., Weber, K., and Mandelkow, E. (1981). *J. Mol. Biol.*, **149**, 285.
68. Meyer, H. E., Eisermann, B., Heber, M., Hoffmann-Posorske, E., Korte, H., Weigt, C., *et al.* (1993). *FASEB J.*, **9**, 776.

69. Boyle, W. J., van der Geer, P., and Hunter, T. (1991). In *Methods in enzymology* (ed. T. Hunter and B. M. Sefton), Vol. 201, p. 110. Academic Press, London.
70. Meyer, H. E., Hoffmann-Posorske, E., Korte, H., and Heilmeyer, L. M. G. (1986). *FEBS Lett.*, **204**, 61.
71. Yano, S., Fukunaga, K., Ushiro, Y., and Miyamoto, E. (1994). *J. Biol. Chem.*, **269**, 5428.
72. Czernik, A. J., Girault, J. A., Nairn, A. C., Chen, J., Snyder, G., Kebabian, J., *et al.* (1991). In *Methods in enzymology* (ed. T. Hunter and B. M. Sefton), Vol. 201, p. 264. Academic Press, London.
73. Tsujimura, K., Ogawara, M., Takeuchi, Y., Imajoh-Ohmi, S., Ha, M. H., and Inagaki, M. (1994). *J. Biol. Chem.*, **269**, 31097.
74. Ogawara, M., Inagaki, N., Tsujimura, K., Takai, Y., Sekimata, M., Ha, M. H., *et al.* (1995). *J. Cell Biol.*, **131**, 1055.
75. Takai, Y., Ogawara, M., Tomono, Y., Moritoh, C., Imajoh-Ohmi, S., Tsutsumi, O., *et al.* (1996). *J. Cell Biol.*, **133**, 141.
76. Sekimata, M., Tsujimura, K., Tanaka, J., Takeuchi, Y., Inagaki, N., and Inagaki, M. (1996). *J. Cell Biol.*, **132**, 635.
77. Snyder, G. L., Girault, J. A., Chen, J. Y., Czernik, A. J., Kebavian, J. W., Nathanson, J. A., *et al.* (1992). *J. Neurosci.*, **12**, 3071.
78. Hagiwara, M., Brindle, P., Harootunian, A., Armstrong, R., Rivier, J., Vale, W., *et al.* (1993). *Mol. Cell. Biol.*, **13**, 4852.
79. Drago, G. A. and Colyer, J. (1994). *J. Biol. Chem.*, **269**, 25073.
80. Nagumo, H., Sakurada, K., Seto, M., and Sasaki, Y. (1994). *Biochem. Biophys. Res. Commun.*, **203**, 1502.
81. Goldstein, M., Lee, K. Y., Lew, J. Y., Harada, K., Wu, J., Haycock, J. W., *et al.* (1995). *J. Neurochem.*, **64**, 2281.
82. Weeks, J. R., Hardin, S. E., Shen, J., Lee, J. M., and Greenleaf, A. L. (1993). *Genes Dev.*, **7**, 2329.
83. O'Brien, T., Hardin, S., Greenleaf, A., and Lis, J. T. (1994). *Nature*, **370**, 75.
84. Epstein, R. J., Druker, B. J., Roberts, T. M., and Stiles, C. D. (1992). *Proc. Natl. Acad. Sci. USA*, **89**, 10435.
85. Bangalore, L., Tanner, A. J., Laudano, A. P., and Stern, D. F. (1992). *Proc. Natl. Acad. Sci. USA*, **89**, 11637.
86. DiGiovanna, M. P. and Stern, D. F. (1995). *Cancer Res.*, **55**, 1946.
87. Lang, E., Szendrei, G. I., Lee, V. M., and Otvos, L. Jr. (1992). *Biochem. Biophys. Res. Commun.*, **187**, 783.
88. Greenberg, S. G., Davies, P., and Binder, L. I. (1992). *J. Biol. Chem.*, **267**, 564.
89. Otvos, L. Jr., Feiner, L., Lang, E., Szendrei, G. I., Goedert, M., and Lee, V. M. (1994). *J. Neurosci. Res.*, **39**, 669.
90. Pope, W. B., Lambert, M. P., Leypold, B., Seupaul, R., Sletten, L., Krafft, G., *et al.* (1994). *Exp. Neurol.*, **126**, 185.
91. Johansson, M. W., Larsson, E., Lüning, B., Pasquale, E. B., and Ruoslahti, E. (1994). *J. Cell Biol.*, **126**, 1299.
92. Coghlan, M. P., Pillay, T. S., Tavare, J. M., and Siddle, K. (1994). *Biochem. J.*, **303**, 893.
93. Liao, J., Lowthert, L. A., Ku, N. O., Frenandez, R., and Omary, M. B. (1995). *J. Cell Biol.*, **131**, 1291.
94. Nakazawa, K., Mikawa, S., Hashikawa, T., and Ito, M. (1995). *Neuron*, **15**, 697.

195. Harlow, E. and Lane, D. (ed.) (1988). *Antibodies: a laboratory manual.* Cold Spring Harbor Laboratory Press, New York.
196. Langone, J. J. (ed.) (1989). *Methods in enzymology*, Vol. 178. Academic Press, London.
197. Lew, J. and Wang, J. H. (1995). *Trends Biochem. Sci.*, **20**, 33.
198. Vallano, M. L., Buckholz, T. M., and DeLorenzo, R. J. (1985). *Biochem. Biophys. Res. Commun.*, **130**, 957.
199. Link, W. T., Dosemeci, A., Floyd, C. C., and Pant, H. C. (1993). *Neurosci. Lett.*, **151**, 89.
100. Guan, R. J., Khatra, B. S., and Cohlberg, J. A. (1991). *J. Biol. Chem.*, **266**, 8262.
101. Sihag, R. K. and Nixon, R. A. (1990). *J. Biol. Chem.*, **265**, 4166.
102. Roder, H. M. and Ingram, V. M. (1991). *J. Neurosci.*, **11**, 3325.
103. Xiao, J. and Monteiro, M. J. (1994). *J. Neurosci.*, **14**, 1820.
104. Hisanaga, S.-I., Kusubata, M., Okumura, E., and Kishimoto, T. (1991). *J. Biol. Chem.*, **266**, 21798.
105. Eggert, M., Radomski, N., Linder, D., Tripier, D., Traub, P., and Jost, E. (1993). *Eur. J. Biochem.*, **213**, 659.
106. Hennekes, H., Peter, M., Weber, K., and Nigg, E. A. (1993). *J. Cell Biol.*, **120**, 1293.

10

The study of mRNA–cytoskeleton interactions and mRNA sorting in mammalian cells

GIOVANNA BERMANO and JOHN E. HESKETH

1. Introduction

It is vital for a wide range of cell functions that newly synthesized proteins are delivered to the correct place in the cell at the appropriate time. There is abundant evidence for targeting motifs in proteins themselves which play a role in protein delivery (e.g. nuclear import signals), and there is increasing evidence that, for a least some proteins, the localization and targeting of their mRNAs plays a significant role in determining the site of protein synthesis within cells (1–3). Thus, in *Drosophila* embryos, the localization of certain mRNAs functions to ensure correct development by determining the site of synthesis, and in this way, the distribution of proteins critical in development (4). mRNA localization has also been demonstrated to occur in mammalian somatic cells (2), and again its role appears to be to determine or restrict protein distribution. This is likely to be important where there are closely similar isoforms and protein targeting alone could perhaps not segregate isoforms (e.g. actin, creatine kinase, metallothionein) (5–7), where the mRNA and protein are unstable (e.g. c-myc) (8) or where inappropriate deposition of the protein might cause disruption of cell organization (e.g. myelin basic protein, vimentin) (9, 10).

In parallel to this work on mRNA localization, a large body of evidence has accumulated showing that a major proportion of cell polysomes and poly(A) mRNA are associated with the cytoskeleton (1, 2, 11, 12). Furthermore, some specific mRNAs are found associated with the cytoskeleton; e.g. c-myc (8, 13) and vimentin (10). More recently, localization and transport of RNA has been found to involve cytoskeletal components (14–17).

The combined use of recombinant DNA techniques, cell transfection, and assays of mRNA localization has enabled the study of which regions of different mRNAs are required for both association with the cytoskeleton and mRNA localization. In both cases the required signals are found within the 3′

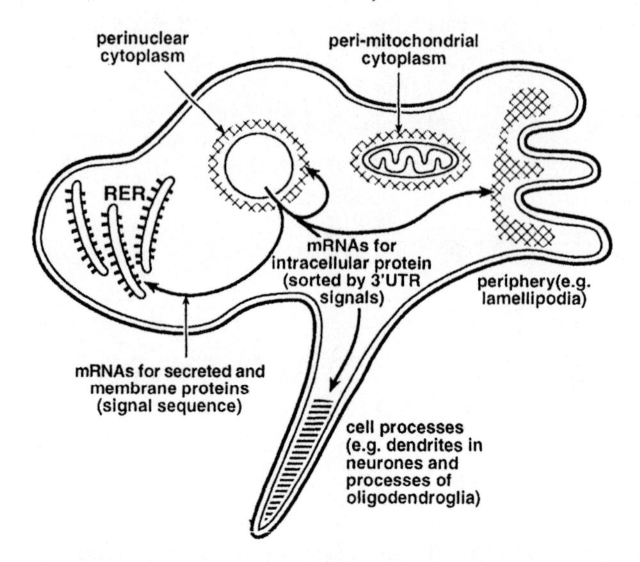

Figure 1. mRNA localization domains and localization signals in an imaginary composite cell. Certain mRNAs exhibit localization in discrete parts of the cytoplasm; examples of known mRNA localization domains are shaded. mRNA coding for intracellular proteins are sorted by signals within their 3′ UTRs. In contrast, mRNAs coding for membrane and secretory proteins are associated with the endoplasmic reticulum (RER) due to the signal sequence in the nascent polypeptide chain. It is unknown if additional transport mechanisms deliver mRNAs to the vicinity of the RER.

untranslated region (3′ UTR) of the mRNA (2); this is the case in mammalian cells, in *Xenopus*, and in *Drosophila*, and thus it is a highly conserved mechanism. This targeting of mRNAs coding for intracellular proteins appears to be distinct from that which produces the translation of mRNAs coding for secreted and membrane proteins in the endoplasmic reticulum compartment.

Thus a picture of the cell is emerging in which the protein synthetic apparatus is highly organized (2) and in which some mRNAs are sorted according to signals in their 3′ UTR (*Figure 1*) and transported and/or retained in different parts of the cytoplasm by differential interactions with the cytoskeleton via RNA–protein interactions (*Figure 2*). Potentially, therefore, there may be a variety of multiple mRNA–cytoskeleton interactions so that some mRNAs are retained in the perinuclear cytoplasm whilst others are transported and then anchored at appropriate sites. Furthermore, the cytoskeleton consists of different filament systems, and these may play different roles in the mRNA localization, based on different mRNA–cytoskeleton interactions. Increased knowledge of these mRNA–cytoskeleton interactions should make important contributions to our understanding of cell organization, particularly organization of protein synthesis, and how the cell sorts mRNAs to regulate the site of mRNA translation.

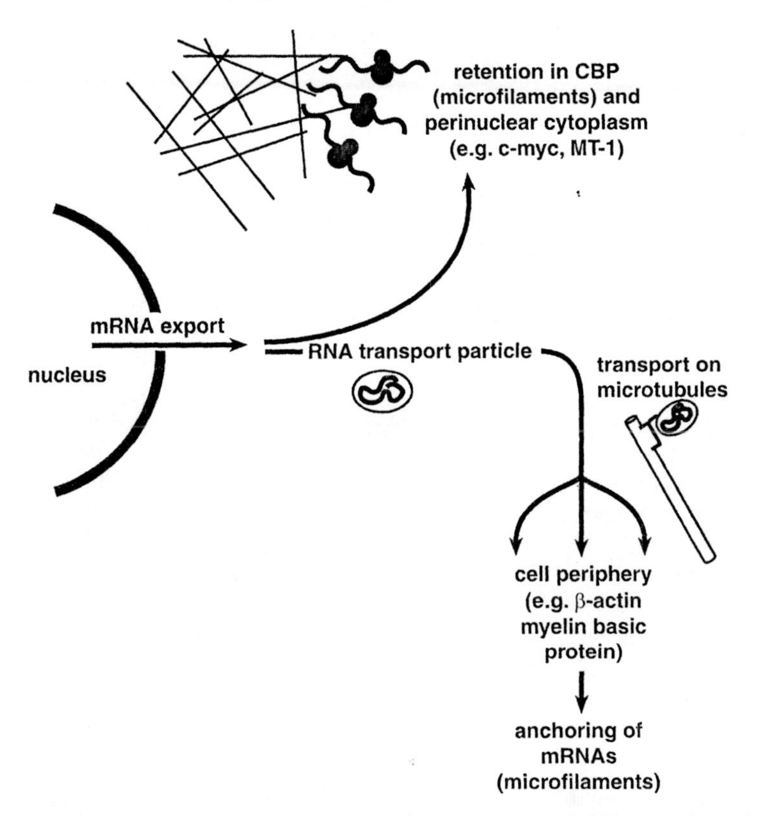

Figure 2. mRNA sorting and the cytoskeleton. The cell can sort mRNAs to direct them to different cell compartments. This involves differential interactions with the cytoskeleton and in some cases transport in RNA-containing particles.

In situ hybridization and sucrose gradient analysis have shown that 75% of mRNAs and polyribosomes are retained in the cell matrix (11, 12). 35% of polyribosomes are released from the cell matrix by treatments that release actin and vimentin (18). Thus, mRNA–cytoskeleton interactions are quantitatively important and make a major contribution to the organization of the protein synthetic apparatus. However, as yet it is unclear which mRNAs are associated with the cytoskeleton and for what functional reasons. Better fractionation techniques are required before we can begin to use differential display type techniques to pull out whole populations of mRNAs associated with the cytoskeleton. Meanwhile, with present techniques of biochemical fractionation and *in situ* hybridization, it is possible to address whether a specific mRNA of interest is associated with the cytoskeleton, whether it is localized, whether the targeting of the mRNA involves the 3′ UTR, and which proteins interact with the mRNA and the cytoskeleton. In this chapter we describe how widely-available techniques can be used to address such questions and so assist the reader in determining the extent and mechanism of mRNA–cytoskeleton interactions. The protocols described should be used in accordance with both general and local safety procedures.

2. Biochemical approaches to mRNA–cytoskeleton interactions

2.1 mRNAs and polysomes in the cell matrix

Treatment of cells in culture with low concentrations of non-ionic detergents solubilizes the plasma membrane and releases soluble cytosolic components, leaving an insoluble network of cytoskeletal filaments known as the cell matrix. Many studies have shown that the cell matrix contains ribosomes, polyribosomes, and mRNAs, and on the basis of such data it was proposed that those mRNAs recovered in the cell matrix were associated with the cytoskeleton (11). However, the cell matrix also contains fragments of rough endoplasmic reticulum (RER) (19) and so at least some of the mRNAs and polyribosomes present in the cell matrix may be associated not with the cytoskeleton but with the RER (2, 20). Indeed, polysomes in the cell matrix or fractions derived from it have been found to contain several mRNAs encoding membrane proteins such as glucose transporter 1 (GLUT 1) and β-microglobulin. Thus, recovery of polysomes or a specific mRNA in the cell matrix does not necessarily mean that they are associated with the cytoskeleton. Some further treatment of the cell matrix is required to separate components of the protein synthetic apparatus associated with the cytoskeleton from those on the RER and so to allow isolation of polysomes and mRNAs specifically associated with the cytoskeleton rather than those that are free in the cytosol or present on the RER.

2.2 Identification of polysomes and mRNAs associated with the cytoskeleton

Two biochemical approaches are available to assess whether a mRNA is specifically associated with the cytoskeleton.

(a) The first is to assess whether or not disruption of the cytoskeleton leads to loss of a particular mRNA from the cell matrix. This approach involves pre-treatment of cells with drugs such as cytochalasins (to depolymerize actin filaments), separation of cytosolic and cell matrix fractions by simple non-ionic detergent treatment (*Protocol 1*, steps 1–6), followed by isolation of RNA and its subsequent analysis by standard mRNA detection techniques such as Northern hybridization.

(b) The second is to produce a fraction enriched in cytoskeletal components but not in cytosolic or RER components. This approach involves treatment of the cell matrix to cause depolymerization of cytoskeletal components and release of these components together with mRNAs and polysomes, followed either immediately, or after separation of polysomes, by isolation and analysis of RNA (*Protocol 1*, steps 1–10).

2.3 Cytoskeletal-bound polysomes

Treatment of the cell matrix with salt concentrations above 100 mM causes breakdown of the cytoskeleton (20) and release of cytoskeletal components, polysomes, and RNA (20, 21). This effect provides the basis for methods to separate polysomes and mRNAs associated with the cytoskeleton, in particular a fractionation procedure (*Figure 3*) to produce 'free' (FP), membrane-bound (MBP), and cytoskeletal-bound (CBP) polysomes and their constituent mRNAs (22).

The procedure is described in *Protocol 1*. The cells are washed and then an initial extraction is carried out with a buffer containing non-ionic detergent and a low salt concentration (e.g. 0.05–0.5% Nonidet P-40 and 25 mM KCl) in order to release cytosolic components but to retain the cell matrix intact. The insoluble cell matrix is then washed before treatment with buffer containing 130 mM KCl; this treatment causes release of actin together with vimentin and cytokeratin 19. In some cell lines 0.05% Nonidet P-40 is also required at this stage. After a further wash, the actin-depleted matrix is treated with both Triton and deoxycholate in order to solubilize the endoplasmic reticulum; deoxycholate is critical for solubilization of phospholipid-rich membranes such as the RER. This sequential detergent and salt extraction procedure thus produces three fractions, one cytosolic in origin, one cytoskeletal, and one derived from the RER, as assessed by markers such as lactic dehydrogenase, actin, and choline incorporation. RNA can be extracted directly from these fractions and analysed. However, it is more usual to separate polysomes from ribonucleoprotein particles by centrifugation of the fractions layered onto a sucrose cushion and then isolate and analyse polysomal RNA, which allows study of those mRNAs that are being translated in polysomes associated with the cytoskeleton. This latter information is not readily available from *in situ* hybridization studies. Our experience is that clearer indications of mRNA enrichment can be obtained by studying the polysomal RNA.

The fractionation procedure described in *Protocol 1* has been used to isolate CBP from a number of cell lines (fibroblasts, hepatoma cells, CHO and ascites cells, MPC-11 cells). The concentration of Nonidet P-40 required for the full release of cytosolic components varies from cell line to cell line; e.g. we have found 0.05% to be sufficient in 3T3 and Ltk^- fibroblasts (8, 13), 0.1% in hepatoma cells (23), but 0.5% is required in Krebs II ascites cells (22). For a new cell type or cell line it is necessary to determine experimentally the detergent concentration required for release of cytosolic components; this is done by measuring lactic acid dehydrogenase activity (24). In addition, it is

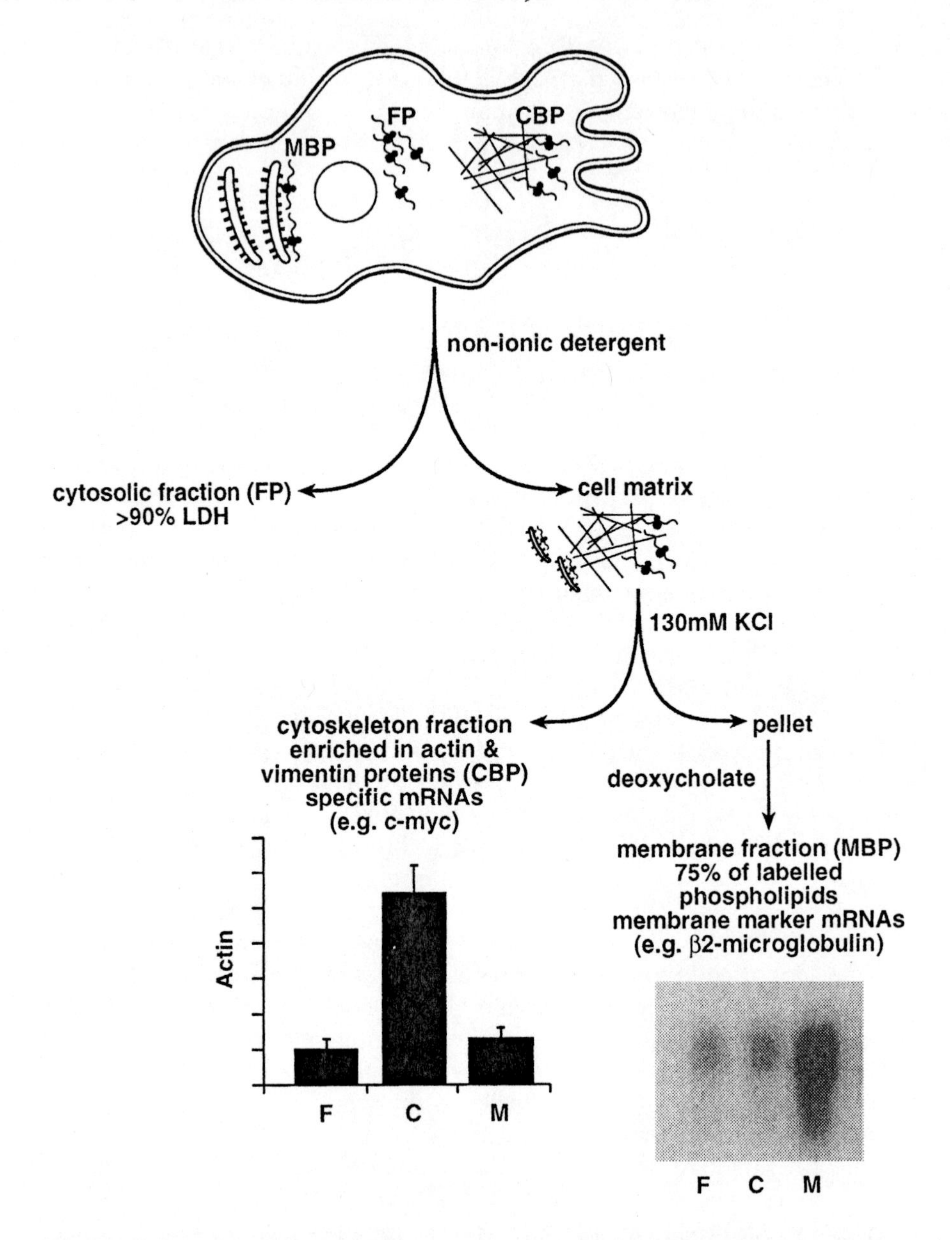

Figure 3. Schematic representation of cell fractionation procedure to produce cytoskeletal-bound polysomes. The diagram illustrates two important points: the importance of using markers such as lactic dehydrogenase (LDH), actin, and marker mRNAs (β2-microglobulin) to characterize fractions; the retention of RER in the cell matrix. FP, free polysomes; CBP, cytoskeletal-bound polysomes, MBP, membrane-bound polysomes.

very important to monitor the fractions produced with cytoskeletal and RER markers: cytoskeletal proteins such as actin and vimentin can be measured by ELISA assay, and the abundance of a mRNA encoding a membrane or secreted protein expected to be found in the RER (e.g. a glucose transporter mRNA) can be assessed by Northern hybridization. In the various cell lines we have studied it has been found consistently that c-myc mRNA is present at a twofold enrichment in CBP. Although this is not a definitive marker in itself, once the procedure has been verified for a given cell line, the measurement of the abundance of c-myc mRNA together with that encoding a membrane or secreted protein allows one to monitor the fractionation of CBP and MBP.

Protocol 1 releases from the cell matrix polysomes which are enriched in particular mRNAs such as that encoding c-myc. It also causes a large increase in the release of actin. Furthermore pre-treatment of cells with cytochalasin B reduces the yield of polysomes released from the matrix in this fraction indicating that, although the fraction contains both actin and intermediate filament (IF) proteins, the major interaction is probably through actin. On such a basis these polysomes are regarded as a population of CBP, distinct from FP and MBP. However, there are limitations to this approach. First, although CBP are likely to be primarily polysomes associated with the actin microfilaments, potentially they may well be a mixed population including some polysomes associated with IF. The extent of this interaction is unclear since, although some cytokeratins and vimentin are released along with actin (the fraction is relatively enriched in both proteins), a major proportion of vimentin is unaffected by the salt treatment and is then solubilized by the double detergent treatment. Thus, any IF-associated mRNAs and polysomes (if they exist) could appear in either the CBP or MBP fraction. Secondly, any polysomes or mRNAs associated with microtubules would be expected to be found largely in FP due to microtubule depolymerization in the initial exposure to ice-cold buffer. Thirdly, the release of mRNAs and polysomes into the cytosolic fraction is dependent on extraction conditions. It is quite likely that the FP fraction consists wholly or partly of mRNAs that are not truly free in the cytosol but 'loosely bound' to the cytoskeleton and more easily released than those recovered primarily in CBP. Fourthly, the fractionation procedure is relatively crude, and there is a degree of cross-contamination between fractions such that mRNA enrichment in CBP is only two- to threefold compared to FP. In summary, by this approach RNA–cytoskeletal interactions are defined operationally within the constraints of the method, and it is likely that this detects only a limited subpopulation. However, fractionation does provide an approach for assessing whether a particular mRNA of interest is enriched in polysomes associated with microfilaments. Different approaches need to be developed (disrupting agents and co-localization studies) to address mRNA interactions with microtubules and intermediate filaments.

Protocol 1. Cell fractionation to produce 'free', cytoskeletal-bound, and membrane-bound polysomes for RNA isolation

Equipment and reagents

- Benchtop and microcentrifuges
- DEPC-treated PBS: 10 mM phosphate buffer, 2.7 mM KCl, 137 mM NaCl pH 7.4, made from tablets (Sigma Chem. Co.,), 0.05% diethylpyrocarbonate (DEPC)
- 10% Nonidet P-40
- 10% sodium deoxycholate
- Solution A: 10 mM Tris pH 7.6, containing 0.25 M sucrose, 25 mM KCl, 5 mM $MgCl_2$, and 0.5 mM $CaCl_2$; 0.05% DEPC
- Solution B: 10 mM Tris pH 7.6, containing 0.25 M sucrose, 130 mM KCl, 5 mM $MgCl_2$, and 0.5 mM $CaCl_2$; 0.05% DEPC
- Solution C: 10 mM Tris pH 7.6, containing 40% sucrose, 130 mM KCl, 5 mM $MgCl_2$, and 0.5 mM $CaCl_2$; 0.05% DEPC
- Solution D: 4 M guanidinium thiocyanate, 25 mM sodium citrate pH 7.0, 0.5% Sarkosyl, 0.1% 2-mercaptoethanol

Method

1. Grow cells in large Petri dishes to 70% confluence. We routinely use 90 mm dishes and use between 5–15 dishes to produce 20 μg RNA from each fraction.
2. Before harvesting, remove the culture medium and rinse the cells twice with PBS.
3. Add 0.7 ml of solution A to each dish and remove cells from the dish into the solution A by scraping the dish with a rubber policeman. Pool the cell suspensions from all dishes and rinse each dish with a further 0.7 ml of solution A.
4. Centrifuge the cell suspension at 1000 *g* for 5 min at 4 °C. Gently pour off the supernatant fluid, leaving behind the loose pellet of cells.
5. Resuspend the cell pellet in 1 ml of ice-cold solution A containing 0.05–0.5% Nonidet P-40 (the concentration of detergent required to release cytosolic components should be ascertained previously by measurements of lactic dehydrogenase activity) (24). Cell resuspension should be achieved by gently drawing the pellet and buffer through a wide-bore sterile pipette tip (typically one used for a 1000 μl automatic pipette) six times. The suspension is then left on ice for 10 min.
6. Centrifuge the suspension at 1000 *g* for 5 min at 4 °C. The supernatant fluid is then carefully removed by pipette and transferred to a sterile microcentrifuge tube and stored on ice. It is the cytosolic fraction containing 'free' polysomes (FP); the pellet is the cell matrix.
7. Wash the pellet. Add 1 ml of solution A, resuspend the pellet gently, and centrifuge the suspension at 1000 *g* for 5 min at 4 °C.
8. Resuspend the pellet in 1 ml solution B and leave on ice for 10 min. Centrifuge the suspension at 1500 *g* for 10 min at 4 °C. Remove the

supernatant fluid with a pipette, transfer to a sterile microcentrifuge tube, and store on ice. This is the cytoskeletal fraction containing cytoskeletal-bound polysomes (CBP).

9. Resuspend the pellet in 0.9 ml solution B containing 0.5% Nonidet P-40 and 0.5% deoxycholate and leave on ice for 10 min. Centrifuge the suspension at 2000 *g* for 10 min at 4°C. Remove the supernatant fluid with a pipette, transfer to a sterile microcentrifuge tube, and store on ice. It is the membrane fraction containing membrane-bound polysomes (MBP).
10. Polysomes are then isolated from all three fractions. Carefully layer the three fractions on top of a 10–15 ml sucrose cushion (solution C) in three separate centrifuge tubes. Centrifuge the samples at 4°C for 17 h at 22 500 *g* (e.g. a Beckman Ti70 rotor at 17 000 r.p.m.).
11. Pour off the cushion and leave tubes to drain for 5–10 min. A small opalescent pellet of polysomes is visible at the bottom of the tube. Resuspend the pellet in solution D and continue RNA extraction according to the procedure of Chomczynski and Sacchi (25).

3. Localization of mRNA by *in situ* hybridization (ISH)

In situ hybridization (ISH) techniques have been applied increasingly to localize gene expression at the cytological level, because they allow specific nucleic acid sequences to be detected in morphologically preserved cells and tissue sections. These techniques offer an alternative approach, complementary to biochemical fractionation, to address the association of mRNAs with the cytoskeleton, as well as being the major tool to investigate mRNA localization. A full description of intracellular distribution of specific mRNAs is important for understanding the spatial organization of the protein synthetic apparatus and the role of the cytoskeleton.

ISH has shown that, in mammalian cells, RNAs for different proteins exhibit characteristic patterns of intracellular distribution (1–5). An example of localization of a specific mRNA is shown in *Figure 4*. ISH can be combined with other techniques such as immunocytochemistry to study co-distribution of mRNAs with cellular components such as the cytoskeleton (see Section 4). Since these techniques are widely used for a variety of purposes, they are continuously being improved in order to optimize conditions for the preservation of cellular morphology, RNA retention, and preservation of the native configuration of RNAs and, consequently, increased sensitivity and specificity. Since its original development by Pardue and Gall in 1969 (26), when radioactive nucleotides were used to synthesize probes and autoradiography was the only means of detecting hybridized sequences, ISH has

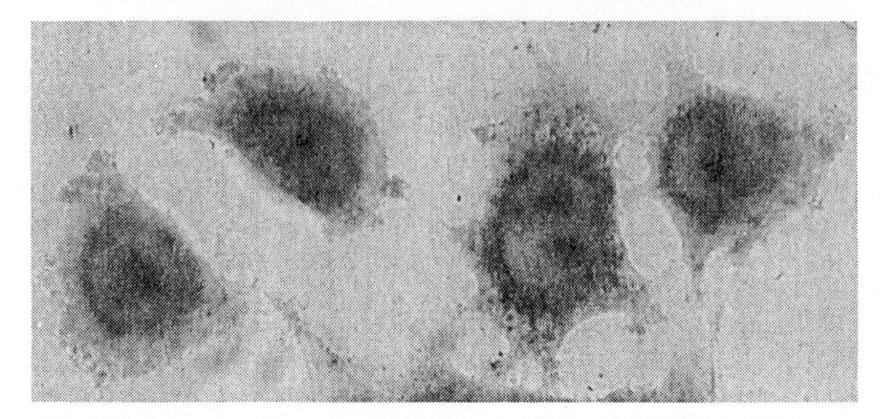

Figure 4. Perinuclear c-myc mRNA detected by *in situ* hybridization: an example of mRNA localization. The picture shows the distribution of c-myc transcripts in cells stably transfected with the c-myc coding sequences and c-myc 3′ UTR (8). DIG-labelled riboprobes prepared according to *Protocol 2,* including alkaline hydrolysis, and *in situ* hybridization was carried out using *Protocol 3.*

been developed to use a variety of non-radioactive probes. The advantage of radiolabelled probes is their ability to detect very low levels of transcripts, whereas the major limitations are very long exposure times for autoradiography and relatively poor resolution, depending on the radioisotope used. The more recent use of non-radioactively labelled nucleotides has considerably improved ISH with both a shortening of the development time and excellent histological resolution. For most mRNA localization studies the resolution of non-radioactive probes is necessary; we will compare the advantages and disadvantages of different probes and different methods of detection.

3.1 Non-radioactive hybridization methods

There are two types of non-radioactive hybridization method (*Figure 5*):

(a) Direct method. The detectable molecule is bound directly to the nucleic acid probe so that probe–target hybrids can be visualized under a microscope directly after the hybridization reaction (e.g. fluorescein-labelled nucleotides).

(b) Indirect method. The probe contains a tag (introduced chemically or enzymatically) that can be detected by affinity cytochemistry (e.g. digoxigenin and biotin labelling of nucleic acids which can subsequently be detected by anti-digoxigenin antibodies or streptavidin-linked reagents).

Since fluorescein is a direct label, no immunocytochemical visualization procedure is necessary and the background is low. However a drawback of

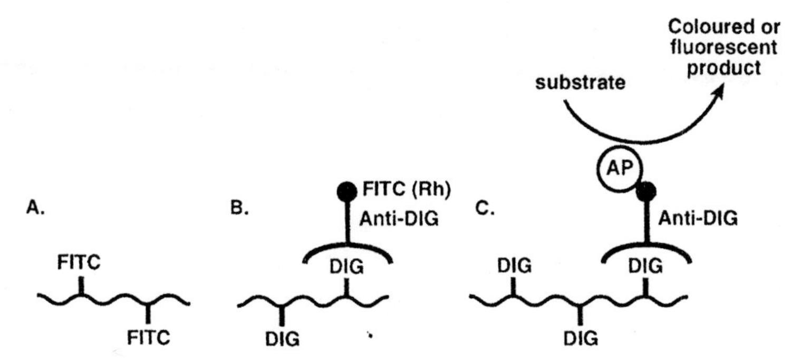

Figure 5. Schematic diagram illustrating direct (A), indirect fluorescent (B), and indirect enzyme-linked (C) detection methods available for *in situ* hybridization. FITC, fluorescein isothiocyanate.

this, and other, direct methods is that they can be less sensitive than indirect ones which include antibody or high affinity multiple binding reactions to amplify the signal. Use of digoxigenin (DIG)-based indirect detection methods has proved to be the most appropriate for rare transcripts. Since DIG is synthesized only in plants of the genus *Digitalis*, background problems in cells of other organisms are avoided. The hybridized DIG-labelled probes may be detected either directly with high affinity anti-DIG antibodies that are conjugated to fluorescein or rhodamine, or with antibodies linked to an enzyme-linked detection system in order to give a further increase in signal amplification. Alternatively, unconjugated anti-DIG antibodies and conjugated secondary antibodies may be used in order to increase the signal. Anti-DIG antibodies linked to peroxidase or colloidal gold are available, and they could allow detection at the electron microscope level. Biotin can be used in the same way as DIG, and usually signal amplification is introduced by reaction with streptavidin or avidin, having a high binding capacity for biotin, linked to an appropriate detection system.

3.2 Choosing the right probe for ISH

A variety of different types of probe, such as RNA probes (riboprobes), oligonucleotides, and single- or double-stranded DNA probes, can be used in ISH. In order to choose the kind of probe most appropriate for a given study, it is important to consider that the strength of the potential interaction between probe and target is critical. Hybrid stability is affected by hybridization conditions such as temperature. In general, since the strength of the bond decreases from RNA–RNA to DNA–RNA, in order to detect RNA, a RNA probe is probably the first choice. Indeed in the detection of RNA transcripts, RNA probes have the advantages over DNA probes of giving higher signal-to-noise ratios and higher stability of RNA hybrids.

RNA probes can be labelled with different molecules, for example DIG, biotin, and fluorescein. DIG-labelled probes have been successfully used in localization studies of several mRNAs. Endogenous β-actin and myosin heavy chain mRNAs have been localized in cultured cells and muscle fibres respectively (5, 27). In transfected cells these types of probe have been used to detect c-myc, β-globin, and metallothionein transcripts (7, 28, 29). DIG-labelled RNA probes are produced by *in vitro* transcription of DNA with DIG RNA labelling mix (see *Protocol 2*). The DNA to be transcribed should be cloned into the polylinker sites of appropriate transcription vectors which contain promoters for SP6, T7, or T3 RNA polymerases adjacent to the polylinker. To synthesize the transcripts, the DNA template must be linearized at a restriction enzyme site just downstream of the cloned insert to avoid transcription of undesirable sequences. The method for DIG labelling of RNA probes is shown in *Protocol 2*. General molecular methods are described in Sambrook *et al.* (30).

The procedure described in *Protocol 2* incorporates one modified nucleotide (in this case DIG-UTP, but also biotin- or fluorescein-UTP can be used) at approximately every 20–25th position in the transcripts. Since the nucleotide concentration does not become limiting in the following reaction, 1 μg of linear plasmid DNA can produce approximately 10 μg of full-length labelled RNA transcript in a two hour incubation. Larger amounts of labelled RNA can be synthesized by scaling up the reaction components.

Before using the labelled riboprobes for ISH, it is good practice to check the extent of label incorporation and the integrity of the probe. In the case of DIG labelling, incorporation is assessed by a dot blot assay. A small amount of fresh probe, together with serial dilutions of a DIG-labelled RNA standard, are spotted onto a nylon membrane (e.g. Genescreen) and the RNA cross-linked to the filter by UV light. The amount of DIG present is then estimated using anti-DIG antibodies linked to alkaline phosphatase and detection similar to that used in the ISH itself (see *Protocol 3*). The integrity and quality of non-radioactive probes can be checked by agarose or urea/polyacrylamide gel electrophoresis and subsequent staining with ethidium bromide. If problems are suspected, the *in vitro* transcription products can be labelled with [^{32}P]UTP, separated by urea/polyacrylamide gel electrophoresis, and analysed in detail by autoradiography (30, 31).

Protocol 2. DIG-labelled RNA probes

Reagents

- 10 × concentrated RNA labelling mix: 10 mM each of ATP, CTP, and GTP; 6.5 mM UTP; 3.5 mM DIG-UTP pH 7.5 (Roche)
- RNase-free DNase, 10 U/μl (Roche)
- SP6, T7, or T3 RNA polymerase, 20 U/μl (Roche)
- 10 × concentrated transcription buffer: 400 mM Tris–HCl pH 8.0, 60 mM $MgCl_2$, 100 mM dithiothreitiol (DTT), 20 mM spermidine (Roche)
- RNase inhibitor, 20 U/μl (Roche)

Method

1. Linearize 1 μg of template DNA with 10 U of appropriate restriction enzyme and 3 μl of 10 × restriction buffer in a final volume of 30 μl. Incubate at 37 °C for 2 h. Check linearization by electrophoresing a small amount on a 1% agarose gel and staining with ethidium bromide.
2. Purify the linearized template DNA by phenol:chloroform extraction, followed by ethanol precipitation (30). Resuspend the pellet in 0.05% DEPC-treated H_2O.
3. Add the following to a 1.5 ml microcentrifuge tube on ice:
   - 1 μg of purified, linearized plasmid DNA
   - 2 μl of 10 × concentrated DIG RNA labelling mix
   - 2 μl of 10 × concentrated transcription buffer
   - 1 μl of RNase inhibitor
   - 2 μl of RNA polymerase (SP6, T7, or T3)
   - enough RNase-free H_2O to make a total reaction volume of 20 μl.
4. Mix gently and incubate for 2 h at 37 °C.
5. Add 2 μl DNase I, RNase-free, and incubate for 15 min at 37 °C to remove the template DNA.
6. Add 2 μl of 0.2 M EDTA to stop the polymerase reaction.
7. Precipitate the labelled RNA transcript by adding 2.5 μl of 4 M LiCl and 75 μl pre-chilled 100% ethanol. Mix well and leave for at least 30 min at –70 °C or 2 h at –20 °C.
8. Centrifuge for 15 min at 12000 *g*, discard the supernatant, and wash the pellet with 50 μl of 70% ice-cold DEPC-treated ethanol (v/v).
9. Dry pellet and dissolve it in 100 μl of DEPC-treated H_2O for 30 min at 37 °C.
10. Add 1 μl RNase inhibitor in order to protect the RNA transcripts.
11. Store the probe solution at –80 °C if not used immediately. Avoid repeat freezing and thawing of the probe.

Alternatively, oligodeoxynucleotides can be used for ISH techniques. The use of non-radioactive labelled oligonucleotides has two advantages: first, they are readily available through automated synthesis. Secondly, the fact that they are small and single-stranded gives them good penetration properties, a factor which is considered critical for successful ISH. In contrast, their small size can be a disadvantage because they usually cover less target than conventional probes. However, their good penetration properties largely compensate for the smaller target and, additionally, the fact that they are single-stranded excludes the possibility of renaturation which can be a problem in double-stranded DNA probes. They can be used for both direct or indirect methods of labelling as, either during synthesis or by end-labelling, they can be labelled directly with a fluorochrome tag, or with biotin or DIG which are visualized indirectly. Oligonucleotides are often end-labelled, and this can be a disadvantage. End-labelling reduces their sensitivity, as there is only one labelled molecule per hybrid formed compared to the larger number present with internal labelling.

In the case of metallothionein transcripts both DNA oligonucleotide and riboprobes have proved effective in ISH. After induction of metallothionein by zinc, metallothionein-1 mRNA was detected and localized using a biotin-labelled oligonucleotide probe (29), whereas later studies in transfected cells used a DIG-labelled riboprobe (7). However, direct comparison of the use of DNA oligonucleotides or RNA probes for the detection of Tac (a subunit of the interleukin 2 receptor) mRNA demonstrated the greater sensitivity of RNA probes (32).

Other types of DNA probes, both single- and double-stranded, have also been used for ISH. Different methods are available to label the DNA, including random primed labelling, nick translation, and polymerase chain reaction (PCR). In the random primed labelling reaction, the template DNA is linearized, denatured, and annealed to a primer. The Klenow enzyme synthesizes new DNA along the single-stranded substrate, introducing one of the labelled nucleotides (DIG, biotin, or fluorescein). The size of the probe obtained can vary between 200–1000 bp. In the nick translation reaction the size of the probe is instead slightly smaller, varying between 200–500 bp. In this case the DNA template can be supercoiled or linear, and after the DNA has been nicked with DNase I, the $5' \rightarrow 3'$ exonuclease activity of DNA polymerase I extends the nicks to gaps; then the polymerase replaces the excised nucleotides with labelled ones. In probes prepared by PCR, the length of the amplified probe is precisely defined by the PCR primers, so PCR allows easy production of optimally sized hybridization probes. In PCR, two oligonucleotide primers hybridize to opposite DNA strands and flank a specific target sequence. Then a thermostable polymerase elongates the two primers, and a repetitive series of cycles results in an accumulation of the target sequence. If, instead of two primers, just one primer is used, a single-stranded DNA probe can be produced. In this case, the chances of renaturation are eliminated, the specificity of the signal increased, and more specific controls to characterize the signal can be used (as in the case of RNA probes) (33).

All three DNA labelling methods allow great flexibility with regard to the length of the labelled fragments. In particular, in nick translation and random primed labelling, a heterogeneous population of probe strands are produced, many of which have complementary regions, which could lead to signal amplification in the hybridization experiment. cDNA and PCR-generated

Table 1. Comparison of properties of different probes for mRNA detection[a]

	Riboprobe	**dsDNA probe**	**ssDNA probe**	**Oligonucleotide**
Affinity	+++	++	++	++
Penetration	++	+	++	+++
Sensitivity	+++	++	++	+
Preparation	++	++	++	+++
Storage	+	++	++	++
Possible controls	+++	++	+++	++

[a] The different properties are scored from + to +++, with +++ representing those most suitable for ISH.

DNA probes have been used successively for localization of myosin, desmin, and vimentin mRNAs during muscle development and β-actin mRNA in fibroblasts (5, 34–36).

Since a number of competing reactions occur during ISH with double-stranded probes, single-stranded DNA or RNA probes provide some advantages. The probe is not exhausted by self-annealing in solution, and large concatenates which prevent penetration into cells or tissue sections are not formed. In the case of single-stranded DNA probes, since DNA–RNA hybrids are more stable than DNA–DNA hybrids, hybridization conditions can be designed in which DNA–RNA hybrid formation is favoured rather than DNA–DNA hybrid formation. A summary and comparison of the properties of the different probes is given in *Table 1*.

3.3 Key features of an ISH protocol

ISH involves a series of critical procedures which are essential for specific hybridization and successful detection of transcripts. They are discussed in detail below in order to provide the information needed for the researcher to decide whether certain steps should be included or modified in a specific protocol. *Protocol 3* describes ISH methodology in current use in our laboratory for examining localization of several different mRNAs in transfected eukaryotic cells. Similar protocols have proved effective for studying endogenous mRNAs (5, 16).

3.3.1 Sample preparation

For eukaryotic cell lines, cells should be grown on coverslips or chamber slides under conditions providing optimal growth. For this reason, glass or plastic should be chosen as appropriate for the particular cells under study (e.g. plastic for myoblasts, glass for fibroblasts). Coverslips or slides can be treated with either collagen or acetylation solution if necessary in order to facilitate cell adhesion.

3.3.2 Fixation and permeabilization

Preservation of the cell morphology is one of the most important issues in ISH, and therefore adequate fixation of biological material is very important. Fixation has to produce a balance between maintenance of cell structure and loss of hybridization. Different fixatives such as 4% paraformaldehyde, 3:1 ethanol:acetic acid, and glutaraldehyde have been used in ISH but unfortunately a fixation protocol which can be used for all samples has not yet been found, and it must be optimized for different applications. In addition, the conditions of fixation such as temperature and duration of the treatment can be varied to optimize the procedure.

The RNA target sequences are surrounded by proteins, and extensive cross-linking of these proteins during fixation can mask the target nucleic acid. Therefore probe penetration is a critical factor in successful ISH. This is usually achieved by employing a permeabilization treatment and by using a probe of sufficiently small size to allow penetration. A permeabilization treatment is the second key step in most ISH protocols. With tissue sections this usually involves a treatment of the sample with proteinase K, but with cultured cells this can be replaced by a treatment with non-ionic detergent.

Probe size can be an important factor in penetration. In the case of riboprobes, it is considered that probes should be kept to approximately 200 nucleotides in length or shorter in order to allow good penetration. If necessary, longer RNA probes generated *in vitro* may be reduced in length by alkaline hydrolysis. Incubate the riboprobe, prepared according to *Protocol 2*, with an equal volume of hydrolysis buffer (80 mM $NaHCO_3$, 20 mM $NaCO_3$, 20 mM 2-mercaptoethanol) at 60°C for a length of time specified by the following formula:

$$T = (L_0 - L_f) / (0.11 \times L_0 \times L_f)$$

where L_0 = original length (kb), L_f = required length (kb). The longer the incubation time is, the shorter the mean fragment length. At the end of the incubation time, add an equal volume of stop buffer (200 mM sodium acetate pH 6.0, 10 mM dithiothreitol, 1% (w/v) acetic acid). Precipitate the hydrolysis products by addition of two volumes of 100% ethanol, wash, air dry pellet, and finally resuspend the RNA in a smaller volume of DEPC-treated H_2O.

In general oligonucleotide probes do not present a penetration problem. With cDNA or single-strand DNA probes any penetration problem is usually circumvented by permeabilization treatments.

Preparations can be treated with Triton X-100, sodium dodecyl sulfate, or other detergents if lipid membrane components have not been extracted by other procedure such as fixation, dehydration, or endogenous enzyme inactivation procedures.

3.3.3 Endogenous enzyme inactivation

When an enzyme is used as the label in the final detection step, it is important to inactivate any endogenous activity which may be present in the sample. In the case of alkaline phosphatase this is often unnecessary, since residual alkaline phosphatase activity is usually destroyed by the hybridization conditions. However, if necessary, levamisole (1 mM) can be added to the substrate solution during the final detection reaction. In the case of peroxidase detection, the inhibition of endogenous activity is achieved by treating the cells with 1% H_2O_2 in methanol for 30 min immediately after the fixation step.

3.3.4 Pre-hybridization

A pre-hybridization incubation is usually included in the protocol to prevent non-specific staining. The pre-hybridization mixture contains all the components of the hybridization mixture except for the probe.

3.3.5 Denaturation of probe and hybridization

Heat denaturation is the most popular way to denature probes because of its simplicity and greater effectiveness. Variations in time and temperature should be evaluated to find the best conditions for denaturation. Denaturation is particularly important when double-stranded DNA probes are used.

Several parameters such as temperature, formamide concentration, probe length, and base mismatch can influence hybridization conditions. Temperature is particularly important, as hybridization depends on the ability of denatured DNA or RNA to reanneal with complementary strands in an environment just below their melting point. Formamide has been shown to reduce the melting temperature of DNA–RNA duplexes as a function of its concentration. In this case hybridization can be performed at lower temperatures in the presence of formamide. The thermal stability of the probe is also influenced by its length. Even if maximal hybridization rates are obtained with long probes, short probes are better because the probe has to diffuse into the dense matrix of a cell. Not only the length, but also the probe concentration, affects hybridization. In fact, the rate of reannealing is dependent on probe concentration; the higher the concentration of the probe, the higher the reannealing rate. Mismatching is particularly important for oligonucleotide probes, as it greatly influences hybrid stability, reducing both hybridization rate and thermal stability of the resulting duplexes (37).

3.3.6 Post-hybridization washes

The labelled probe can hybridize non-specifically to sequences which are partially homologous to the probe sequence. Such hybrids are less stable than perfectly matched ones and they can be dissociated by performing washes of various stringency. The stringency of the washes is thus an important way to improve the specificity of hybridization; this can be achieved by changing formamide concentration, salt concentration, and temperature.

Protocol 3. *In situ* hybridization in eukaryotic cells and DIG detection

Reagents

- 4% paraformaldehyde in DEPC-treated PBS (see *Protocol 1*)
- 0.2% Triton X-100 in 4% paraformaldehyde in DEPC-treated PBS
- 1 × SSC: 15 mM sodium citrate, 150 mM NaCl pH 7.4
- 50% formamide in 2 × SSC
- *In situ* hybridization mix: 50% formamide, 0.3 M NaCl, 20 mM Tris pH 7.4, 5 mM EDTA pH 8, 10 mM NaH_2PO_4 pH 8, 10% dextran sulfate, 1 × Denhardt's (0.02% polyvinyl chloride, 0.02% pyrrolidone, 0.02% BSA), 0.5 mg/ml tRNA
- DNase-free RNase A (Sigma Chem. Co.)
- Wash buffer: 0.4 M NaCl, 10 mM Tris pH 7.5, 5 mM EDTA
- Buffer 1: 100 mM Tris pH 7.5, 150 mM NaCl
- Buffer 2: 1% blocking reagent (Roche) in buffer 1
- Anti-DIG-AP, Fab fragment, 750 U/ml (Roche)
- Buffer 3: 100 mM Tris pH 9.5, 100 mM NaCl, 50 mM $MgCl_2$
- Nitroblue tetrazolium chloride (NBT) (Roche)
- 5-Bromo-4-chloro-3-indolyl-phosphate, 4-toluidine salt (BCIP) (Roche)
- Buffer 4: 10 mM Tris pH 8, 1 mM EDTA

A. *In situ hybridization*

1. Grow the cells on chamber slides (Nunc, Gibco BRL) to 70% confluence.
2. Wash the cells directly on the chamber slide three times with DEPC-treated PBS to eliminate any residual cell culture medium.
3. Fix the cells with 4% paraformaldehyde in DEPC-treated PBS for 10 min on ice.
4. Wash the cells twice with DEPC-treated PBS for 5 min at room temperature.
5. Treat the cells with DEPC-treated 70% ethanol for between 30 min and 2 h on ice to extract lipid membrane components and partially dehydrate the sample.
6. Wash the cells twice with DEPC-treated PBS for 5 min at room temperature.
7. Incubate the cells with 0.2% Triton X-100 in 4% paraformaldehyde in DEPC-treated PBS for 10 min on ice.
8. Wash the cells twice with DEPC-treated PBS for 5 min at room temperature.
9. Re-fix the cells with 4% paraformaldehyde in DEPC-treated PBS for 5 min on ice.
10. Wash the cells twice with DEPC-treated PBS for 5 min at room temperature.

11. Pre-hybridize the cells with 50% formamide in 2 × SSC for 10 min at room temperature.
12. Hybridize the cells with 80 μl of probe diluted 1:10 with ISH mix per slide. The diluted probe was heat inactivated at 80 °C for 2 min before use. Hybridization was performed overnight at 55 °C in humidified atmosphere.

B. *Washes*

1. Wash slides with 5 × SSC for 30 min at 55 °C to allow coverslips to slip off.
2. Wash slides with 50% formamide in 2 × SSC for 30 min at 55 °C.
3. Wash slides twice with wash buffer for 10 min at 37 °C.
4. Incubate slides with 20 μg/ml DNase-free RNase in wash buffer for 30 min at 37 °C to remove unbound riboprobe.
5. Wash slides in wash buffer for 5 min at 37 °C.
6. Wash slides twice with 2 × SSC for 10 min at 37 °C.

C. *DIG detection*

1. Wash slides in buffer 1 for 2 min at room temperature.
2. Incubate slides in buffer 2 for 30 min at room temperature in order to reduce background.
3. Wash slides in buffer 1 briefly at room temperature.
4. Incubate slides with 7.5 U/ml anti-DIG-AP, Fab fragments in buffer 1 for 30 min at room temperature, using 160 μl per each slide and protecting with coverslip.
5. Remove unbound antibody conjugate by washing slides twice with buffer 1 for 15 min at room temperature.
6. Wash slides in buffer 3 for 2 min.
7. Incubate slides at room temperature with freshly prepared colour substrate solution (45 μl NBT and 35 μl BCIP in 10 ml buffer 3) in the dark.
8. When the desired staining intensity is reached, stop the reaction by immersing the slides in buffer 4 for 5 min, and then briefly wash the slides in PBS.
9. Drain the slides and mount them with Aquamount and coverslips. Seal the coverslips with nail varnish.
10. Examine the slides under a microscope.

3.3.7 Immunological detection

When using indirect detection systems, it is very important to use a blocking step prior to the immunological part of the protocol to reduce background staining. In a typical procedure, the target cell preparation is incubated with blocking reagent for 30–60 minutes, washed briefly with PBS, and then incubated with the appropriate antibody-conjugate solution for a minimum of 30 minutes. The samples are then washed well two or three times to remove unbound antibody.

The antibody-conjugate can carry an enzyme which produces a coloured or fluorescent product during incubation with an appropriate substrate. The most common enzymes are peroxidase or alkaline phosphatase. Incubation times are often two to three hours but, if extra sensitivity is required, the reaction can be allowed to continue overnight. Excessive incubation times can lead to loss of resolution and specificity. At first it may be necessary to monitor the reaction to determine the optimal incubation time.

3.3.8 Controls

It is essential to have adequate controls to be confident of the specificity of the staining pattern and to eliminate potential problems both at the hybridization stage and during the detection step. Several different controls should be used routinely.

When RNA probes or single-stranded DNA probes (antisense sequences to the target) are used in ISH procedures, an ideal control is to use a sense probe generated from the complementary DNA strand by using an alternative RNA polymerase or PCR primer. Such sense probes do not have a sequence complementary to the target mRNA and therefore should not hybridize.

The most common way to check for non-specific hybridization or to check that the hybridization is to RNA and not to DNA is to treat cells with RNase to destroy endogenous mRNA transcripts. The preparations are incubated in DNase-free RNase (750 μg/ml) in 2 × SSC at 37 °C for 30 min after *Protocol 3A*, step 8.

A second control to check for non-specific hybridization is to process in parallel cells which are known not to contain the target RNA. If the probe used is specific for the target mRNA, no signal should be present. This control is particularly useful when studying transfected cells.

Finally, in order to check the detection step, hybridization should be carried out without addition of any probe. No staining should be apparent in this case at the end of the procedure.

3.4 Advantages and disadvantages of the different detection systems

The ability to detect specifically RNA transcripts by ISH depends not only on the choice of probe but also on the method of detection. The different methods can be broadly divided into three classes, colorimetric, fluorescent, and electron-dense product detection.

The advantages of colorimetric detection methods, such as those based on peroxidase and alkaline phosphatase, are the low diffusion, and thus good localization properties of the product, as well as high sensitivity of the assay, and stability of the precipitates. In the case of peroxidase, the diaminobenzidine (DAB)/imidazole reaction produces a dark brown colour which contrasts with the blue alkaline phosphatase product. Colorimetric methods can be detected by light microscopy, and this allows a broad definition of cellular regions (i.e. localization in the peripheral cytoplasm, cell extensions and processes, perinuclear cytoplasm). However, it is desirable to have sufficiently increased resolution to be able to correlate the spatial distribution of mRNAs with specific cellular components. As it appears that the spatial organization of mRNA is defined by structures which are not resolved by light microscopy (5), a higher level of resolution is necessary and confocal or electron microscopy is required.

An increasingly wide range of fluorescent detection methods are becoming available. The green fluorophore fluorescein, the red fluorochromes rhodamine and Texas Red, and the blue fluorochrome AMCA (amino-methylcoumarin-acetic acid) are all available directly linked to antibodies. Other methods depend on the production of a fluorescent compound as the result of peroxidase or alkaline phosphatase reactions (e.g. that producing Vector Red), and thus potentially give greater sensitivity. The use of fluorochromes as a detection system has several advantages over colorimetric detection: the greater sensitivity of fluorescent microscopy, the ability to excite up to three different immunofluorophores at the same time allowing the identification of three different transcripts (or transcripts and proteins) in co-localization studies, and the use of highly sensitive image acquisition systems and confocal microscopy. The latter offers improved facilities for quantification and for precise co-localization.

Detection of RNA transcripts by electron microscopy requires a detection system which generates electron-dense products. Both enzyme-linked peroxidase and colloidal gold detection systems have been used successfully. For example, the combination of electron microscopy and ISH has been used to demonstrate a close apposition of actin and vimentin mRNAs to cytoskeletal filaments (10). Following a RT-PCR procedure, anti-DIG antibodies linked to colloidal gold were used for localization of RNA transcripts in rat liver by electron microscopy (38).

3.5 Interpretation and limitations of ISH

The interpretation of ISH results in terms of mRNA localization is not necessarily straightforward. There may be ambiguities over what to define as localization and how to interpret the distribution in terms of cell shape.

Our appreciation of the mechanism and extent of mRNA localization is still relatively rudimentary, and as yet the nature of mRNA localization domains is poorly understood. Thus, the exact nature of localization and the size of localization domains may be different in different cell types (e.g. in a highly polarized cell such a neuron compared with a fibroblast) and between different transcripts.

Furthermore, if a transcript is found in only one part of the cell, it is clearly localized. However, if a transcript is found throughout the cell but at a distinctly higher concentration in one particular region, it can also be considered to be localized.

In many cell populations there are variations in cell shape, cell size, stage of the cell cycle, and the level of expression of the mRNA under study. These variations can make it difficult to assess mRNA localization or how that localization is influenced by a particular experimental treatment. Indeed, the major problem in the interpretation of localization patterns is the comparison of the patterns within a cell population where there is variation in cell shape and level of gene expression. The solution to this problem is the use of some form of quantification or image analysis to remove subjective bias from the analysis (see Section 3.6). Whether ISH is used to detect endogenous mRNAs in a mixed population of cells or to detect a product of a transgene after transfection, there is the problem that cells have different shapes even if they are derived from a single clone. Cells which are not in the same stage of the cell cycle have different shapes and cytoskeletal organizations, and thus may be expected to exhibit different localization patterns. This makes interpretation of data from mixed cell populations even more difficult. One solution to this problem is the use of synchronized cells in order to have cells at the same stage of the cell cycle and so be able to relate localization patterns to cell shape and cell cycle or growth.

It is well known that there are variations in the thickness of cytoplasm across the cell; for example the cytoplasm is particularly thick around the nucleus. Thus, one should always bear in mind that increased staining in a particular cytoplasmic region may not be due to higher concentration of transcripts but rather to a greater thickness of cytoplasm. Quantification may indicate that this effect is unlikely to explain the localization. Unequivocal interpretation is best achieved by using confocal microscopy which allows the electronic analysis of a series of cytoplasmic slices of equal thickness.

In some cases lack of sensitivity can be a limitation to the use of ISH. To date the successful use of ISH to study localization of mRNAs or their association with the cytoskeleton has been largely limited to the study of transcripts, either endogenous mRNAs expressed at relatively high levels, or products from transfected genes. Recently, however, Braissant and Wahli (39) have shown that it is possible to study the expression of genes with transcript levels ranging from very low (20–30) to high (several thousands) copy numbers per cell. They optimized a simplified *in situ* hybridization protocol which uses non-radioactively labelled probes to detect abundant and rare mRNAs in tissue sections. By processing cryosections immediately after sectioning and by increasing the hybridization time to 40 hours, they were able to detect the distribution gradients of mRNAs present at less than 30 copies within the cells.

The sensitivity of fluorescent ISH has increased as more powerful microscope systems have become available. By modifying fluorescence ISH and digital imaging microscopy, it has proved possible to identify specific nucleic acid sequences and reveal sites of RNA processing, transport, and cytoplasmic localization, to the extent that a quantitative approach which can identify RNA molecules in regions of low concentration has been developed (40, 41). Such methods have detected single β-actin mRNA molecules. Multiple probes specific to β-actin mRNA were used to generate high-intensity point sources that result from hybridization to individual mRNAs. Then the light intensity for each point source was quantitated by acquiring digital images from a series of focal planes through the cell and processing these images with a constrained deconvolution algorithm which gave a final image consisting of numerous distinct points of light throughout the cytoplasm. At present this type of study requires highly specialized microscope/image analysis equipment.

3.6 Quantification of localization signal

Subjective, visual analysis of mRNA distribution patterns needs, in most cases, to be backed up by image analysis and quantification of the staining pattern to obtain an unbiased, objective assessment of localization. This is particularly important when comparing the extent of localization of different transcripts and where there is variation in the level of expression and/or cell shape (see Section 3.5).

The simplest way to determine if a transcript is localized is for two or more individuals to assess, *independently* and *blind*, localization in a high number of cells in randomly selected fields (about 600 from at least two different experiments) (5). The experimenters count the number of cells showing localization and then calculate the percentage of cells which show a particular distribution pattern. For example, this system has been used successfully for the assessment of actin localization in chicken embryo fibroblasts and of the ability of the β-actin 3′ UTR to localize a reporter sequence (5, 42, 43).

A slightly more complicated method of quantification requires an image analysis system and a software package which allows measurements of intensity in different areas of the cytoplasm or across the cell to produce a profile of staining intensity. In this way it is possible to identify subcellular regions with a higher staining concentration. This method has been used for the analysis of transcripts exhibiting a perinuclear localization (7, 8, 28, 44). Intensity values were collected in the perinuclear and peripheral cytoplasmic regions of the cell, several measurements were taken for each cell, and 20 or more cells were analysed in each experiment. *Protocol 4* gives an outline of this method of quantification, which uses a simple image analysis system. *Figure 6* is a schematic representation of it.

Confocal microscopy also offers the possibilities of quantification. Even more sophisticated methods of quantification were used recently by Femino *et al.* (41) and, as described above, these allow the detection of single mRNA molecules and their quantification. By this study and others, it becomes more evident that the use of fluorescence ISH provides better definition and improves quantification in conjunction with the use of confocal microscopy and digital imaging microscopy.

Protocol 4. Quantification

Equipment

- Image analysis system connected to PC with image processing and analysis software
- Microscope

Method

1. Acquire several images (20–30) of cells by randomly choosing different fields in the same experiment.
2. For each cell, collect intensity values in at least two different regions of the cell by drawing five or six small boxes in each region and quantifying the staining (see *Figure 6*).
3. For the same preparation collect background measurements by selecting areas between cells.
4. Subtract the mean background measurement from all the intensity values for each cell preparation.
5. Perform the same measurements for appropriate controls.
6. If appropriate, express the corrected intensity values as ratios between the staining intensity in two distinct regions.
7. Perform statistical analysis.

4. Co-localization studies

ISH combined with light or confocal microscopy can provide information as to whether a mRNA is localized within a cell and in what part of the cytoplasm it is localized; used alone it is limited and tells us little or nothing about the interactions of that mRNA with the cytoskeleton. For example, an mRNA which is localized could theoretically be associated with organelles, with a specific membrane compartment, or with the cytoskeleton. Thus, ISH is essentially a first step which allows us to build up a description of mRNA

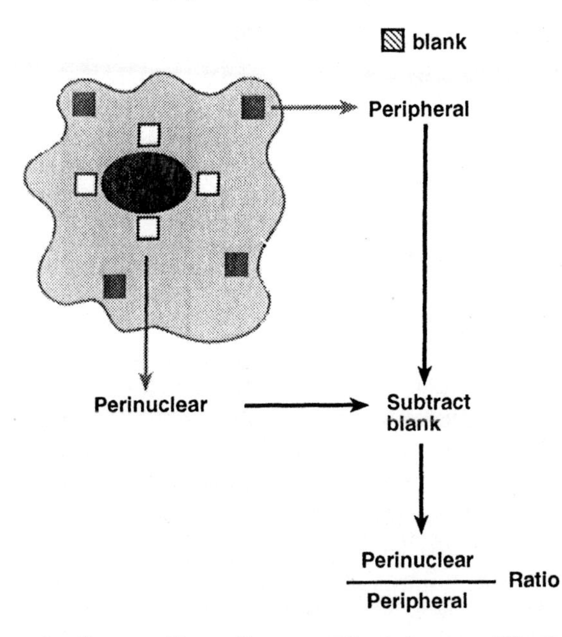

Figure 6. Schematic diagram illustrating a method for quantification of perinuclear mRNA localization. Multiple measurements are made in perinuclear and peripheral cytoplasm. After correction of the mean values for background staining (subtraction of a blank measurement taken outside the cell) a perinuclear/peripheral ratio can be calculated. By choosing appropriate areas this approach can be used to study localization in different cytoplasmic domains.

distribution within the cell. Parallel biochemical and localization studies can provide evidence that a mRNA is both localized and associated with the cytoskeleton. For example, c-myc mRNA is found in polysomes released from the cell matrix along with actin, and in the perinuclear cytoplasm (8), suggesting the mRNA is associated with microfilaments around the nucleus. However, in order to investigate directly the role of cytoskeletal components in biochemical interactions which lead to localization, the ISH approach needs to be directly combined with other approaches.

Combined with electron microscopy, ISH can show whether or not an mRNA is found in close proximity to a particular cell structure. In this way actin mRNA was found close to microfilaments and vimentin mRNA close to intermediate filaments (10). However, this approach also has its limitations. One must be very careful with the fixation procedure to avoid potential artefacts. This method also requires laborious, complex analysis to be certain of mRNA localization and association with specific structures (38).

A potentially informative approach is to combine ISH with immunohistochemistry followed by confocal analysis. By using antibodies directed against cytoskeletal components, it is possible to carry out co-localization of specific mRNAs with cytoskeletal components (16); this requires the antibody detection and the *in situ* detection system to be linked to different fluorochromes, such as fluoroscein and rhodamine. Although such dual labelling studies can be carried out using standard light microscopy, it is far preferable to use confocal microscopy, as this provides higher resolution and greater precision needed for co-localization studies. The presence of an mRNA close to specific cell components can be studied using either electron microscopy or immunohistochemistry combined with ISH; however, these approaches cannot demonstrate a direct interaction.

The demonstration that such interactions lead to localization requires disruption of the cytoskeleton. Agents such as colchicine and nocodazole can be used to produce microtubule depolymerization, and cytochalasin B, D, and lantrunculin can be used to disrupt microfilaments. Incubation of cells with such agents followed by ISH has the potential to show whether or not microtubules or microfilaments interact with the mRNAs or polysomes. Since there are no agents which produce specific disruption of intermediate filaments, this approach is not available for studying potential IF–mRNA interactions. Disadvantages of this approach are the possible non-specific effects of the agents and the complex interrelationships between the different cytoskeletal components, which leads to one filament system depending on others to maintain their integrity. For example, loss of microtubule organization may influence IF, and thus any effect of microtubule disruption on mRNA localization may reflect an association of that mRNA not with microtubules but with IF.

The increasing availability of mutant cell lines expressing altered cytoskeletal organization offers new and potentially more specific systems for investigating the effects of cytoskeletal disruption on mRNA localization.

5. Localization signals in mRNAs

There is now strong evidence that certain mRNAs are localized in different regions of cells, and that signals in particular sections of transcripts are involved in localization. More and more examples are appearing in the literature indicating an important role in localization for the 3′ untranslated region (3′ UTR) of mRNA (2, 3). A common approach in all these localization studies has been the combination of the use of chimeric genes introduced into cells by transfection or microinjection and the determination of transcript localization using ISH or biochemical fractionation. In one method a reporter sequence is attached to the region of transcript under investigation, the construct introduced into the cell, and the transcript localization studied. In the second method the gene of interest with part of the mRNA sequence removed or exchanged is introduced into the cells. An application of the first method is shown in *Figure 7*; fractionation, ISH, and

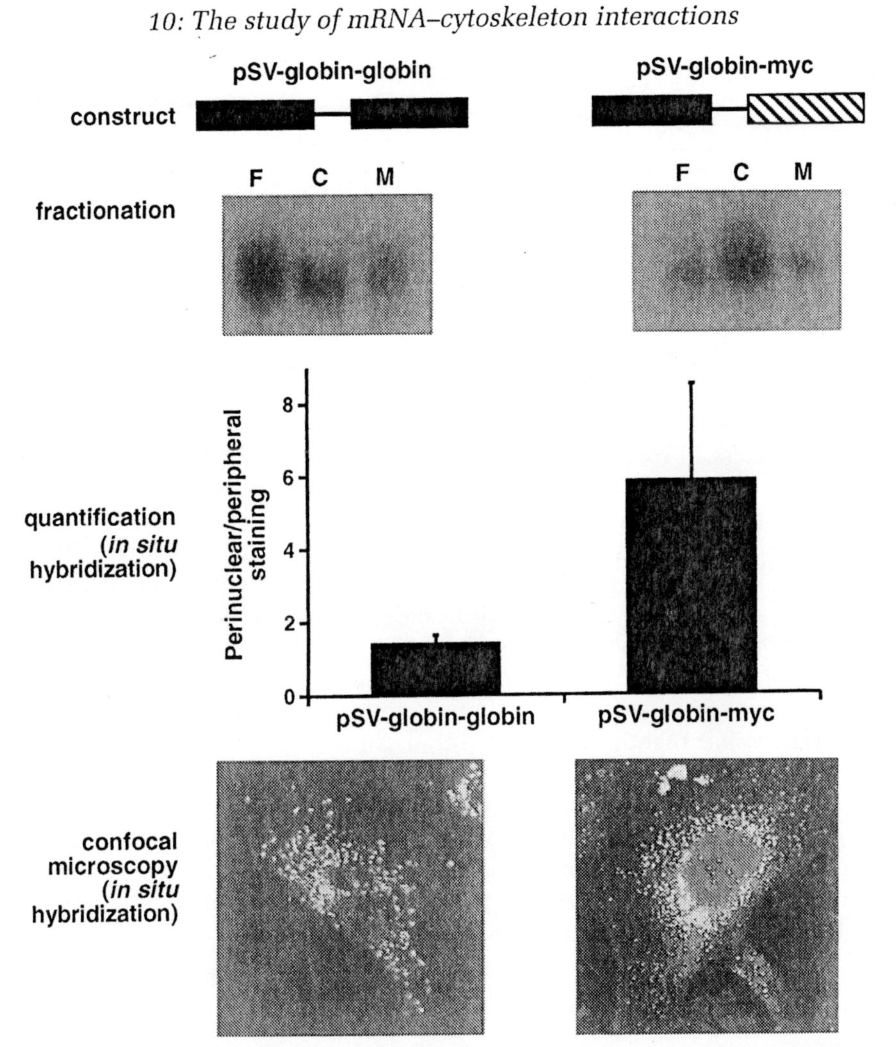

Figure 7. The use of reporter gene systems in transfected cells to investigate the role of 3′ UTR sequences. Cells were transfected with either a control β-globin construct with the globin 3′ UTR (globin–globin) or a chimeric construct with the c-myc 3′ UTR linked to globin coding sequences (globin–myc). Fractionation using *Protocol 1* showed a shift of globin transcripts from FP (F) to CBP (C) and increased perinuclear localization of globin transcripts as indicated by quantification of standard *in situ* hybridization (*Protocol 3*) and by confocal microscopy (8, 28).

quantification techniques were used to study the role of c-myc 3' UTR in localizing a reporter gene, in this case β-globin sequence.

5.1 The use of reporter genes to study localization signals

Genes of interest can be introduced into cells either transiently or stably, depending on the cell type and experiment. Stable introduction requires the presence of a selectable marker in the plasmid DNA. Using this approach, the cells containing the gene of interest can be readily obtained by growth in the appropriate selective medium, e.g. with the use of the neomycin resistance gene and selection with geneticin (G418). In this case, in fact, once the DNA has passed the cell membrane, it is free to enter the nucleus, to be transcribed and to become integrated into the host genome to generate a permanently transfected cell line. In the case of transiently transfected cells, the mRNA distribution is studied only 24–48 hours after transfection, and after a few days the foreign DNA is excluded from the cell and the transcript is no longer present. Transiently transfected cells usually exhibit higher expression of the transgene than stably transfected cells, possibly because in the transient transfectants either the DNA is highly transcribed or more than one DNA copy is present in the same cell. As a result of this high expression, the ISH signal is greater in transiently transfected cells and a less sensitive procedure is required. However, because the expression is so high, there is a risk of saturating the localization system so that any potential localization of the transcript is not evident. ISH of transiently transfected cells is carried out so soon after transfection that residual DNA is present in the medium, and this can remain attached to the cells after fixation and cause interference with the ISH signal. DNase treatment can reduce, but not always abolish, this problem.

When using modified genes to study localization, it is important to consider that it is possible that the new mRNA transcript may have a different stability. In this case any apparent change in localization may be secondary to the altered stability, or the altered stability may prevent a change in localization being evident by ISH. It is, therefore, very important to check the stability of the different transcripts in studies of mRNA localization which involve transfection of modified gene constructs with different sequences or different region of the same 3′ UTR (e.g. 28, 44).

Protocol 5. Stable transfection

Reagents

- LipofectAMINE reagent, 2 mg/ ml (Gibco BRL)
- Geneticin (Gibco BRL)
- Opti-MEM I reduced serum medium (Gibco BRL)

Method

1. In a 35 mm tissue culture plate, seed ~ 1–3 × 10^5 cells in 2 ml of the appropriate complete growth medium.
2. Incubate the cells at 37 °C in a CO_2 incubator until the cells are 50–80% confluent. This will usually take 18–24 h, but the time will vary among cell types.
3. For each transfection prepare solution A: 1–2 μg of DNA in 100 μl

Opti-MEM, and solution B: 5–10 μl of LipofectAMINE in 100 μl Opti-MEM.

4. Combine the two solutions, mix gently, and incubate at room temperature for 45 min to allow DNA–liposome complexes to form.
5. While complexes form, rinse the cells once with 2 ml Opti-MEM.
6. For each transfection, add 0.8 ml of Opti-MEM to the tube containing the complexes. Mix gently and overlay the diluted complex solution onto the rinsed cells.
7. Incubate the cells with the complexes for a minimum of 5 h at 37 °C in a CO_2 incubator.
8. Following incubation, add 1 ml of growth medium containing twice the normal concentration of serum without removing the transfection mixture.
9. Replace the medium with fresh, complete medium at 18–24 h following the start of transfection.
10. For stable expression, at 48 h after transfection, move cells into the selective medium for the reporter gene transfected. For example for transfection using DNA carrying the neomycin gene, geneticin antibiotic (G418) is added to the medium in a concentration ranging from 0.4–0.8 mg/ml.
11. Culture cells in G418 for two to three weeks, after which time cells not expressing the transgene should die. Select clones during this period or allow a mixed population to be generated. Maintain cell stocks in medium containing G418.

5.2 Different reporter genes

Different reporter genes have been used to study localization signals: β-globin mRNA coding sequences, β-galactosidase protein (β-Gal), and more recently the green fluorescent protein (GFP). The different reporters have specific characteristics which can be used to address different questions.

Originally, β-globin sequences were used as a reporter gene by our group to study localization signals in the c-myc 3′ UTR. It was chosen for three reasons: first, it was not expressed in a wide range of cells; secondly, its transcripts are of the high stability; thirdly, after transfection, the globin transcript did not show any particular localization pattern and was found at highest abundance in FP rather than CBP, thus indicating little or no association with the cytoskeleton. Thus, globin coding sequences present a stable, non-localized transcript present throughout the cytoplasm. Since the first studies on c-myc 3′ UTR (8), globin has been used as a reporter gene to look at the role of vimentin and myosin heavy chain 3′ UTRs in localizing the corresponding transcripts (44, 45). In both cases the 3′ UTRs were able to localize globin transcript to the perinuclear region of the cell. In the case of myosin heavy chain, this has also been found in cardiomyocytes using a β-galactosidase reporter (46) and in CHO cells using a GFP reporter (authors' unpublished data).

β-Galactosidase as a reporter gene can be used in two different ways: its transcript can be detected by ISH or its protein by histochemistry. Histochemical detection of β-Gal protein can be used to determine mRNA localization as the β-Gal reporter protein has been shown to diffuse only a short distance from its site of synthesis (four subunits are required for activity), making it an excellent marker for mRNA localization (42, 43). The detection of β-Gal activity instead of the mRNA is an advantage, as the protein is more stable compared to the mRNA, and protein detection is more straightforward than ISH techniques.

Green fluorescent protein (GFP) has been used widely as a reporter in protein targeting studies by fusing GFP to the target protein. The distribution of GFP cannot be used to study mRNA localization, as β-Gal has been used, because GFP diffuses rapidly from its site of synthesis throughout the cell. However, ISH can be used to detect GFP mRNA, and in mRNA localization studies this reporter presents the advantage that the protein is visible directly under UV light, allowing the selection of cells which have been positively transfected and the estimation of transfection efficiency. In our laboratory this has proved a useful approach with transiently transfected cells, but the toxicity of GFP makes it difficult to obtain stable transfectants.

6. mRNA transport

The approaches discussed in the previous sections can all provide a static picture of localization and can provide evidence that a specific mRNA is associated with the cytoskeleton or is localized. Indeed, these approaches have provided a large body of evidence to support the hypothesis that, in mammalian cells, there is a sorting and targeting of transcripts to particular locations and that this targeting involves the cytoskeleton and signals in the 3′ UTR of the mRNAs.

To examine the dynamics of mRNA localization one must study mRNA transport. At present two alternative methods present themselves for the study of individual mRNAs. One is to microinject labelled RNA into cells. The RNA of interest (or a part containing the appropriate localization signal) is synthesized *in vitro* using standard transcription methods (see *Protocol 2*) but including uridine tagged with a fluorescent label such as fluorescein in the incubation to produce fluorescent RNA. An alternative with a short RNA sequence would be chemical synthesis. This method has proved effective in the study of myelin basic protein mRNA transport (14, 15). Combination of such an approach with the use of cytoskeleton-disrupting drugs allows the role of microtubules and microfilaments in mRNA transport to be studied.

Potentially the site of microinjection may be critical, and this should be considered when such experiments are designed. For example, if localization requires the transcript to attach to some factor in the nucleus, then injection into the nucleus may be essential.

The second approach to mRNA transport is the use of inducible promoters and cell transfection. In these experiments an inducible promoter such as the fos promoter (responds to serum) or the tetracycline promoter system (47) is included in the vector containing the gene coding for the RNA of interest. The vector is introduced into an appropriate cell line by transfection (see *Protocol 5*) and culture conditions modified to produce a short pulse of expression of the transgene. ISH can then be used to follow the distribution of mRNA over time and thus mRNA transport.

There is increasing evidence for localized mRNAs being transported in particles containing various components of the protein synthetic apparatus. The movement of such particles can be studied in living cells by the use of the membrane-permeable nucleic acid stain STYO 14 (17).

7. The biochemistry of RNA–cytoskeleton interactions: RNA-binding proteins

Ultimately, mRNA sorting, targeting, and localization must depend on mRNA–protein interactions which result in some mRNAs being transported or associated with certain cytoskeletal components in certain parts of the cell. In all cases discovered to date, the signal for localization resides in the 3′ UTR of the mRNA (1–3). In order to achieve localization these signals must interact with specific proteins. Little is, at present, known about such proteins. Theoretically, the RNA could interact directly with the cytoskeleton or indirectly through linker proteins or other components of the protein synthetic apparatus, such as initiation factors. High resolution confocal microscopy has shown that those mRNAs which are transported in particles contain translation factors and ribosomes (14–16).

The first step in analysing the proteins involved in localization is to identify the proteins that interact specifically with the 3′ UTR localization signal. To date, this has usually been achieved using standard techniques such as gel retardation and UV cross-linking assays. Of the two the UV cross-linking is generally regarded as the less problematical. It involves an initial incubation of labelled RNA and protein extract, a cross-linking step to join RNA and protein, an RNase digestion to remove excess RNA, and finally SDS gel polyacrylamide electrophoresis to separate the proteins according to molecular weight. Proteins which have bound the RNA of interest retained a small labelled RNA fragment because they have protected that RNA from digestion, and they can be detected as labelled bands in the gel. Compared to gel retardation assays this method has the advantages that it gives an indication of the number of proteins binding and also allows estimation of their approximate molecular weight. In all RNA–protein binding assays it is vital to reduce or eliminate the possibility that the binding is non-specific and is due to a specific interaction. This is usually achieved by competition experiments in which up to 100 times molar excess unlabelled RNA is included in the incubation. For example, an RNA corresponding a known localization signal will compete for any binding protein and thus reduce or abolish the binding, whereas an unrelated RNA or a sequence containing a mutation known to abolish localization should have no effect.

These techniques are being used to identify proteins which bind to known localization and transport elements in 3′ UTRs (48, 49) as well as 3′ UTR-binding proteins of unknown function (31) but which have a potential role in mRNA–cytoskeleton interactions.

Protocol 6. UV cross-linking to detect RNA–cytoskeleton interactions

Equipment and reagents

- UV cross-linker (e.g. Stratalinker)
- Facilities for handling radioactivity [^{32}P]
- Facilities for quantifying bound radioactivity (e.g. Canberra Packard Instantimager)
- [^{32}P]UTP, 800 Ci/ mmol (Dupont Ltd.)
- Plasmids: the RNA sequences to be studied should be present in a vector containing appropriate polymerase promoter sites (e.g. Bluescript plasmids which contain T3 and T7 polymerase sites)
- Reagents for labelling transcripts (see *Protocol 2*)
- RNase inhibitor, 40 U/μl (Promega)
- 5% polyacrylamide (Severn Biotech Ltd.)
- Incubation buffer: 10 mM Hepes pH 7.2, 0.5 μg tRNA, 100 mM KCl, 3 mM $MgCl_2$, 5% (v/v) glycerol, 1 mM dithiothreitol
- Heparin, 50 mg/ml (Sigma Chem. Co.)
- RNase A (Sigma Chem. Co.)
- RNase T1 (Sigma Chem. Co.)
- 10% SDS (sodium dodecyl sulfate)
- 10% mercaptoethanol
- 10% SDS–polyacrylamide

A. *Protein extract*

1. Prepare protein extract to be studied. This might be a total cytoplasmic extract or a cytoskeletal fraction of some sort, e.g. a 130 mM KCl extract as prepared in *Protocol 1*. If necessary include protease inhibitors. Store in aliquots at –80 °C.

B. *RNA transcripts*

1. Linearize 10 μg plasmid with 40 U of appropriate restriction enzyme as described in *Protocol 2*.
2. Extract linearized plasmid with phenol:chloroform and precipitate the DNA with ethanol (30).
3. Synthesize capped RNA transcripts using a transcription kit (e.g. Pharmacia Transprobe T or Stratagene mcapRNA Transcription kit). Mix 1 μg of linearized DNA and 15 U of polymerase and kit buffer (with

NTPs and capping analogues) in a final volume of 25 μl according to the manufacturer's instructions. To produce labelled transcripts add 20 μCi [^{32}P]UTP (800 Ci/mmol). All radioactive work should be carried out behind a protective screen such as a Perspex barrier. Incubate at 37 °C for 30 min.

4. Unlabelled transcripts are prepared in parallel with the addition of unlabelled UTP from the transcription kit in the place of radioactive UTP.
5. Purify RNA transcripts using Chroma Spin columns (Clontech), followed by phenol:chloroform extraction and ethanol precipitation (30).
6. Resuspend RNA in 20 μl DEPC-treated water and add 0.5 μl RNase inhibitor (RNasin).
7. Check integrity of the transcripts by running them in a non-denaturing 5% (w/v) polyacrylamide gel at 120 V for 0.5–3 h (according to size), or in a denaturing urea–polyacrylamide gel.
8. Measure by scintillation counting the amount of radioactivity in a small aliquot of RNA. From the specific activity of the UTP used, its activity date, and the number of Us in the RNA synthesized, it is then possible to calculate the amount of labelled probe synthesized. By direct comparison one can then also calculate the amount of unlabelled transcripts synthesized in parallel incubations.
9. Aliquot transcripts and store at –80 °C.

C. *Cross-linking assay*

1. Add labelled RNA (10^5 c.p.m.) and 10–15 μg of protein extract to incubation buffer so that the final volume is 10 μl. In the case of a competition experiment add the unlabelled RNA immediately before addition of the labelled RNA. Incubate the mixture for 25 min at room temperature (22 °C).
2. Add 50 μg heparin and incubate for a further 5 min.
3. Irradiate the mixture for 2 × 960 mJ on ice.
4. Add 4 μg RNase A, 12 U of RNase T1, and incubate for 15 min at 37 °C.
5. Add 10% SDS, 10% mercaptoethanol, and heat at 80 °C for 5 min.
6. Load onto a 10% SDS–polyacrylamide gel with 5% stacking gel. Run molecular weight standards in parallel. Electrophorese for 1 h at 25 mA. Dry the gel.
7. Visualize and quantify any bound RNA using the Packard Instantimager or by autoradiography and densitometry. Estimate the molecular weight of the RNA-binding proteins by comparison of mobility of radioactive bands with the known marker proteins.

8. Future developments

Both our knowledge of mRNA–cytoskeleton interactions and the techniques available to investigate them are advancing rapidly. The development of confocal and digital microscopy coupled to sophisticated image analysis has allowed more accurate localization and co-localization studies (14–16) to the extent that single mRNAs have been detected (41). These techniques have also shown that mRNAs, such as those coding for actin and myelin basic protein, are transported in RNA-containing particles along with other components of the protein synthetic apparatus, such as initiation factors, aminoacyl tRNA synthetases, and ribosomal subunits (15, 17). Use of ISH with sophisticated microscopy presents major possibilities to describe localization events, along with use of co-localization and other approaches to define mRNA–cytoskeleton interactions. On the other hand, more precise biochemical techniques are required to produce fractions enriched in cytoskeletal-bound mRNAs or localized mRNAs to such a degree that differential screening techniques can be used to characterize different subcellular pools of mRNA.

Four outstanding questions remain: what proportion of mRNAs (and which ones) are localized and/or associated with the cytoskeleton; what are the functional consequences of a loss of mRNA localization; what are the detailed mechanisms which allow some mRNAs to be sorted to the perinuclear cytoplasm and others to the cell periphery of cytoplasmic extensions; in what form of complex are mRNAs localized or transported?

With regard to mechanism and mRNA transport complexes, major efforts should now be made to accurately characterize RNA particles, to describe the mechanism and site of their formation, and how RNA transport is related to nuclear RNA export. It will also be important to dissect out the relationship between localization of mRNAs, their cytoskeletal-association, their stability, and their translation. At the heart of much of this work is the search for the proteins involved in mRNA–cytoskeleton or mRNA particle complexes. As knowledge of RNA localization signals becomes more extensive, it will allow affinity and yeast two/three hybrid methods (50) to be used to detect and isolate proteins which bind to localization signals or which form part of mRNA–cytoskeleton complexes.

The ability to identify and define localization signals in 3′ UTRs combined with recombinant DNA technology provides the basis to modify genes to alter the localization of the transcripts derived from them without altering the coding sequence of the protein. Introduction of such constructs into cell lines of no or low expression (e.g. those derived from 'knock-out' mice or ES cells) then presents the opportunity to study the functional effects of modified mRNA localization.

The presence of localization mechanisms in yeast (40) and *Drosophila* (4) will present in the future strong genetic methods to dissect out mRNA–

cytoskeleton interactions and discover gene products with possible mammalian homologues having important roles in localization and transport of RNA.

Acknowledgements

J. E. Hesketh is supported by SOAEFD, G. Bermano by Muscular Dystrophy Group. We thank Gill Campbell for the work with these protocols over many years.

References

1. Bassell, G. and Singer, R. H. (1997). *Curr. Opin. Cell Biol.*, **9**, 109.
2. Hesketh, J. E. (1996). *Exp. Cell Res.*, **225**, 219.
3. Nasmyth, K. and Jansen, R.-P. (1997). *Curr. Opin. Cell Biol.*, **9**, 396.
4. St Johnston, D. (1995). *Cell*, **81**, 161.
5. Sundell, C. L. and Singer, R. H. (1990). *J. Cell Biol.*, **111**, 2397.
6. Wilson, I. A., Brindle, K. M., and Fulton, A. M. (1995). *Biochem. J.*, **308**, 599.
7. Mahon, P., Partridge, K., Beattie, J. H., Glover, A., and Hesketh, J. E. (1997). *Biochim. Biophys. Acta*, **1358**, 153.
8. Hesketh, J. E., Campbell, G. P., Piechacyzk, M., and Blanchard, J.-M. (1994). *Biochem. J.*, **298**, 143.
9. Ainger, K. D., Avossa, F., Morgan, S. J., Hill, C., Barry, E., Barbarese, E., *et al.* (1993). *J. Cell Biol.*, **123**, 431.
10. Singer, R. H., Langevin, G. L., and Lawrence, J. B. (1989). *J. Cell Biol.*, **108**, 2343.
11. Lenk, R., Ransom, L., Kaufmann, Y., and Penman, S. (1977). *Cell*, **10**, 67.
12. Taneja, K. L., Lifshitz, L. M., Fay, F. S., and Singer, R. H. (1992). *J. Cell Biol.*, **119**, 1245.
13. Hesketh, J. E., Campbell, G. P., and Whitelaw, P. F. (1991). *Biochem. J.*, **274**, 607.
14. Barbarese, E., Koppel, D. E., Deutscher, M. P., Smith, C. L., Ainger, K., Morgan, F., *et al.* (1995). *J. Cell Sci.*, **108**, 278.
15. Ainger, K., Avossa, D., Diana, A. S., Barry, C., Barbarese, E., and Carson, J. H. (1997). *J. Cell Biol.*, **138**, 1077.
16. Bassell, G. J., Zhang, H., Byrd, A. L., Femino, A., Singer, R. H., Taneja, K. L., *et al.* (1998). *J. Neurosci.*, **18**, 251.
17. Knowles, R. B., Sabry, J. H., Martone, M. E., Deerinck, T. J., Ellisman, M. H., Bassell, G. J. *et al.* (1996). *J. Neurosci.*, **16**, 7812.
18. Hesketh, J. E. and Pryme, I. F. (1988). *FEBS Lett.*, **231**, 62.
19. Dang, C. V., Yang, D. C. H., and Pollard, T. D. (1983). *J. Cell Biol.*, **96**, 1138.
20. Hesketh, J. E. and Pryme, I. F. (1991). *Biochem. J.*, **277**, 1.
21. Bird, R. and Sells, B. (1986). *Biochim. Biophys. Acta*, **868**, 215.
22. Vedeler, A., Pryme, I. F., and Hesketh, J. E. (1991). *Mol. Cell. Biochem.*, **100**, 2397.
23. Hovland, R., Campbell, G. P., Pryme, I. F., and Hesketh, J. E. (1995). *Biochem. J.*, **310**, 193.
24. Moldeus, P., Hogberg, J., and Orrhenius, S. (1978). In *Methods in enzymology*, Vol. III, p. 60.

25. Chomczynski, P. and Sacchi, N. (1987). *Anal. Biochem.*, **162**, 156.
26. Pardue, M. L. and Gall, J. G. (1969). *Proc. Natl. Acad. Sci. USA*, **64**, 600.
27. Aigner, S. and Pette, D. (1990). *Histochemistry*, **95**, 11.
28. Veyrune, J.-L., Campbell, G. P., Wiseman, J. W., Blanchard, J.-L., and Hesketh, J. E. (1996). *J. Cell Sci.*, **109**, 1185.
29. Mahon, P., Beattie, J. H., Glover, L. A., and Hesketh, J. E. (1995). *FEBS Lett.*, **373**, 76.
30. Sambrook, I., Fritsch, E. F., and Maniatis, T. (ed.) (1989). *Molecular cloning: a laboratory manual.* Cold Spring Harbor Laboratory Press, NY.
31. Zehner, Z. E., Shepherd, R. K., Gabryszuk, J., Fu, T.-F., Al-Ali, M., and Holmes, W. M. (1997). *Nucleic Acids Res.*, **25**, 3362.
32. White, K. N. and Fu, L. (1995). *Methods Mol. Cell. Biol.*, **5**, 222.
33. Tautz, D., Hulskamp, M., and Sommer, R. J. (1992). In *In situ hybridization: a practical approach* (ed. D. G. Wilkinson). IRL Press, Oxford.
34. Pomeroy, M. E., Lawrence, J. B., Singer, R. H., and Billings-Gagliardi, S. (1991). *Dev. Biol.*, **143**, 58.
35. Cripe, L., Morris, E., and Fulton, A. B. (1993). *Proc. Natl. Acad. Sci. USA*, **90**, 2724.
36. Morris, E. J. and Fulton, A. B. (1994). *J. Cell Sci.*, **107**, 377.
37. Non-radioactive *in situ* hybridization application manual. Boehringer Mannheim.
38. Cohen, N. S. (1996). *J. Cell. Biochem.*, **61**, 81.
39. Braissant, O. and Wahli, W. (1998). *Boehringer Mannheim Biochem.*, **1**, 10.
40. Long, R. M., Singer, R. H., Meng, X., Gonzales, I., Nasmyth, K., and Jansen, R.-P. (1997). *Science*, **277**, 383.
41. Femino, A. M., Fay, F. S., Fogarty, K., and Singer, R. H. (1998). *Science*, **280**, 585.
42. Kislauskis, E. H., Li, Z., Taneja, K. L., and Singer, R. H. (1993). *J. Cell Biol.*, **123**, 165.
43. Kislauskis, E. H., Xiaochun, Z., and Singer, R. H. (1994). *J. Cell Biol.*, **127**, 441.
44. Wiseman, J. W., Glover, A., and Hesketh, J. E. (1997). *Int. J. Biochem. Cell Biol.*, **29**, 1013.
45. Wiseman, J. W., Glover, A., and Hesketh, J. E. (1997). *Cell Biol. Int.*, **21**, 243.
46. Goldspink, P. H., Sharp, W. W., and Russell, B. (1996). *J. Cell Sci.*, **110**, 2969.
47. Gossen, M., Freunlieb, S., Bender, G., Muller, G., Hillen, W., and Bujard, H. (1995). *Science*, **268**, 1766.
48. Hoek, K. S., Kidd, G. J., Carson, J. H., and Smith, R. (1998). *Biochemistry*, **37**, 7021.
49. Ross, A. F., Oleynikov, Y., Kislauskis, E. H., Taneja, K. L., and Singer, R. H. (1997). *Mol. Cell. Biol.*, **17**, 2158.
50. SenGupta, D. J., Zhang, B., Kraemer, B., Pochart, P., Fields, S., and Wickens, M. (1996). *Proc. Natl. Acad. Sci. USA*, **93**, 8496.

11

Cell shape control and mechanical signalling through the cytoskeleton

WOLFGANG H. GOLDMANN, JOSÉ LUIS ALONSO, KRZYSZTOF BOJANOWSKI, CLIFFORD BRANGWYNNE, CHRISTOPHER S. CHEN, MARINA E. CHICUREL, LAURA DIKE, SUI HUANG, KYUNG-MI LEE, ANDREW MANIOTIS, ROBERT MANNIX, HELEN McNAMEE, CHRISTIAN J. MEYER, KEIJI NARUSE, KEVIN KIT PARKER, GEORGE PLOPPER,THOMAS POLTE, NING WANG, LI YAN, and DONALD E. INGBER

1. Introduction

For the past twenty years, our group has explored the possibility that cells and tissues use a form of architecture, known as 'tensegrity', to control their shape and to transduce mechanical signals into changes in biochemistry and gene expression (1–5). The importance of the tensegrity paradigm is that it predicts that the form and function of living cells are controlled through changes in mechanical interactions between cells and their underlying extracellular matrix (ECM) adhesions that, in turn, alter cytoskeletal and nuclear structure inside the cell (1–6).

Tensegrity structures gain their stability from continuous tension and local compression, much like a tent fabric stiffened by internal tent poles and external pegs (5). In contrast to conventional engineering models, which viewed the cell as a viscous cytosol surrounded by an elastic membrane, the tensegrity model assumes that the cell is a discrete (porous) mechanical network that requires prestress (internal tension) to fully stabilize itself. This model recognizes that the molecular elements of the cell are dynamic. However, it predicts that, at any instant in time, the entire cell will behave as if it were 'hard-wired' by a continuous series of molecular struts, cables, and ropes that stretch from specific adhesion receptors on the cell surface to physically couple to discrete contacts on the surface of the nucleus and from there to the chromatin and genes within (2–4, 7). In this type of structure, external mechanical signals would not be transmitted equally across all points on the cell surface, rather mechanical stresses should be preferentially transferred across specific transmembrane receptors that physically link the cytoskeleton to the ECM or to junctions on other cells (2, 6). Furthermore, changes in the balance of forces transmitted through the cytoskeletal network should result in short- and long-range changes in molecular arrangements in the cytoskeleton and nucleus as well as associated alterations in thermodynamic and kinetic parameters that may influence cellular biochemistry (2, 4, 6). The tensegrity model, therefore, led to many testable predictions relating to how cells structure themselves as well as how they sense and respond to external mechanical signals.

Since this theory was first published, we have set out to systematically test these predictions. Because of the key roles of cell microarchitecture and mechanical forces in this model, this required that we develop entirely new experimental techniques that would allow us to control cell deformation independently of cell binding to growth factors or ECM as well as methods for applying controlled mechanical stresses to specific cell surface receptors. In this chapter, we describe these methods and briefly review new insights into cytoskeletal signalling and cell regulation that have emerged from these studies.

2. Application of techniques

2.1 Control of cell shape by varying extracellular matrix density

One of the fundamental predictions of the tensegrity model is that cell form and function are determined through *mechanical* interactions with the cell's ECM adhesions. The corollary to this is that cell shape, mechanics, and function should change if cell–ECM binding interactions are varied so as to prevent or promote cell distortion. Indeed, in early studies, we were able to show that the growth and differentiation of specialized cells (e.g. capillary endothelial cells, primary hepatocytes) can be controlled by altering the ability of the ECM substrate to resist cell tractional forces (8–10).

This was accomplished by varying the density of purified ECM molecules, such as fibronectin (FN), coated on otherwise non-adhesive, bacteriological plastic dishes, as described in *Protocol 1*. In these studies, the cells had to be cultured in serum-free medium to focus on effects of varying cell–ECM binding interactions, independently of exogenous matrix proteins that are found in high concentrations in serum (e.g. FN, vitronectin). When cells are plated on these dishes, cell spreading and growth increase in parallel as the ECM coating density is raised (8–10) (*Figure 1*). When spreading is restricted and growth suppressed, differentiation (e.g. tube formation by capillary cells, secretion of liver-specific proteins by hepatocytes) is concomitantly turned on (9, 10). This system has also been used to demonstrate shape-dependent control of cell contractility and mechanics in vascular smooth muscle cells (11). We are now using this as a model to analyse how changes in cell binding to ECM and concomitant alterations in cell shape modulate the intracellular

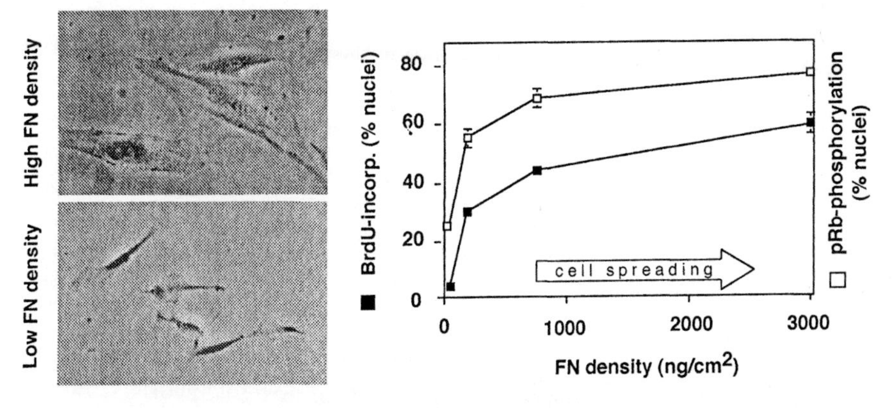

Figure 1. Modulation of cell shape. (*Left*) Phase-contrast view of human capillary endothelial cells plated on bacteriological dishes coated with high (*top*) and low (*bottom*) FN density. (*Right*) Cells plated on progressively higher FN density exhibit an increase in pRb phosphorylation and S phase entry as measured by nuclear BrdU incorporation. pRb phosphorylation was measured using *in situ* assay (12).

signalling cascade that controls this contractile response as well as cell cycle progression.

Protocol 1. Control of cell shape and function by varying ECM coating density

Equipment and reagents

- Bacteriological Petri dishes (Falcon Labware)
- 96-well plates (Immunolon II, Dynatech)
- Confluent cell monolayers
- Culture medium
- Lyophilized FN (Collaborative Biomedical Products, or Organon Teknika-Cappel, or Calbiochem)
- Carbonate buffer pH 9.4
- PBS
- 1% bovine serum albumin (BSA; Fraction V, Sigma)
- Trypsin-EDTA
- Serum-free media

Method

1. Two days prior to experiments, re-feed confluent cell monolayers with conventional culture medium containing low (0.5–1%) serum and no growth supplements in order to induce the cells to enter quiescence. Note: this may be difficult with transformed cells or certain cell lines. We have found that lovastatin provides an even more efficient method to synchronize cells in early G1 (12).
2. At least one day prior to experiment, resuspend purified ECM component, such as lyophilized FN in sterile distilled water (final concentration of 5 μg/ml) and store at 4°C. Allow 30–60 min for material to go into solution; do not pipette, agitate, or swirl. Note: similar results

Protocol 1. *Continued*

have been obtained with FN from multiple commercial sources. Store aliquots of lyophilized ECM molecules at –70 °C.

3. Dilute FN in carbonate buffer pH 9.4 to a final concentration that will add the required FN per well or dish in the following coating volumes: 100 μl/well in 96-well plates; 500 μl/24 mm dish, 5 ml/60 mm dish, 20 ml/100 mm dish, 30 ml/150 mm dish.
4. Incubate dishes overnight at 4 °C. Dishes can be stored for longer times in the cold prior to use. Cover tightly with Parafilm to minimize evaporation.
5. On the day of experiment, aspirate coating solution, wash twice with PBS, once with basal medium, and then block non-specific binding sites by incubating dishes in medium containing 1% BSA at 37 °C for at least 30 min.
6. While ECM-coated plates are sitting in medium containing 1% BSA, dissociate quiescent cell monolayers by brief exposure to trypsin-EDTA, collect by centrifugation, wash in medium containing 1% BSA to neutralize the trypsin, and plate cells on ECM-coated dishes in chemically-defined, serum-free medium. Note: use of serum to stop the trypsin will interfere with ECM-dependent control of cell shape and function; soybean trypsin inhibitor can also be added to ensure complete trypsin inactivation.
7. If the primary focus is on ECM-dependent control of cell shape and function, plate cells at a low density to minimize cell–cell contact formation in chemically-defined, serum-free medium containing 1% BSA. The exact composition of the medium will be dictated by the cell of choice. In certain cell types, such as capillary endothelial cells, higher plating densities are utilized when studies focus on control of ECM-dependent modulation of multicellular differentiation (e.g. capillary tube formation) (9).

2.2 Analysis of focal adhesion formation and integrin signalling

The results obtained with cells cultured on varying ECM densities demonstrated close coupling between cell spreading and growth (8–10), as demonstrated in past studies using other model systems (13). However, increasing the FN coating density does more than promote cell spreading, it also induces local clustering of cell surface ECM receptors, called integrins (14, 15). For this reason, it was necessary to develop methods to dissect out the biological effects induced by integrin clustering, independent of any associated cell shape change. We accomplished this by allowing round cells plated on a low FN density or suspended spherical cells to bind to small microbeads (4.5 μm diameter) coated with a high density of FN, synthetic RGD-containing peptides, or specific anti-integrin antibodies (*Protocol 2*). Under these conditions, we could demonstrate that cells formed focal adhesion complexes (FACs) containing clustered integrins, talin, and vinculin, directly at the site of bead binding within 5–15 minutes after stimulation (16). In contrast, beads coated with a control ligand, acetylated-low density lipoprotein (AcLDL), bound to cell surfaces via transmembrane metabolic (scavenger) receptors but did not induce recruitment of any of these FAC proteins.

The rapid and synchronous induction of FAC formation obtained with this bead technique permitted an analysis of integrin signalling not possible with conventional methods. For example, we extended this work by using immunofluorescence microscopy to demonstrate that many signal transducing molecules that are turned on in response to binding to growth factor receptors as well as integrins (e.g. $pp60^{c\text{-}src}$, $pp125^{FAK}$, phosphatidylinositol-3-kinase, phospholipase C-γ, Na^+/H^+ antiporter, protein tyrosine phosphorylation) were also recruited to the cytoskeletal framework of the FAC in an integrin-specific manner and at similar times (17) (*Figure 2*). Finally, we took advantage of the fact that the beads we used were paramagnetic to develop a means to physically isolate these FACs away from the cell and remaining cytoskeleton (16, 17) (*Figure 3*; also see *Protocol 3*). Western blot analysis confirmed that FACs isolated using RGD-beads were enriched for $pp60^{c\text{-}src}$, $pp125^{FAK}$, phospholipase C-γ, and the Na^+/H^+ antiporter when compared with intact cytoskeleton or basal cell surface preparations that retained lipid bilayer. Isolated FACs were also greatly enriched for the high affinity fibroblast growth factor receptor. Most importantly, isolated FACs continued

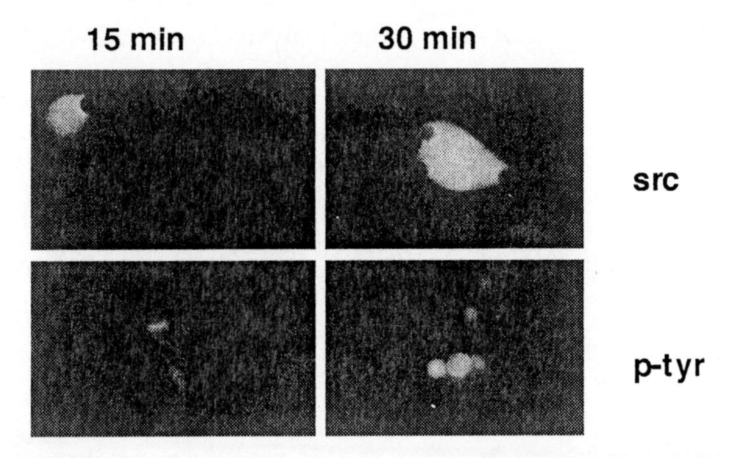

Figure 2. Recruitment of signalling molecules to the FAC in response to cell binding to FN-coated microbeads for 15 min (*left*) and 30 min (*right*). Positive immunostaining for src- and tyrosine-phosphorylated proteins is concentrated in a crescent-like pattern along the cell–bead interface.

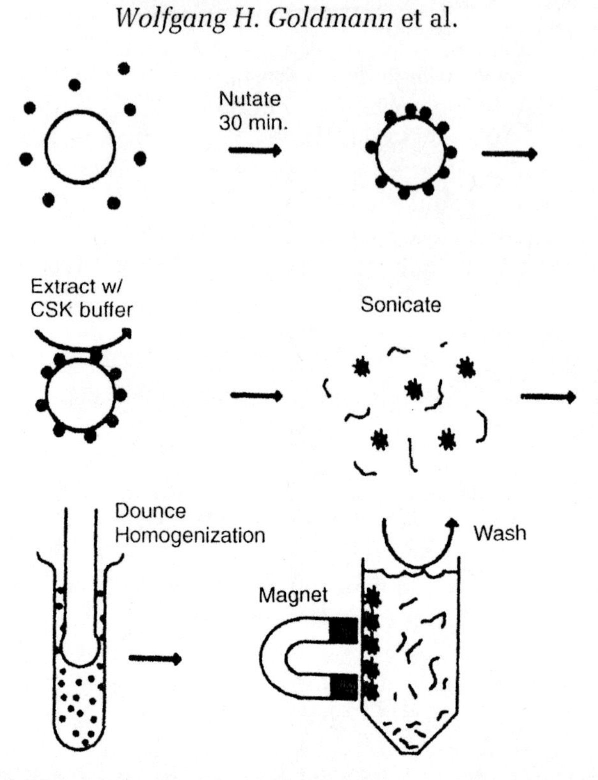

Figure 3. FAC isolation. Cells bound to RGD-coated beads were magnetically pelleted, extracted in CSK buffer, sonicated, and homogenized to remove nuclei and structures not intimately associated with the beads.

to exhibit multiple chemical signalling activities *in vitro*, including protein tyrosine kinase activities ($pp60^{c\text{-}src}$ and $pp125^{FAK}$) as well as the ability to undergo multiple sequential steps in the inositol lipid synthesis cascade (17, 18). This work, in combination with similar work from other laboratories (19), led to the realization that the FAC represents a major site for signal integration between growth factor and integrin pathways at the cell surface.

Protocol 2. Rapid induction of focal adhesion formation

Equipment and reagents

- Glass coverslips or slides (LabTek 8-well; Nalge NUNC International)
- Serum-free, chemically-defined medium
- Tosyl-activated magnetic microbeads (4.5 μm diameter; Dynal Inc.)
- 0.1 M carbonate buffer pH 9.4
- PBS
- 0.1% BSA
- Cytoskeleton stabilizing buffer (CSK–TX): 50 mM NaCl, 150 mM sucrose, 3 mM $MgCl_2$, 20 μg/ml aprotinin, 1 μg/ml leupeptin, 1 μg/ml pepstatin, 1 mM phenylmethylsulfonyl fluoride, 10 mM piperazine-*N*,*N*′-bis(2-ethanesulfonic acid) pH 6.8
- Triton X-100
- 4% paraformaldehyde

Method

1. Culture cells in serum-free, chemically-defined medium on glass coverslips or slides coated with a low FN density (e.g. 25–50 ng/cm^2) to hold cells in a round form and prevent spreading, using *Protocol 1*. This coating concentration may vary between different cell types and should be determined empirically.
2. Coat tosyl-activated magnetic microbeads (4.5 μm diameter) with ECM molecule, synthetic RGD peptide (Peptide 2000), specific antibodies, or control ligands, such as AcLDL (Biomedical Technologies Inc.) at 50 μg/ml in 0.1 M carbonate buffer pH 9.4 for 24 h at 4°C (15, 16). Coated beads are washed twice in PBS, incubated in medium containing 1% BSA for at least 30 min, and then stored at 4°C in PBS. Note: microbeads from Dynal or other suppliers that are pre-coated with secondary antibodies or other cross-linking agents may be used in a similar manner.
3. Add coated microbeads to cells (20 beads/cell) and allow to incubate for 5–30 min at 37°C.
4. To identify cytoskeletal-associated FAC proteins, incubate cells for 1 min in ice-cold cytoskeleton stabilizing buffer (CSK–TX) which maintains the integrity of the cytoskeleton (20). Incubate for 1 min in the same buffer supplemented with 0.5% Triton X-100 (CSK+TX) to remove membranes and soluble cytoplasmic components. Note: removal of soluble cytoplasmic contents also greatly increases the signal-to-noise ratio of FAC protein staining and may be advantageous for any study that focuses on morphological analysis of the FAC.
5. Fix detergent-extracted cells in 4% paraformaldehyde/PBS, wash with PBS, and incubate with primary antibodies diluted in 0.2% Triton X-100/0.1% BSA in PBS. Visualize primary antibodies using the appropriate affinity-purified anti-IgGFc antibodies conjugated to fluorescein or rhodamine (Organon Teknika-Cappel).

Protocol 3. Isolation of focal adhesion complexes

Equipment and reagents

- See *Protocols 1* and *2*
- Polypropylene tubes (Costar)
- 15 ml conical tubes (Falcon)
- Rotator (Nutator)
- Side pull magnetic separation unit (Advanced Magnetics)
- XL 2005 cell disruptor (Heat Systems)
- Dounce homogenizer (Wheaton)
- 1% BSA/DMEM
- CSK–TX and CSK+TX buffer (see *Protocol 2*)
- RIPA buffer: 1% Triton X-100, 1% deoxycholate, 0.1% SDS, 150 mM NaCl, 50 mM Tris pH 7.2, 0.1 mM AEBSF

Protocol 3. *Continued*

Method

1. Disperse quiescent cells with trypsin-EDTA as described in *Protocol 1*, wash with 1% BSA/DMEM, and place cells in polypropylene tubes.
2. Suspend approx. 10^6 cells/ml in defined medium without growth factors, add an equal volume of medium containing RGD-coated magnetic microbeads (2×10^7 beads/ml; coated as in *Protocol 2*), and place on a rotator for 30 min at 37 °C. RGD-coated beads are utilized because they exhibit less non-specific clumping in suspension than FN-beads, and thus allow greater binding efficiency. However, similar results have been obtained with FN-coated beads.
3. Use a side pull magnetic separation unit to collect microbeads and bound cells. This and all subsequent procedures are carried out at 4 °C.
4. Resuspend cell/bead pellet in ice-cold CSK–TX buffer and transfer these to 15 ml conical tubes.
5. Re-pellet using magnetic separator. Resuspend bead pellets in cold CSK+TX buffer (with detergent), incubate on ice for 5 min, sonicate for 10 sec (output setting, 4; output power, 10%), and homogenize (20 strokes) in a (100 μm) Dounce homogenizer.
6. Pellet microbeads magnetically and wash five times with 10 ml CSK+TX buffer. RIPA buffer is used to remove protein from beads for biochemical analysis.

2.3 Geometric control of cell shape and function

This finding that integrin clustering alone is sufficient to activate internal signal transduction pathways that can influence cell behaviour complicated the cell shape story. However, the cells that bound to these beads which induced integrin clustering and early signalling events, including expression of immediate early growth response genes (21), never entered S phase (8). Thus, we then set out to make cell shape or distortion an independent variable. In other words, we attempted to devise a system in which cells could bind to optimal densities of ECM and soluble growth factors, but yet could be restricted in their spreading by purely physical means.

To control cell deformation in this manner, we adapted a soft lithography-based micropatterning technique that was originally developed in Dr George Whitesides' laboratory at Harvard as an alternative manufacturing approach for the microchip industry (22, 23). This method allowed us to create ECM-coated adhesive islands with defined shape, size, and position on the micron scale that were separated by non-adhesive, polyethylene glycol (PEG)-coated boundary regions that do not support protein adsorption and hence, prevent cell adhesion. When cells are plated on these substrates they adhere and

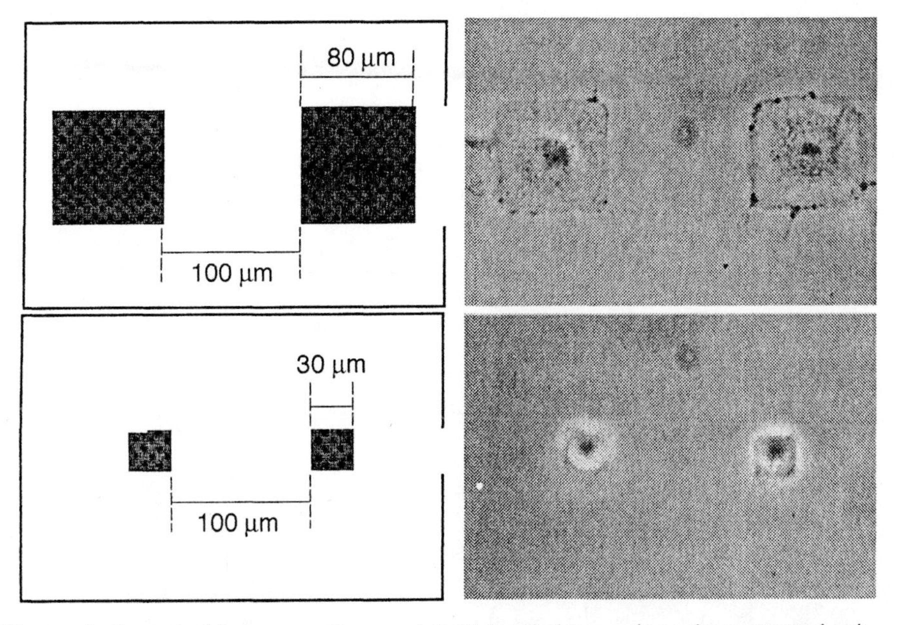

Figure 4. Control of human capillary endothelial cell shape using micropatterned substrates. Phase-contrast view (*right*) of CE cells plated on patterns containing FN-coated square islands 80 μm and 30 μm wide, respectively. The design of the pattern geometry is shown in the left panels, with the shaded area indicating the protein-adsorbing areas.

change their shape to fit the size and form of their container (i.e. of the patterned adhesive island) (24, 25) (*Figure 4*).

Methods for generating micropatterned substrates are described in *Protocols 4–8*. In brief, it involves a single photolithography step to generate a master etched pattern in silicon using a computer program commonly used for integrated circuit designs (23) (*Figure 5*). An elastomer (polydimethylsiloxane; PDMS) is then poured over the surface of this master and polymerized to form a 'rubber stamp' that retains the precise surface features of the micropattern down to less than 1 μm resolution. The outward facing features of the stamp are coated with a chemical 'ink' composed of alkanethiols. When the inked stamp is brought into contact with a gold-coated surface, the alkanethiols adhere tightly to the gold (Au) and self-assemble to form a space-filling monolayer that precisely fills the island form. A solution containing the same alkanethiol conjugated on its tail to a PEG blocking group is then poured over the slide surface. These alkanethiols self-assemble to fill all the spaces between the islands, creating a fully chemically-defined surface monolayer. However, because of the PEG blocking groups, ECM molecules will only adsorb to the surfaces of the islands when added in solution. This results in fabrication of highly adhesive, ECM-coated islands of defined shape, size, and position surrounded by non-adhesive, PEG-coated

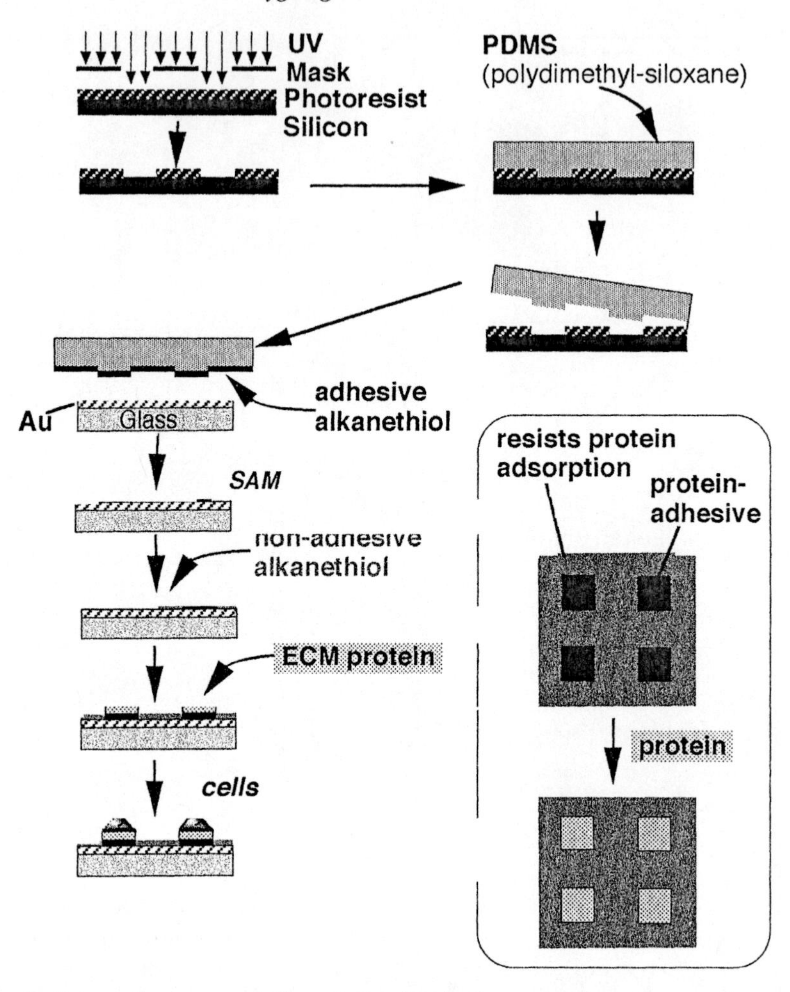

Figure 5. Schematic representation of the fabrication procedure for creating micropatterned substrates using microcontact printing. See text for details.

barrier regions. When cells are plated on these substrates they only adhere to the islands, spreading until they reach the surrounding non-adhesive boundary. By changing the size of the island, cells can be held in fully spread, moderately spread, or fully retracted form; by changing island form, cell shape can be controlled as well.

Using this micropatterning approach, we showed that capillary endothelial cells can be switched between growth, differentiation, and apoptosis programs simply by varying cell geometry (25, 26). For example, cells on large islands (50 μm diameter) that promoted cell extension supported growth, whereas cells on small islands (10 μm diameter) that remained fully retracted underwent a suicide program. To explore this mechanism, we allowed single cells to

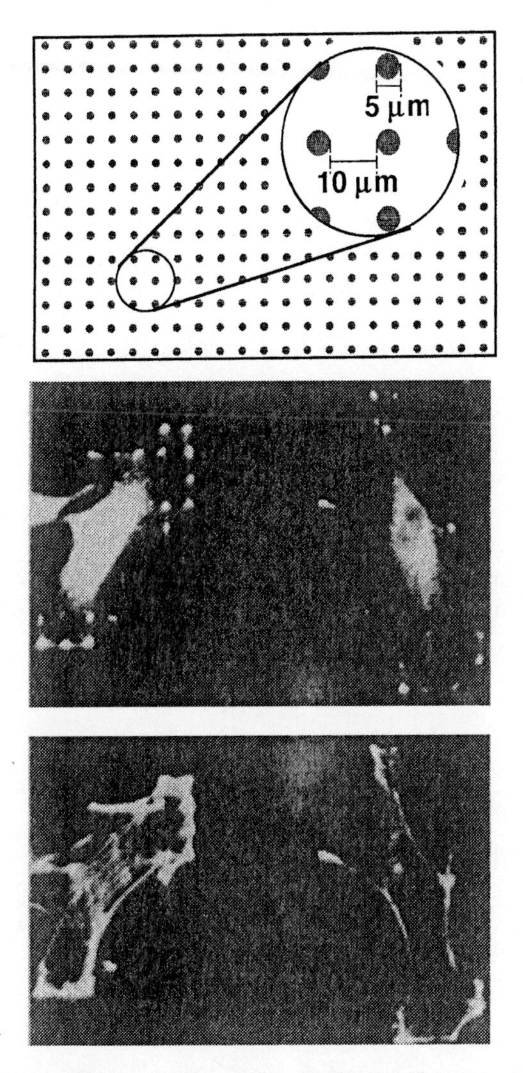

Figure 6. Micropatterns of 5 μm circles coated with fibronectin separated by 10 μm non-adhesive spaces (*top*), and immunofluorescence micrographs of adherent CE cells stained with anti-vinculin antibody (*middle*), and with FITC phalloidin (*bottom*) to visualize FACs and actin filaments, respectively.

spread across multiple, closely spaced, smaller adhesive islands that were the size of individual focal adhesions (3–5 μm diameter). Though the total area of these tiny islands was smaller than a single small (10 μm diameter) island, cells adherent to these patterned substrates spread as well as on a large island, stretching from island to island across non-adhesive areas (*Figure 6*). By increasing the spacing between islands, cell spreading could be increased by a factor of ten without altering the total cell–ECM contact area. These observations revealed that cell shape—the degree to which the cell physically extends or retracts—governs whether cells grow or die (24). Most recently, using the same approach, we found that the motility of cells constrained on square adhesive islands is also controlled geometrically. Lamellipodia preferentially extend from the corners of these cells when stimulated with motility factors.

We are also using this micropatterning technique to expedite detection of focal adhesion-associated antigens in cells. Focal adhesion proteins commonly appear as small, spear-like streaks at the base of the cell when analysed by immunofluorescence microscopy. However, this is a highly subjective method and definitive characterization of a positively stained structure requires additional techniques, such as electron microscopy or interference reflection microscopy. By using substrates containing multiple focal adhesion-sized adhesive islands oriented in a regular pattern, we have overcome this limitation. Positive staining on these tiny islands represents antigens that are definitively concentrated in regions of cell anchorage to ECM and, hence, in focal adhesions. We are using this technique as a screening method, in conjunction with monoclonal antibody generation methods, to identify antibodies to novel focal adhesion proteins that may have structural or signalling roles relevant for cellular regulation.

Protocol 4. Generating a patterned silicon master using photolithography

Equipment and reagents

- Mask aligner (Karl Suss model)
- Silicon wafers
- Microscope
- Desiccator
- Trichloroethylene
- Acetone
- Methanol
- Hexamethyldisilazane (Shipley)
- 1813 positive photoresist (Shipley)
- (Tridecafluoro-1,1,2,2,-tetrahydro-octy)-1-trichlorosilane (United Chemical Technologies)

Method

1. Clean silicon wafers in a clean room (preferably Class A) by sonicating for 5 min successively in trichloroethylene, acetone, and methanol.
2. Heat wafers at 180 °C for 10 min to dry thoroughly.
3. Spin coat (1500 g for 40 sec) the wafers with approx. 1–2 μm hexamethyldisilazane followed by a 1.3 μm thick layer of Shipley 1813 positive photoresist (1500 g for 40 sec).
4. Bake the resist at 105 °C for 3.5 min.
5. Expose the wafer on a mask aligner through a photomask with features etched in chrome deposited on quartz (Advance Reproductions Corp.) for 5.5 sec at 10 mW/cm^2. The patterns on the

photomask are normally created on a glass plate coated with photosensitive emulsion using a pattern generator. The final film is made by contact printing of the emulsion plate onto the chromium-coated quartz plate.

6. Develop the features by immersing in Shipley 351 for 45 sec, then rinse with distilled water, and dry with a stream of nitrogen. The proper development of the features should be checked under a microscope using a red filter in front of the light source to avoid unwanted exposure of the photoresist.
7. Place the wafers in a desiccator under vacuum for 2 h. Use a vial containing 1–2 ml (tridecafluoro-1,1,2,2-tetrahydro-octy)-1-trichlorosilane that reacts with the silicon areas not covered by the photoresist and makes them hydrophobic. Rinse only with water and avoid all contact with organic solvents.

Protocol 5. Moulding elastomeric microstamps

Equipment and reagents

- Petri dishes
- Razor blade
- Polydimethylsiloxane (PDMS) (Sylgard 184, Dow Chemical Co.)

Method

1. Polydimethylsiloxane (PDMS) pre-polymer is made by mixing ten parts of monomer Silicone Elastomer-184 and one part of initiator Silicone Elastomer Curing Agent-184 thoroughly in a plastic container and degassing it under vacuum for approx. 1 h until air bubbles no longer rise to the top.
2. Pour the pre-polymer in a Petri dish that contains the patterned silicon wafer, and cure for at least 2 h at 60 °C. Often small air bubbles form in the PDMS after it is poured on the master stamp. Cover the dish and gently tap it to allow the bubbles to diffuse out of the pre-polymer. Typically, stamps are 0.5–1 cm tall.
3. Peel the PDMS from the wafer and cut the stamps to the desired size with a razor blade. During curing, a layer of PDMS forms underneath the wafer and holds it to the dish. Invert the dish and gently press on the reverse side until the cured PDMS dewets from the surface of the dish. Invert the dish and use a dull edge to trace the contour of the PDMS so as to lift it off the dish. Often the PDMS remains attached to the wafer. Carefully cut the layer of PDMS found under the wafer and gently peel the two surfaces away from each other.

Protocol 6. Metal coating of glass substrates

Equipment and reagents

- Microscope slides (Fisher, No. 2)
- Electron beam evaporator
- Titanium (Aldrich, 99.99% purity)
- Gold (Materials Research Corporation, 99.99% purity)

Method

1. Load microscope slides on a rotating carousel in an electron beam evaporator (most of these are home-built).
2. Perform evaporation at pressure $< 1 \times 10^{-6}$ Torr. Occasionally, during the evaporation of titanium, the pressure increases above 1×10^{-6} Torr, but decreases after allowing the chamber to stabilize for approx. 2 min. Note: a sputter coat system also may be used to prepare these coatings. However, we recommend using an evaporator to coat the substrates because most sputter coaters are single source and, thus, they are impractical for coating two different metals (Ti and Au) on a single substrate. In addition, sputtering gives less homogeneous films that require an additional annealing step. Sputtering systems also generally produce films with higher quantities of metal oxides and other impurities that could interfere with the generation of the self-assembled monolayer (SAM) surface.
3. Allow the metals to reach evaporation rates of 1 Å/sec.
4. Allow 400–500 Å of each metal to evaporate before opening the shutters and exposing the glass slides to 15 Å of titanium (99.99% purity) followed by 115 Å of gold (99.99% purity).

Notes on storage: typically, gold-coated substrates become 'mottled' after four to five weeks and are no longer deemed suitable for experiments; streaks with heterogeneous transparency develop (they are obvious to the naked eye). This may be caused by rearrangements in the thickness of the gold layer related to impurities present on the glass before evaporation of the gold. Gold substrates that are stamped immediately after evaporation are generally more stable over time (approximately three months), perhaps because the SAM acts as a resist against impurities.

Protocol 7. Stamping micropatterned adhesive substrates

Equipment and reagents

- Q-tips
- Forceps
- Pasteur pipette
- Gold-coated substrate
- PDMS stamp
- 2 mM solution of hexadecanethiol (HS-$(CH_2)_{15}CH_3$) (Aldrich) in ethanol
- 2 mM PEG-terminated alkanethiol: tri(ethylene glycol)$HS(CH_2)_{11}(OCH_2CH_2)_3OH$ (synthesized at Dr G. Whitesides' laboratory)

Method

1. Lay metal-coated substrate on clean flat surface, with gold facing upward. Take care not to scratch the surface with sharp forceps, or to place the substrate upside down. If there is dust visible on the substrate, blow gently with pressurized air or nitrogen.
2. Rinse the PDMS stamp with ethanol and remove the ethanol vigorously with a stream of pressurized air or nitrogen for at least 10 sec. If any dust remains on the stamp, repeat this procedure.
3. Dip a Q-tip into a 2 mM solution of hexadecanethiol in ethanol and gently paint a layer of the solution onto the PDMS stamp. Use a stream of air or nitrogen to gently evaporate the ethanol off the stamp.
4. Gently place the stamp face down onto the gold-coated substrate and allow it to adhere; this step may require gentle pressure. Let the fully adhered stamp remain on the substrate for at least 10 sec. Observe the light reflected from the micropatterned substrate at an angle to ensure that the stamp has fully adhered to the substrate. A pink colour will ensure that full adhesion has occurred. Both under-stamping and over-stamping results in a loss of this interference pattern. Make sure that no patches of non-adhesion remain.
5. Use forceps to gently peel away the stamp from the substrate, making sure not to move the stamp against the substrate or to let the stamp re-adhere to the substrate.
6. Return to step 2 to continue stamping more substrates. After all substrates are stamped, proceed with step 7.
7. Use a Pasteur pipette to deliver an ethanol solution containing 2 mM PEG-terminated alkanethiol [tri(ethylene glycol)$HS(CH_2)_{11}(OCH_2CH_2)_3OH$] dropwise onto each substrate until the liquid covers it entirely. This usually requires approx. 0.5–1 ml per square inch of substrate. Incubate with the PEG-alkanethiol for 30 min to fill in the remaining bare regions of gold and complete the self-assembled monolayer surface coating. Note: always stamp the hexadecanethiol, rather than the PEG-alkanethiol. Stamping the PEG-alkanethiol results in less efficient pattern transfer, incomplete formation of the self-assembled monolayer, and non-specific adsorption of proteins in the 'non-adhesive' PEG-coated barrier regions.
8. With forceps—cleaned with ethanol and blown dry—grasp the corner of the substrate and rinse with a stream of ethanol on both sides of the pattern for 20 sec. Place the substrate on a clean surface and rinse the forceps with ethanol. Grasp the substrate again in a different location and rinse with ethanol to wash the area previously masked by the forceps.

Protocol 8. *Continued*

9. Blow off ethanol from the substrate with pressurized air or nitrogen.
10. Place the stamped substrates into containers, taking care not to allow the patterned surface to rub against any coarse surfaces. Store under nitrogen gas in a cool, dark location. Place the containers in a Ziplock bag filled with nitrogen.

Notes on storage: typically, alkanethiols kept in ethanolic solutions for more than three months become oxidized and form significant amounts of disulfides. Disulfides of PEG are detected by TLC as spots with an R_f of approximately 0.15, while the thiol has an R_f of 0.25 using CH_2Cl_2:CH_3OH (98:2) as the eluent. By NMR, disulfides can be distinguished from alkanethiols by the presence of a triplet of peaks (from the methylene group adjacent to the sulfur atom) at approximately 2.6 p.p.m. instead of a quartet at 2.5 p.p.m. Although disulfides are known to form SAMs with interfacial properties similar to those formed with alkanethiols, their assembly is 75 times slower.

Protocol 8. Culturing cells on micropatterned substrates

Equipment and reagents

- Petri dishes
- Microscope
- PBS
- ECM protein (25 μg/ml)
- 1% BSA dissolved in PBS
- Trypsin-EDTA
- 1% BSA/DMEM
- Serum-free media

Method

1. Prepare a solution of phosphate-buffered saline (PBS) containing the ECM protein (25 μg/ml) to be coated on the adhesive islands.
2. Place a small droplet (0.25 ml) of ECM solution onto a bacteriological Petri dish or another disposable hydrophobic surface that is non-adhesive for cells in the absence of serum. Typically, 0.25 ml of solution per square inch of substrate is sufficient. Float each substrate, with patterned side face-down, on the drops. Let sit for 2 h at room temperature.
3. After 2 h, add a large amount (5–15 ml) of 1% BSA dissolved in PBS directly to the dish in order to dilute the ECM protein solution and block further coating. Remove the substrates, flip the slide so that micropattern side is facing-up, and place directly into serum-free medium containing 1% BSA. Note: when adding the BSA/PBS solution, the slides can sink onto the dish and adhere to it; since the substrates face the bottom of the dish, the pattern may be damaged.

To avoid this, gently add the BSA solution around the edges of the substrate so that the slides remain afloat.

4. Dissociate quiescent cell monolayers by brief exposure to trypsin-EDTA, wash, and resuspend in 1% BSA/DMEM, as described in *Protocol 1*. Remove one-half volume of 1% BSA/DMEM from Petri dishes containing the micropatterned substrates and replace with an equal volume of medium containing cells at a low plating density (e.g. 7500 cells/cm^2) plus any required medium supplements at twice the concentration. Note: if serum-containing medium must be utilized, initially plate cells for at least 1 h in serum-free medium (this may only be necessary for certain cell types). Subsequent medium changes should minimize drying, which can non-specifically adsorb medium proteins onto the substrate and thus, compromise shape control. We commonly remove 75–90% of the old medium and replace it with new at each refeeding rather than remove all fluid during these medium changes. Note: a low cell plating density is utilized to optimize plating of single cells on individual islands and because the actual surface area available for cell adhesion on the patterned slides is a fraction of that of a regular culture dish. The cell plating densities and medium conditions will vary between different cell types and thus, should be determined empirically.
5. Visualize cells by any standard microscopic technique when micropatterning is performed on glass slides. Immunofluorescence staining of cytoskeletal and FAC components may be performed as described in *Protocol 2*.

2.4 Analysis of transmembrane mechanical signalling

The tensegrity model predicts that ECM regulates cell shape and function based on its ability to balance cytoskeletal tension that is transmitted across discrete transmembrane adhesion receptors on the cell surface. To analyse the molecular mechanism by which cells transfer external mechanical signals across the cell surface and to the cytoskeleton, we developed a magnetic twisting device (magnetic cytometry system) in which controlled mechanical stresses can be applied directly to specific subclasses of receptors on the surface of living cells (27, 28). In this technique, cultured cells are allowed to bind to small ferromagnetic beads coated with specific receptor ligands, as described in *Protocol 2*. Shear stress (torque) is applied to the membrane receptor-bound ferromagnetic microbeads by magnetically twisting the beads. This is accomplished by first magnetizing the beads by applying a very brief, but strong, homogeneous magnetic field in the horizontal direction. Then a weaker, but sustained magnetic field is applied in the vertical direction. This does not remagnetize the beads; instead, the beads realign along the new

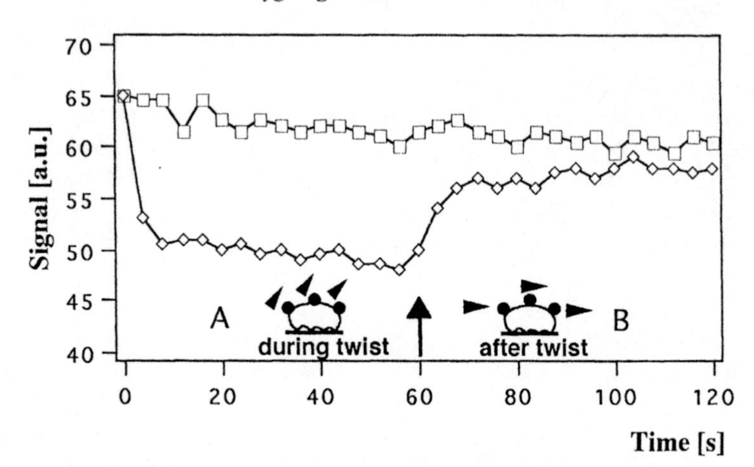

Figure 7. Data obtained using magnetic twisting cytometry showing the magnetic signal in the absence (squares) and presence (diamonds) of the twisting field. Application of a constant twisting stress (40 Gauss) for 60 sec in the vertical direction results in an instantaneous decrease in the signal. The difference in signal (on/off) at 60 sec is used to calculate the apparent stiffness, *E*. The rate of recovery after 60 sec multiplied by *E* determines the apparent viscosity (see *Equation 1*).

applied field lines much like a compass needle. Thus, the cellular response to applied stress can be measured simultaneously by quantifying changes in the rate and degree of magnetic bead rotation (i.e. angular strain) using an in-line magnetometer (*Figure 7*). This method is described in *Protocol 9*. Note: ferromagnetic beads must be used in these studies because, in contrast to the paramagnetic beads we utilized in our FAC isolation procedure, these materials must maintain their magnetization when an applied magnetic field is removed.

The values of the material properties measured using this technique will vary depending on the cell surface receptor to which that force is applied. Thus, this method allows one to measure micromechanical behaviour of different molecular supporting frameworks in the cell. When force is applied through cell surface integrins, the resulting shear stress is transmitted through the cell, causing the intracellular cytoskeleton to deform without producing large scale changes in cell shape (27). Using this technique, we showed that integrin receptors efficiently mediate mechanical force transfer across the cell surface through the FAC and to the cytoskeleton in endothelial cells and vascular smooth muscle cells, although different receptor subtypes varied in their response (integrin $\beta 1 > \alpha Vb3 > \alpha 5 > \alpha 2 > \alpha V$) (11, 27–33). Other adhesion receptors, such as cadherins and selectins, mediated mechanical force transfer to the cytoskeleton to a moderate degree, whereas transmembrane receptors not involved in adhesion, such as metabolic scavenger receptors and histocompatibility antigens, displayed minimal mechanical coupling (27, 31, 34). In addition, we consistently found that the stiffness (i.e. the ratio of stress to strain) of the cytoskeleton increased in direct proportion to the applied stress and thus, that the cytoskeleton behaves as a tensionally-integrated or 'tensegrity' structure (27).

Additional support for this model came from the finding that the mechanical stiffness at the cytoskeleton depended directly on the level of pressure (pretension) in the cytoskeleton (11, 30, 35).

We recently showed that the same magnetic twisting method can be used to analyse mechanochemical signalling in living cells. For this application, the magnetometer is used to verify the level of applied mechanical twisting stresses rather than to record experimental data. Changes in cellular biochemistry are measured simultaneously using conventional biochemical and molecular biological techniques. For example, magnetically twisting beads bound to cell surface integrins in adherent endothelial cells were found to induce recruitment of total poly(A) mRNA and ribosomes to the focal adhesions that form at the site of bead binding, as detected using high resolution *in situ* hybridization and immunofluorescence microscopy (36). Recruitment of the protein translation machinery was shown to be integrin-specific and dependent on the level of applied mechanical stress. These data suggest that altering the balance of mechanical forces, specifically across integrins, changes the intracellular biochemistry and, in this case, induces the formation of a microcompartment at the focal adhesion specialized for protein synthesis. Again, these results add experimental support for the predictions made by the tensegrity model. We are now using this method of localized stress application in cells expressing GFP-labelled cytoskeletal proteins (e.g. actin, tubulin, vinculin) to analyse directly how the cytoskeleton bears and distributes mechanical loads.

Protocol 9. Magnetic twisting cytometry

Equipment and reagents

- Spherical ferromagnetic beads obtained commercially (4.5 μm diameter, CFM-40-10 carboxylated-ferromagnetic beads; Spherotech) or from independent laboratories (1.4 μm or 5.5 μm diameter beads; Dr W. Moller, Gauting, Germany)
- Magnetic twisting device
- In-line magnetometer
- 96-well Removawells (Immunolon II)
- 100 000 centipoise, PSAV100K (Petrarch Systems Inc.)

Method

When carboxylated-Spherotech beads are utilized, ligands are conjugated at 50 μg/ml using a carbodiimide (EDC; Sigma E-6383) reaction, as described by the supplier. The other beads are coated with adhesive ligands or antibodies at the same concentration, using methods described for paramagnetic Dynal microbeads in *Protocol 2*. Beads may also be

Protocol 9. *Continued*

obtained pre-coated with secondary antibodies (e.g. goat anti-mouse IgG Fc domain), to which can be added primary antibody in sterile PBS at 10–50 μg/ml (this concentration needs to be determined empirically for each antibody to ensure optimal binding). Store beads in sterile PBS at 4°C. Note: do not magnetically separate during washing steps because these beads will remain magnetized and thus, will be difficult to dissociate.

1. Plate cells (3×10^4/well) on FN-coated bacteriological plastic dishes (96-well Removawells) and culture for 6–10 h in chemically-defined medium before bead addition. The beads are quenched in chemically-defined medium containing 1% BSA for at least 30 min before being added to the cells.
2. Add approx. 1–10 beads/cell (this should be determined experimentally to obtain approx. 1–4 bound beads/cell) and incubate in chemically-defined medium containing 1% BSA for 15–30 min. At this time, the cells are washed in serum-free medium to remove unbound beads.
3. Place the individual well containing cultured cells and bound beads within a plastic vial, and then place it into the holder of the magnetic twisting device. Rotate at a constant rate (5 r.p.m.) to reduce ambient noise. The vial prevents the circulating water used for temperature control from getting into the culture well.
4. Apply a brief (10 μsec) but strong (1000 Gauss) magnetic pulse in the horizontal direction (parallel to the culture surface), using one pair of the magnetic coils of the device. After several seconds, apply a much weaker magnetic twisting field (0–40 Gauss) in the vertical direction and twist the beads for 1 min. The vertical field can be altered to assess the effects of force duration and frequency as well as the form of the stress regimen (e.g. square wave versus sinusoidal).
5. Use an in-line magnetometer to detect changes in the magnitude of the bead magnetic vector in the horizontal direction. Note: the torque of the applied twisting field is proportional to the twisting field, bead magnetization, and the sine of the angle between the twisting field vector and the bead magnetization vector (27). In the absence of force transmission across the cell surface, the spherical beads turn in place by 90° into complete alignment with the twisting field, and the remaining field vector immediately drops to zero. In contrast, transmission of force to the cytoskeleton results in increased resistance to deformation and decreased bead rotation. Angular strain (bead rotation) is calculated as the arc cosine of the ratio of remanent field after 1 min twist to the field at time 0.

6. Determine the apparent stiffness (ratio of stress to strain) and viscosity using the following relation:

$$E = \frac{\sigma}{\phi} \quad \text{and} \quad \eta = E\tau \qquad [1]$$

where E is the apparent stiffness; σ is the effective stress; ϕ is the angular rotation of the microbead; η is the apparent viscosity; and τ is the time constant of recovery after stress release. In the absence of the applied twisting field, the magnetic signal exhibits only a small decrease or relaxation in signal that is due to thermal motion and membrane movement. However, when a constant twisting field (e.g. 30 Gauss) is applied in the vertical direction for 1 min, the magnetic signal of the beads decreases almost instantaneously (to zero when the beads are free in solution). By subtracting the relaxation signal from the control, the remanent field signal almost reaches a plateau after 1 min. This plateau value obtained 1 min after stress application (representing magnetic torque being balanced by the elastic component of the cell) is used to calculate the apparent stiffness of the cell according to the above relation (*Figure 7*).

7. Quantitate the energy stored elastically in the cell by turning off the twisting field for 1 min and measuring the extent of recovery of the bead magnetic signal. The difference in the signal that remains after 2 min relative to the remanent field signal is a readout of permanent deformation.

8. Calibrate the effective applied stress by placing microbeads in a standard solution of known viscosity (e.g. 100 000 centipoise, PSAV100K) (28, 29) and measuring angular strain. The effective stress equals:

$$\sigma = cH_a \frac{B_{twist}}{B_{relax}} \; ; \; \phi = \cos^{-1}\left(\frac{B_{twist}}{B_{relax}}\right) \qquad [2]$$

where c is a constant dependent on the magnetic property of the bead:

$$c = v \times 1/T$$

where v is the standard viscosity; $1/T$ is the slope of *tan* $(90° - \phi/2)$ versus time; H_a is the applied twisting field; B_{twist} is the remaining twisting signal at 60 sec; B_{relax} is the remanent relaxation signal at 60 sec. Note: it is assumed that the angular rotation of the bead is a direct readout of the angular strain of the cell.

For studies on mechanochemical transduction, controlled stresses are applied in a similar manner, and biochemical or morphological changes are measured inside the cells using conventional analytical techniques. For example, cells plated on small ECM-coated coverslips can be bound to magnetic beads, twisted, fixed, and then analysed using *in situ* hybridization or immunofluorescence microscopy (36). It is also possible to use this technique to twist large numbers (millions) of suspended cells bound to magnetic beads in order to measure biochemical changes inside the cells (e.g. cAMP levels). Larger numbers (10–20) of beads bound per cell may be utilized for these studies.

2.5 Detection of long-range mechanical signal transfer

The tensegrity model of integrated cell shape control assumes that transmembrane ECM receptors, cytoskeletal filaments, and nuclear scaffolds are 'wired' together in such a way that mechanical stimuli can change the organization of molecular assemblies deep in the cytoplasm and nucleus. This is contrary to many existing models which view the cell as a viscous cytoplasm surrounded by an elastic membrane that bears most of the mechanical load. We recently developed a microsurgical approach to demonstrate that living cells and nuclei are indeed hard-wired such that a mechanical tug on cell surface receptors can immediately change the organization of molecular assemblies in the cytoplasm and nucleus (37). Specifically, when integrins were pulled by micromanipulating bound ECM-coated microbeads or micropipettes, cytoskeletal filaments reoriented, nuclei distorted, and nucleoli redistributed along the axis of the applied tension field. These effects were specific for integrins, independent of cortical membrane distortion, and mediated by direct linkages between the cytoskeleton and nucleus.

We extended these studies to analyse long-distance stress transfer and connectivity in the cytoskeleton and the nucleus, by using a very fine microneedle (0.5 μm diameter) to 'harpoon' the cytoplasm at precise distances from the nucleus and then rapidly pulling it away from the nucleus, toward the cell periphery (37). Cells remained viable during this procedure. We used digitized image analysis and real-time video microscopy to determine the position of multiple phase-dense structures within the cytoplasm and the nucleus (e.g. mitochondria, vacuoles, nucleoli, nuclear envelope) before and after pulling the micropipette away from the nucleus using the micromanipulator. By harpooning the cytoplasm with uncoated micropipettes and applying stresses directly to the cytoskeleton in cells treated with different cytoskeleton-modulating drugs, we were able to show that actin microfilaments mediated force transfer to the nucleus at low strain; however, tearing of the actin gel resulted in greater distortion (37). In contrast, intermediate filaments effectively mediated force transfer to the nucleus under both conditions, and cell tearing resulted when intermediate filaments were disrupted either by pharmacological means (37) or using genetic recombination (38). Finally, by placing the pipette tip closer to the nuclear border (2–4 μm) or within the nucleoplasm itself and then pulling away from the nuclear envelope, we could demonstrate the presence of discrete (localized) sites of mechanical stress transfer between the cytoplasm and nucleus as well as between the nucleoplasm and the nuclear envelope (37) (*Figure 8*).

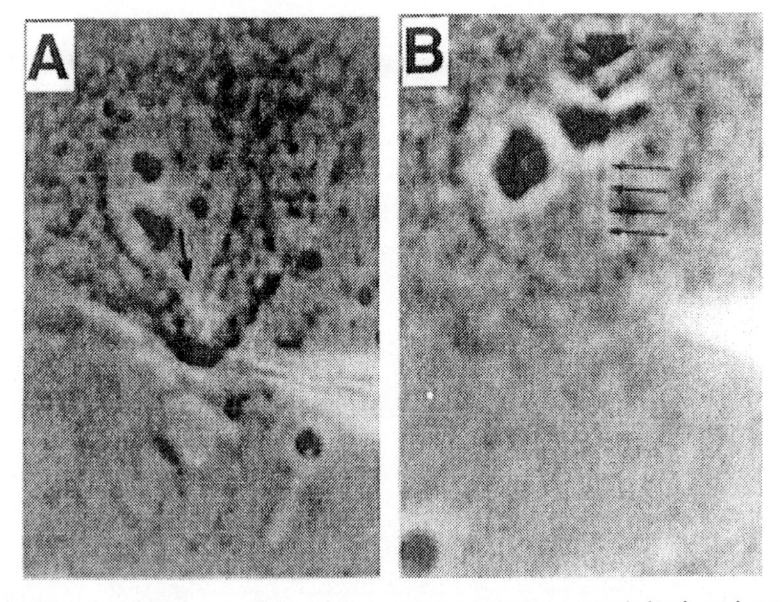

Figure 8. Demonstration of discrete connections between cytoskeletal and nuclear scaffolds. (A) A cell harpooned by a microneedle close to the nuclear border after stress application. The *arrow* indicates a local tongue-like protrusion of the nuclear envelope. (B) Invagination of the nuclear envelope in response to harpooning the nucleoplasm. Four small *arrows* indicate the tensed nucleoplasmic thread stretching from the pipette tip.

We also used the harpooning approach to estimate relative changes in mechanical stiffness of the cytoplasm and nucleus (37). The stiffness (E) of the cytoplasm and nucleus, as in any material, equals stress (s) (force/cross-sectional area) divided by strain (e) (change in length/initial length). The stiffness of the cytoplasm and nucleus cannot be determined directly using the harpooning technique, because only induced strains are measured. However, the ratio of stiffness in the cytoplasm (c) and nucleus (n) can be determined in the following manner (*Figure 9*). The ratio of nuclear to cytoplasmic stiffness (E_n/E_c) will equal the ratio of cytoplasmic to nuclear strain ($\varepsilon_c/\varepsilon_n$) measured in these regions when exposed to the same stress. If the cell responds isotropically and homogeneously to stresses applied over short (micrometre) distances, then the stress tensor (three-dimensional stress field) produced at any point will depend primarily on its location relative to the site of force application. Thus, the ratio of nuclear to cytoplasmic stiffness can be calculated by determining the ratio of induced strains measured in regions of the cytoplasm and nucleus when placed at the same distance from the micropipette (i.e. even if in different regions of the cell). We adapted the harpooning method to quantitate apparent Poisson's ratios as a measure of mechanical connectivity within the cytoplasm and nucleus of living cells.

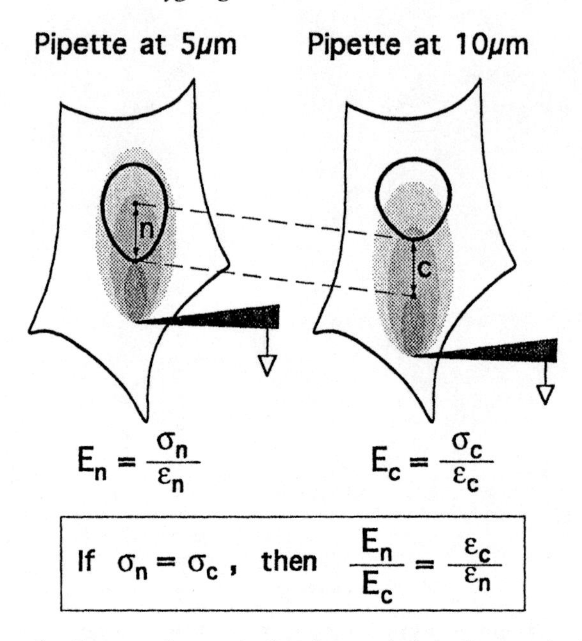

Figure 9. Schematic diagram demonstrating how relative changes in the mechanical stiffness of the cytoplasm and nucleus can be analysed in living cells.

Based on these studies, we found that intermediate filaments and actin filaments acted as molecular guy wires to mechanically stiffen the nucleus and anchor it in place, whereas microtubules acted to hold open the intermediate filament lattice and to stabilize the nucleus against lateral compression. Thus, these results added direct experimental support for our hypothesis that molecular connections between integrins, cytoskeletal filaments, and nuclear scaffolds may provide a discrete path for long-range mechanical signal transfer through cells as well as a mechanism for producing integrated changes in cell and nuclear structure in response to changes in ECM adhesivity or mechanics. Our methods for analysis of long-distance force transfer through the cytoskeleton as well as cytoplasmic and nuclear mechanics are presented in *Protocols 10* and *11*.

Protocol 10. Microsurgical analysis of long-range mechanical signalling

Equipment and reagents

- FN-coated glass coverslips
- Needles (Narishige)
- Pipette puller (Sutter)
- Manual micromanipulator (Leitz)
- Tosyl-activated microbeads (4.5 μm diameter; Dynal)
- Culture media

Method

1. Culture cells for 6–24 h in serum-free, chemically-defined medium containing 1% BSA on FN-coated glass coverslips. For studies involving force application to cell surface integrins, cells should be cultured on a suboptimal FN density (200–400 ng/cm^2) that promotes only a moderate degree of spreading, using the method in *Protocol 1*. Note: long-range distance transfer is difficult to visualize in well-spread cells on a high ECM density because these cells are too stiff, and most of the applied load is borne by their basal adhesions to the rigid ECM-coated substrate.
2. Pull needles with pipette puller and affix them to a manual micromanipulator. The micropipettes should be formed with tips approx. 0.5–1 μm wide along a length of 40–100 μm.
3. Add microbeads (4.5 μm diameter, tosyl-activated) coated with integrin ligands, as described in *Protocol 2*, to cells (1–4 beads/cell) for 10–15 min. Beads should be bound to cells at this time, but not internalized.
4. Wash cells with cell culture medium to remove unbound beads. Transfer the coverslip onto the lid of 35 mm Petri dish filled with DMEM, and place it in the heated chamber (Omega RTD 0.1 stage heating ring at 37 °C) on the stage of an inverted microscope.
5. Position the tip of the uncoated glass micropipette alongside the surface-bound beads using the micromanipulator. Rapidly pull the bead away from the cell (about 5–10 μm/sec), parallel to dish surface.
6. Measure nuclear distortion, movement of the nuclear border, and any other cellular response to stress (e.g. redistribution of phase-dense organelles, such as mitochondria, vacuoles, or secretory granules) in the direction of the applied tension field using inverted phase-contrast, fluorescence, Nomarski, or birefringence microscopy in conjunction with real-time computerized image analysis (we use Oncor *Image Analysis* software).

Protocol 11. Analysis of stress transfer through the cytoskeleton

Equipment and reagents

- Nocodazole (10 μg/ml; Sigma), acrylamide (5 mM; Bio-Rad), or cytochalasin D (0.1–1 μg/ml; Sigma)
- Glass micropipette (0.5 μm diameter)
- See *Protocol 10*

Method

1. To analyse force transfer between the cytoskeleton and nucleus, 'harpoon' the cytoplasm with an uncoated glass micropipette (0.5 μm

Protocol 11. *Continued*

diameter) 10 μm from the nuclear border and then pull the pipette away, first 10 μm and then 20 μm at a rate of 5–10 μm/sec (*Figure 8*).

2. Culture cells in the presence of nocodazole (10 μg/ml), acrylamide (5 mM), or cytochalasin D (0.1–1 μg/ml) to disrupt or compromise microtubules, intermediate filaments, or microfilaments, respectively. Note: doses need to be adjusted for different cell types, and immunofluorescence staining should be carried out in parallel to ensure that the effects are specific and maximal.
3. Measure resultant changes in nuclear deformation induced by the 10 μm and 20 μm pulls simultaneously, using real-time video microscopy in conjunction with computerized image analysis, as described in *Protocol 10*.
4. Calculate nuclear strains in the direction of pull at 10 μm and 20 μm displacements. Nuclear movement is defined as displacement of the rear border of the nucleus in the direction of pull. Negative lateral nuclear strain (nuclear narrowing) is calculated by measuring changes in nuclear width perpendicular to the direction of pull.
5. To demonstrate direct connections between the cytoskeleton and nucleus, apply tension via a pipette placed closer to the nuclear border (2–4 μm). Note: if the cytoplasm pulls away from the nuclear border without deforming the nucleus or by causing it to narrow along its entire length (i.e. sausage-casing type effect), then the two structures are merely interposed with no connectivity. If stress application causes a small region of the nuclear envelope to protrude locally toward the pipette in the region of highest stress (*Figure 8*), then a direct connection must exist.
6. To demonstrate the presence of a distinct filamentous network within the nucleus, harpoon the nucleoplasm and pull inward. Local indentation of the nuclear border again indicates the presence of direct mechanical coupling at a localized site (*Figure 8*).

Protocol 12. Analysis of nuclear and cytoplasmic stiffness and connectivity

Equipment and reagents

- See *Protocols 10* and *11*

Method

1. To determine the ratio of nuclear to cytoplasmic stiffness, calculate strains in the direction of pull within regions of the nucleus and

cytoplasm located at the same distance from a pipette that is pulled 10 μm toward the cell periphery. This is accomplished by placing the pipette 5 μm or 10 μm from the nuclear border in two separate pulling experiments (i.e. in different cells under similar conditions).

2. When the pipette is placed 10 μm from the nuclear border, measure induced strains in the direction of pull in the cytoplasm adjacent to the pipette (0–5.5 μm from the tip), in distal cytoplasm adjacent to the nucleus (5–10 μm away), and in the proximal portion of the nucleus (10–15 μm away).
3. Perform identical measurements in similarly treated cells with a pipette placed 5 μm from the nuclear border to determine strains at the same distances (0–5, 5–10, or 10–15 μm) from the pipette tip and hence, under similar stress. These locations now fall in the cytoplasm adjacent to the nucleus, in the proximal nucleus, and in the distal nucleus, respectively (*Figure 9*).
4. Calculate the ratio of nuclear to cytoskeletal stiffness by determining the ratio of strains measured in the adjacent cytoplasm and proximal nucleus (i.e. 5–10 μm away from pipettes placed 10 μm and 5 μm away from the nuclear border, respectively), according to the equations described in *Figure 9*.
5. To measure apparent Poisson's ratios, harpoon the cells 10 μm from the nuclear envelope and pull the pipette 5 μm away from the nuclear border. Calculate the ratio of the strain (per cent changes in distances between different phase-dense particles in cytoplasm or nucleus) before and after stress application in the region along the axis perpendicular to the direction of pull divided by the strain in the direction of pull. All strains are measured in equal areas (9 μm^2), equally distant (4–5 μm) from both the pipette and the nuclear border, and all displacements should be of equal magnitude. Note: Poisson's ratio calculated in this manner must be viewed as an 'apparent' rather than absolute value, because the ratio is based on a two-dimensional projection of a three-dimensional material in cells adherent to an underlying solid substrate.

2.6 Analysing chromosomal connectivity and mechanics

Our work on mechanical connectivity in living cells suggested the existence of a continuous filamentous network within the nucleus as well as in the cytoplasm (37). To test directly whether discrete networks physically interlink chromatin in the nucleus, we used very fine glass microneedles (tips less than 0.5 μm diameter) to harpoon individual nucleoli within cultured interphase cells or individual chromosomes in mitotic cells. Using this approach, we found that pulling a single nucleolus or chromosome out from interphase or

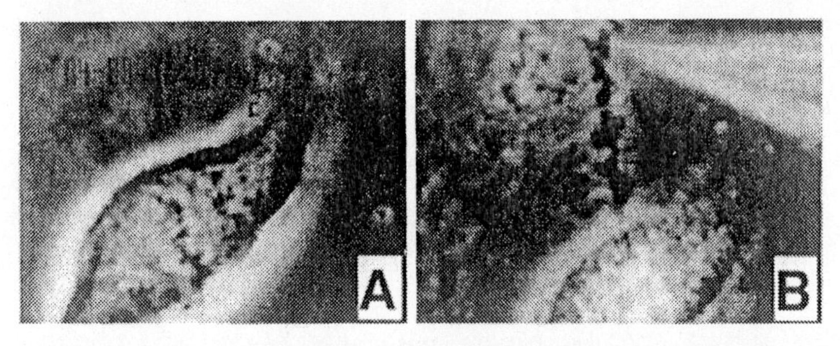

Figure 10. Microsurgical analysis of chromatin structure. A mitotic endothelial cell before (A) and after (B) a single chromosome was harpooned and removed from the cell using a micromanipulator. Note that pulling on one chromosome results in sequential removal of all the other chromosomes, revealing mechanical connectedness in the entire genome.

mitotic cells resulted in sequential removal of the remaining nucleoli and chromosomes, interconnected by a continuous elastic thread (40) (*Figure 10*). Enzymatic treatments of interphase nucleoplasm and chromosome chains held under tension revealed that mechanical continuity within the chromatin was mediated by DNA. Furthermore, when ion concentrations were raised and lowered, both the chromosomes and the interconnecting strands underwent multiple rounds of decondensation and recondensation (39–41). Fully decondensed chromatin strands also could be induced to recondense into chromosomes with pre-existing size, shape, number, and position by adding antibodies against histone H1 or topoisomerase II alpha, but not other nuclear proteins (39–40). These data suggest that DNA, its associated protein scaffolds, and surrounding cytoskeletal networks function as a structurally unified system. Mechanical coupling within the nucleoplasm may co-ordinate dynamic alterations in chromatin structure, guide chromosome movement, and ensure fidelity of mitosis. A typical assay is presented in *Protocol 13*.

Protocol 13. Analysis of chromatin mechanics and dynamics

Equipment and reagents

- See *Protocols 10* and *11*
- Glass coverslips
- Microscope
- 60 mM $MgCl_2$; trypsin; proteinase K

Method

1. Culture cells on glass coverslips in standard serum-containing medium. We have used endothelial cells and fibroblasts (39–41).
2. Transfer the coverslip to the lid of a 35 mm Petri dish covered with 2 ml serum-free medium with Hepes pH 7.4 and position on the stage of a Diaphot inverted microscope. For studies involving oil-immersion, use a cell culture chamber with a glass coverslip bottom, instead of 35 mm dish lid.

3. Fabricate the microneedle (0.5 μm wide tip), using methods in *Protocols 10* and *11*, and position it with the micromanipulator above a nucleolus or the edge of a mitotic plate.
4. Rapidly introduce the microneedle into the cell in a single motion and 'harpoon' an individual nucleolus or chromosome by gently touching it with the tip of the needle.
5. Pull out the nucleolus or chromosome attached to the microneedle through the original micropuncture opening made in the cell membrane by reversing the direction of pipette movement using the micromanipulator. The remaining chromatin strand will follow forming a continuous, elastic 'chain'. Note: the chromatin chain can be held in an extended form for study by either holding the micromanipulator in place or adhering the distal end of the chain to another cell or the surface of the culture dish by gently pressing the pipette against the surface. The connectivity between chromosomes can be studied by using the micropipette to apply a mechanical force directly to chromosome chains.
6. Probe the effect of different ions, chemicals, or biological compounds on chromatin mechanics and dynamics by placing a droplet (2 μl) containing the compound of interest directly above the isolated chain, or by replacing the original medium with the compound at the desired final concentration.
7. To analyse chromosome dynamics, incubate the extended chromosome chain with high salt, such as 60 mM $MgCl_2$ or place a 2 μl droplet of protease, such as the combination of trypsin (5 μg) and proteinase K (50 μg) above the isolated chain. The chromosomes will unfold and become phase lucent within seconds. These dynamic effects are reversible upon the return to the original ionic conditions or after addition of molecules involved in chromosome compaction (e.g. histone H1). Note: caution should be used during the handling of unfolded chromosomes, because excessive fluid turbulence can easily disrupt their pattern, entangle them, or cause non-specific sticking to the substrate. A more thorough description of methods and materials used to study chromosome dynamics *in situ* may be found in our recent publications (39–41).

3. Conclusion

In this chapter, we review a variety of different techniques that we have developed to manipulate and probe cell structure and cytoskeletal signalling. Our initial working hypothesis, based on a model of cellular tensegrity, suggested that cell form and function may be largely controlled based on mechanical interactions between cells and their ECM. Our initial methods for varying cell–ECM interactions in a controlled manner were consistent with this model, because they showed tight coupling between cell spreading and growth. However, we could not discriminate signals triggered by cell shape modulation from signalling induced by local integrin receptor binding and clustering. In fact, by developing a microbead technique for inducing integrin clustering independently of cell spreading we could confirm that integrin binding alone is sufficient to activate many chemical signalling pathways that are involved in cellular control. Furthermore, we showed that much of this signalling occurs on the cytoskeleton at the site of integrin binding to the backbone of the FAC.

Nevertheless, when we finally developed a method to separate local cell–ECM binding events from cell spreading using micropatterned substrates, we discovered that cell shape exerts separate and distinct regulatory signals that act many hours after initial integrin and growth factor receptor signalling events to control cell cycle progression. In addition, we developed various other techniques to probe cell structure which revealed that the cell is indeed 'hard-wired' to respond to mechanical stresses; that cell surface adhesion receptors provide preferred sites for transmembrane mechanical transfer; and that living cells behave as if they are tensegrity structures when they are mechanically stressed. Taken together, these new methods have led to a new view of cell regulation based on tensegrity architecture in which mechanics and chemistry are tightly coupled. Better understanding of how cells control their shape and function will require development of more techniques that incorporate methods for controlling and quantitating changes in cell structure as well as biochemical events.

Acknowledgements

We would like to acknowledge the invaluable technical assistance of Deborah Flusberg, Adi Loebl, and Paul Kim, and thank Ruth L. Capella and Jeanne Nisbet for careful reading and editing of the manuscript. We would also like to thank Dr George Whitesides and the members of his group for their help in developing the micropatterning technique. This work was supported by grants from NIH, NASA, NSF, DOD, and NATO.

References

1. Ingber, D. E., Madri, J. A., and Jamieson, J. D. (1981). *Proc. Natl. Acad. Sci. USA*, **78**, 390.
2. Ingber, D. E. and Jamieson, J. D. (1985). In *Gene expression during normal and malignant differentiation* (ed. L. C. Andersson, C. G. Gahmberg, and P. Ekblom), p. 13. Academic Press, Orlando.
3. Ingber, D. E. (1993). *J. Cell Sci.*, **104**, 613.

4. Ingber, D. E. (1997). *Annu. Rev. Physiol.*, **59**, 575.
5. Ingber, D. E. (1998). *Sci. Am.*, **278**, 48.
6. Ingber, D. E. (1991). *Curr. Opin. Cell Biol.*, **3**, 841.
7. Ingber, D. E. (1993). *Cell*, **75**, 1249.
8. Ingber, D. E. (1990). *Proc. Natl. Acad. Sci. USA*, **87**, 3579.
9. Ingber, D. E. and Folkman, J. (1989). *J. Cell Biol.*, **109**, 317.
10. Mooney, D., Hansen, L., Farmer, S., Vacanti, J., Langer R., and Ingber, D. E. (1992). *J. Cell. Physiol.*, **151**, 497.
11. Lee, K.-M., Tsai, K., Wang, N., and Ingber, D. (1998). *Am. J. Physiol.*, **274**, H76.
12. Huang, S., Chen, C. S., and Ingber, D. E. (1998). *Mol. Biol. Cell*, **9**, 3179.
13. Folkman, J. and Moscona, A. (1978). *Nature*, **273**, 345.
14. Katz, B. Z. and Yamada, K. M. (1997). *Biochimie*, **79**, 467.
15. Schwartz, M. A., Lechene, C., and Ingber, D. E. (1991). *Proc. Natl. Acad. Sci. USA*, **88**, 7849.
16. Plopper, G. and Ingber, D. E. (1993). *Biochem. Biophys. Res. Commun.*, **193**, 571.
17. Plopper, G., McNamee, H., Dike, L. E., Bojanowski, K., and Ingber, D. E. (1995). *Mol. Biol. Cell*, **6**, 1349.
18. McNamee, H., Liley, H., and Ingber, D. E. (1996). *Exp. Cell Res.*, **224**, 116.
19. Miyamoto, S., Teramoto, H., Coso, O. A., Gutkind, J. S., Burbelo, P. D., Akiyama, S. K., *et al.* (1995). *J. Cell Biol.*, **131**, 791.
20. Burr, J. G., Dreyfuss, G., Penman, S., and Buchanan, J. M. (1980). *Proc. Natl. Acad. Sci. USA*, **77**, 3484.
21. Dike, L. E. and Ingber, D. E. (1996). *J. Cell Sci.*, **109**, 2855.
22. Xia, Y., Mrksich, M., Kim, E., and Whitesides, G. M. (1995). *J. Am. Chem. Soc.*, **117**, 9576.
23. Wilbur, J. L., Kim, E., Xia, Y., and Whitesides, G. M. (1995). *Adv. Mater.*, **7**, 649.
24. Singhvi, R., Kumar, A., Lopez, G. B., Stephanopoulos, G. N., Wang, D. I. C., Whitesides, G. M., *et al.* (1994). *Science*, **264**, 696.
25. Chen, C. S., Mrksich, M., Huang, S., Whitesides, G. M., and Ingber, D. E. (1997). *Science*, **276**, 1425.
26. Dike, L. E., Chen, C. S., Mrkisch, M., Tien, J., Whitesides, G., and Ingber, D. E. (1999). *In Vitro*, in press.
27. Wang, N., Butler, J. P., and Ingber, D. E. (1993). *Science*, **260**, 1124.
28. Wang, N. and Ingber, D. E. (1994). *Biophys. J.*, **66**, 2181.
29. Wang, N. and Ingber, D. E. (1995). *Biochem. Cell Biol.*, **73**, 327.
30. Wang, H. (1998). *Hypertension*, **32**, 162.
31. Potard, U. S., Butler, J. P., and Wang, N. (1997). *Am. J. Physiol.*, **272**, C1654.
32. Ezzell, R. M., Goldmann, W. H., Wang, N., Parasharama, N., and Ingber, D. E. (1997). *Exp. Cell Res.*, **231**, 14.
33. Goldmann, W. H., Galneder, R., Ludwig, M., Xu, W., Adamson, E. D., Wang, N., *et al.* (1998). *Exp. Cell Res.*, **239**, 235.
34. Yoshida, M., Westlin, W. F., Wang, N., Ingber, D. E., Rosenzweig, A., Resnick, N., *et al.* (1996). *J. Cell Biol.*, **133**, 445.
35. Pourati, J., Maniotis, A., Spiegel, D., Schaffer, J. L., Butler, J. P., Fredberg, J. J., *et al.* (1998). *Am. J. Physiol.*, **274**, C1283.
36. Chicurel, M. E., Singer, R. H., Meyer, C. J., and Ingber, D. E. (1998). *Nature*, **392**, 730.

37. Maniotis, A. J., Chen, C. S., and Ingber, D. E. (1997). *Proc. Natl. Acad. Sci. USA*, **94**, 849.
38. Eckes, B., Dogic, D., Colucci-Guyon, E., Wang, N., Maniotis, A. J., Ingber, D., *et al.* (1998). *J. Cell Sci.*, **111**, 1897.
39. Maniotis, A. J., Bojanowski, K., and Ingber, D. E. (1997). *J. Cell Biochem.*, **65**, 114.
40. Bojanowski, K., Maniotis, A. J., Plisov, S., Larsen, A. K., and Ingber, D. E. (1998). *J. Cell Biochem.*, **69**, 127.
41. Bojanowski, K. and Ingber, D. E. (1998). *Exp. Cell Res.*, **244**, 286.

A1

List of suppliers

Adam and List Associates, 110 Shames Drive, Westbury, NY 11590, USA.

Amersham

Amersham International plc., Lincoln Place, Green End, Aylesbury, Buckinghamshire HP20 2TP, UK.

Amersham Corporation, 2636 South Clearbrook Drive, Arlington Heights, IL 60005, USA.

Amicon Ltd., distributed by Millipore, Upper Mill, Stone House, Gloucestershire GL20 2BJ, UK.

Anderman

Anderman and Co. Ltd., 145 London Road, Kingston-Upon-Thames, Surrey KT17 7NH, UK.

BDH, Merck Ltd., Hunter Boulevard, Magna Park, Lutterworth, Leicestershire LE17 4XN, UK.

Beckman Instruments

Beckman Instruments UK Ltd., Oakley Court, Kingsmead Business Park, London Road, High Wycombe, Bucks HP11 1J4, UK.

Beckman Instruments Inc., PO Box 3100, 2500 Harbor Boulevard, Fullerton, CA 92634, USA.

Becton Dickinson

Becton Dickinson and Co., Between Towns Road, Cowley, Oxford OX4 3LY, UK.

Becton Dickinson and Co., 2 Bridgewater Lane, Lincoln Park, NJ 07035, USA.

Becton Dickinson, Collaborative Biomedical Products, 2 Oak Park, Bedford, MA 01730, USA..

Becton Dickinson Labware, Falcon, Franklin Lakes, NJ 07417, USA.

Bio

Bio 101 Inc., c/o Statech Scientific Ltd, 61–63 Dudley Street, Luton, Bedfordshire LU2 0HP, UK.

Bio 101 Inc., PO Box 2284, La Jolla, CA 92038–2284, USA.

Bio-Rad Laboratories

Bio-Rad Laboratories Ltd., Bio-Rad House, Maylands Avenue, Hemel Hempstead HP2 7TD, UK.

Bio-Rad Laboratories, Division Headquarters, 3300 Regatta Boulevard, Richmond, CA 94804, USA.

BioWhittaker, Inc., 8830 Biggs Ford Road, Walkersville, MD 21793, USA.

Boehringer Mannheim

Boehringer Mannheim UK (Diagnostics and Biochemicals) Ltd, Bell Lane, Lewes, East Sussex BN17 1LG, UK.

Boehringer Mannheim Corporation, Biochemical Products, 9115 Hague Road, P.O. Box 504 Indianapolis, IN 46250–0414, USA.

Boehringer Mannheim Biochemica, GmbH, Sandhofer Str. 116, Postfach 310120 D-6800 Mannheim 31, Germany.

British Drug Houses (BDH) Ltd, Poole, Dorset, UK.

Calbiochem, PO Box 12087, La Jolla, CA 92039–2087, USA.

Campden Instruments, King Street, Sileby, Loughborough LE12 7LZ, UK.

Chance Propper coverslips, supplied by Philip Harris Lynn Lane, Shenstone, Lichfield, Staffordshire WS14 0EE, UK.

Clark Electroinstruments, Reading, UK.

Clontech Laboratories

Clontech Laboratories Ltd., Wade Road, Basingstoke, Hampshire, RG24 8NE, UK.

Clontech Laboratories, Inc., 1020 E. Meadow Circle, Palo Alto, CA 94303, USA.

Corning, Inc., PO Box 5000, Corning, NY 14830, USA.

Cytoskeleton, Inc., 1650 Fillmore St., Denver, CO 80206, USA.

DAGE-MTI, Inc., 701 N. Roeske Avenue, Michigan City, IN 46360, USA.

David Kopf Instruments, 7324 Elmo Street, PO Box 636, Tujunga, CA 91043, USA.

Difco Laboratories

Difco Laboratories Ltd., P.O. Box 14B, Central Avenue, West Molesey, Surrey KT8 2SE, UK.

Difco Laboratories, P.O. Box 331058, Detroit, MI 48232–7058, USA.

Dow Corning Co., PO Box 0994, Midland, MI 48640, USA.

Du Pont

Dupont (UK) Ltd., Industrial Products Division, Wedgwood Way, Stevenage, Herts, SG1 4Q, UK.

Du Pont Ltd., NEN Life Science Products, PO Box 66, Hounslow TW5 9RT, UK.

Du Pont Co. (Biotechnology Systems Division), P.O. Box 80024, Wilmington, DE 19880–002, USA.

Eppendorf

Eppendorf, Postfach 650670, 2000 Hamburg 65, Germany.

Eppendorf–Netheler–Hinz GmbH, Barkhausenweg 1, D-2000 Hamburg 63, Germany.

Eppendorf North America, Inc., Molecular and Cell Biology Division, 545 Science Drive, Madison, WI 53711, USA.

European Collection of Animal Cell Culture, Division of Biologics, PHLS Centre for Applied Microbiology and Research, Porton Down, Salisbury, Wilts SP4 0JG, UK.

Falcon (Falcon is a registered trademark of Becton Dickinson and Co.).

Fisher Scientific Co., 711 Forbest Avenue, Pittsburgh, PA 15219–4785, USA.

Flow Laboratories, Woodcock Hill, Harefield Road, Rickmansworth, Herts. WD3 1PQ, UK.

Fluka

Fluka-Chemie AG, CH-9470, Buchs, Switzerland.

Fluka Chemicals Ltd., The Old Brickyard, New Road, Gillingham, Dorset SP8 4JL, UK.

Gibco BRL

Gibco BRL (Life Technologies Ltd.), Trident House, Renfrew Road, Paisley PA3 4EF, UK.

Gibco BRL (Life Technologies Inc.), 3175 Staler Road, Grand Island, NY 14072–0068, USA.

Hamamatsu Photonics K.K., Systems Division, 812 Joko-cho, Hamamatsu City, 431-31, Japan.

Arnold R. Horwell, 73 Maygrove Road, West Hampstead, London NW6 2BP, UK.

Hybaid

Hybaid Ltd., 111–113 Waldegrave Road, Teddington, Middlesex TW11 8LL, UK.

Hybaid, National Labnet Corporation, P.O. Box 841, Woodbridge, NJ. 07095, USA.

HyClone Laboratories 1725 South HyClone Road, Logan, UT 84321, USA.

International Biotechnologies Inc., 25 Science Park, New Haven, Connecticut 06535, USA.

Invitrogen Corporation

Invitrogen Corporation 3985 B Sorrenton Valley Building, San Diego, CA. 92121, USA.

Invitrogen Corporation c/o British Biotechnology Products Ltd., 4–10 The Quadrant, Barton Lane, Abingdon, OX14 3YS, UK.

Kodak: Eastman Fine Chemicals 343 State Street, Rochester, NY, USA.

Leitz/Leica, 111 Deer Lake Road, Deerfield, IL 60015, USA.

Life Technologies

Life Technologies Inc., 8451 Helgerman Court, Gaithersburg, MD 20877, USA.

Life Technologies, Gibcvo BRL, Inc., 3175 Staley Road, Grand Island, NY 14072, USA.

Ludl Electronics Products, Ltd., 171 Brady Avenue, Hawthorne, NY 10532, USA.

Mallinckodt-Baker, Inc., 16305 Swingley Ridge Drive, Chesterfield, MO 63017, USA.

MatTeK Corp., 200 Homer Avenue, Ashland, MA 01721, USA.
Merck
Merck Industries Inc., 5 Skyline Drive, Nawthorne, NY 10532, USA.
Merck, Frankfurter Strasse, 250, Postfach 4119, D-64293, Germany.
Millipore
Millipore (UK) Ltd., The Boulevard, Blackmoor Lane, Watford, Herts WD1 8YW, UK.
Millipore Corp./Biosearch, P.O. Box 255, 80 Ashby Road, Bedford, MA 01730, USA.
Molecular Probes, Inc., PO Box 22010, 4949 Pitchford Avenue, Eugene, OR 97402-9165, USA.
Narishige
Narishige Europe, Ltd., Willow Business Park, Willow Way Unit 7, London SE26 4QP, UK.
Narishige USA, Inc., One Plaza Road, Greenvale, NY 11548, USA.
National Cancer Institute, Drug Synthesis and Chemistry Branch, Developmental Therapeutics Program, Division of Cancer Treatment, Bethesda, MD, USA.
New England Biolabs (NBL)
New England Biolabs (NBL), 32 Tozer Road, Beverley, MA 01915–5510, USA.
New England Biolabs (NBL), c/o CP Labs Ltd., P.O. Box 22, Bishops Stortford, Herts CM23 3DH, UK.
Nikon
Nikon Europe B.V., Schipholweg 321, PO Box 222, 1170 AE Badhoevedorp, The Netherlands.
Nikon, Inc., Instrument Group, 623 Steart Avenue, Garden City, NY 11530, USA.
Nikon Corporation, Fuji Building, 2–3 Marunouchi 3-chome, Chiyoda-ku, Tokyo, Japan.
Nunc–Gibco, Life Technologies Ltd., 3 Fountain Drive, Inchinnan Business Park, Paisley PA4 9RF, UK.
Oncogene Research Products, 84 Rodgers Street, Cambridge, MA 02142, USA.
Perkin-Elmer
Perkin-Elmer Ltd., Maxwell Road, Beaconsfield, Bucks. HP9 1QA, UK.
Perkin-Elmer Ltd., Post Office Lane, Beaconsfield, Bucks, HP9 1QA, UK.
Perkin-Elmer-Cetus (The Perkin-Elmer Corporation), 761 Main Avenue, Norwalk, CT 0689, USA.
Pharmacia Biotech Europe Procordia EuroCentre, Rue de la Fuse-e 62, B-1130 Brussels, Belgium.
Pharmacia Biosystems
Pharmacia Biosystems Ltd. (Biotechnology Division), Davy Avenue, Knowlhill, Milton Keynes MK5 8PH, UK.

Pharmacia LKB Biotechnology AB, Björngatan 30, S-75182 Uppsala, Sweden.

Pierce and Warriner, 44 Upper Northgate Street, Chester CH1 4EF, UK.

Pierce Chemical Co., 3747 N. Meridian Road, PO Box 117, Rockford, IL 61105, USA.

Polysciences, Inc., 400 Valley Road, Warrington, PA 18976, USA.

Princeton Instruments, Inc., 3660 Quaker Bridge Road, Trenton, NJ 08619, USA.

Promega

Promega Ltd., Delta House, Enterprise Road, Chilworth Research Centre, Southampton, UK.

Promega Corporation, 2800 Woods Hollow Road, Madison, WI 53711–5399, USA.

Qiagen

Qiagen Inc., c/o Hybaid, 111–113 Waldegrave Road, Teddington, Middlesex, TW11 8LL, UK.

Qiagen Inc., 9259 Eton Avenue, Chatsworth, CA 91311, USA.

Roche Diagnostics Ltd., Bell Lane, Lewes, East Sussex BN7 1LG, UK.

Sarstedt, Inc., PO Box 468, Newton, NC 28658–0468, USA.

Schleicher and Schuell

Schleicher and Schuell Inc., Keene, NH 03431A, USA.

Schleicher and Schuell Inc., D-3354 Kassel, Germany.

Schleicher and Schuell Inc., c/o Andermann and Company Ltd.

Severn Biotech Ltd., Park Lane, Kidderminster, Worcester DY11 6TJ, UK.

Shandon Scientific Ltd., Chadwick Road, Astmoor, Runcorn, Cheshire WA7 1PR, UK.

Sigma Chemical Company

Sigma Chemical Company (UK), Fancy Road, Poole, Dorset BH17 7NH, UK.

Sigma Chemical Company, 3050 Spruce Street, P.O. Box 14508, St. Louis, MO 63178–9916.

Sorvall DuPont Company, Biotechnology Division, P.O. Box 80022, Wilmington, DE 19880–0022, USA.

Stratagene

Stratagene Ltd., Unit 140, Cambridge Innovation Centre, Milton Road, Cambridge CB4 4FG, UK.

Stratagene Inc., 11011 North Torrey Pines Road, La Jolla, CA 92037, USA.

Transduction Laboratories, 133 Venture Court, Lexington, KY 40511–2624, USA.

United States Biochemical, P.O. Box 22400, Cleveland, OH 44122, USA.

Universal Imaging Co., 502 Brandywine Parkway, West Chester, PA 19380, USA.

Upstate Biotechnology, 199 Saranac Avenue, Lake Placid, NY 12946, USA.

Vincent Associates, 1255 University Avenue, Rochester, NY 14607, USA.

Warner Instrument Corporation, 1125 Dixwell Avenue, Hamden, CT 06514, USA.

Wellcome Reagents, Langley Court, Beckenham, Kent BR3 3BS, UK.

Whatman, Inc., 8 Bridewell Place, Clifton, NJ 07014, USA.

Worthington Biochemical Co., 730 Vassar Avenue, Lakewood, NJ 08701, USA.

Carl Zeiss

Carl Zeiss, Inc., One Zeiss Drive, Tornwood, NY 10594, USA.

Carl Zeiss Ltd., PO Box 78, Woodfield Road, Welwyn Garden City, Hertfordshire AL7 1LU, UK.

Zymed Laboratories, 458 Carlton Court, South San Francisco, CA 94080, USA.

Index

Index

Index

Printed in the United States
35601LVS00002B/152